About Island Press

Since 1984, the nonprofit Island Press has been stimulating, shaping, and communicating the ideas that are essential for solving environmental problems worldwide. With more than 800 titles in print and some 40 new releases each year, we are the nation's leading publisher on environmental issues. We identify innovative thinkers and emerging trends in the environmental field. We work with world-renowned experts and authors to develop cross-disciplinary solutions to environmental challenges.

Island Press designs and implements coordinated book publication campaigns in order to communicate our critical messages in print, in person, and online using the latest technologies, programs, and the media. Our goal: to reach targeted audiences—scientists, policymakers, environmental advocates, the media, and concerned citizens—who can and will take action to protect the plants and animals that enrich our world, the ecosystems we need to survive, the water we drink, and the air we breathe.

Island Press gratefully acknowledges the support of its work by the Agua Fund, Inc., Annenberg Foundation, The Christensen Fund, The Nathan Cummings Foundation, The Geraldine R. Dodge Foundation, Doris Duke Charitable Foundation, The Educational Foundation of America, Betsy and Jesse Fink Foundation, The William and Flora Hewlett Foundation, The Kendeda Fund, The Andrew W. Mellon Foundation, The Curtis and Edith Munson Foundation, Oak Foundation, The Overbrook Foundation, the David and Lucile Packard Foundation, The Summit Fund of Washington, Trust for Architectural Easements, Wallace Global Fund, The Winslow Foundation, and other generous donors.

The opinions expressed in this book are those of the author(s) and do not necessarily reflect the views of our donors.

ABOUT THE SOCIETY FOR ECOLOGICAL RESTORATION INTERNATIONAL

The Society for Ecological Restoration (SER) International is an international non-profit organization comprising members who are actively engaged in ecologically sensitive repair and management of ecosystems through an unusually broad array of experience, knowledge sets, and cultural perspectives.

The mission of SER is to promote ecological restoration as a means of sustaining the diversity of life on Earth and reestablishing an ecologically healthy relationship between nature and culture.

The opinions expressed in this book are those of the author(s) and are not necessarily the same as those of SER International. Contact SER International at 285 W. 18th Street, #1, Tucson, AZ 85701. Tel. (520) 622-5485, Fax (270) 626-5485, e-mail, info@ser.org, www.ser.org.

RESTORING WILDLIFE

Restoring Wildlife

Ecological Concepts and
Practical Applications

Michael L. Morrison

SOCIETY FOR ECOLOGICAL RESTORATION INTERNATIONAL

Washington · Covelo · London

Island Press is a trademark of The Center for Resource Economics.

Library of Congress Cataloging-in-Publication Data

Morrison, Michael L.
 Restoring wildlife : ecological concepts and practical applications / Michael L. Morrison. —
2nd ed.
    p. cm. — (The science and practice of ecological restoration)
 Rev. ed. of: Wildlife restoration. ©2002.
 Includes bibliographical references and index.
 ISBN-13: 978-1-59726-492-1 (cloth : alk. paper)
 ISBN-10: 1-59726-492-X (cloth : alk. paper)
 ISBN-13: 978-1-59726-493-8 (pbk. : alk. paper)
 ISBN-10: 1-59726-493-8 (pbk. : alk. paper) 1. Restoration ecology. 2. Habitat (Ecology)
3. Wildlife management. I. Morrison, Michael L. Wildlife restoration. II. Title.
 QH541.15.R45M67 2009
 639.9—dc22

                                                                        2008044974

Printed on recycled, acid-free paper

Manufactured in the United States of America

10 9 8 7 6 5 4 3 2 1

CONTENTS

ACKNOWLEDGMENTS

There are, of course, many people who have helped bring this project to fruition. My first book on wildlife restoration was heavily influenced by Don Falk, John Rieger, and Thomas Scott; that influence carried forward and formed the foundation of this new book. Work subsequent to publishing my earlier book on wildlife restoration substantially contributed to development of this volume, and discussions on restoration with people such as K. Shawn Smallwood, Paul Krausman, Shane Romsos, Mollie Hurt, Kathi Borgmann, Julie Roth, Patricia Manley, and others have been most helpful to me. Perhaps most influential to me were the undergraduate and graduate students in my course at Texas A&M University on wildlife restoration; I especially acknowledge the students who are listed as authors of case studies in chapter 10. Joyce Vandewater did the graphics and artwork for the entire volume; her work is excellent and substantially enhanced the presentation. Angela Hallock assisted in many ways, including gathering of the copyright permissions, compiling the literature, and other important tasks. Melissa Rubio assisted with final preparation of the text. Also I appreciate the skillful (and patient) guidance of this project by Barbara Dean at Island Press; many of her editorial suggestions greatly streamlined and enhanced the final presentation. Barbara Youngblood, also of Island Press, assisted in many ways to help see this work to completion. Finally I thank my wife, Venetia, and daughter, Madeleine, for keeping me firmly grounded.

Chapter 1

Introduction: Restoring and Preserving Wildlife

I began my previous book on wildlife restoration (Morrison 2002) with a quote from Michael Gilpin: "The restorationist who sees himself or herself as the practical arm of some abstract, academic discipline is not likely to give great thought to the theoretical aspects of the restoration process. . . . He will probably view his failures as a result of personal ignorance of the received wisdom" (1987, 305). I used this quote to emphasize the fundamental role that ecology plays in restoration. I study wildlife, and the profession of wildlife biology began with well-meaning people doing the best they could to manage animals based on personal experience, knowledge passed down to them by other experienced people, and a good amount of best guess. The wildlife profession has moved increasingly toward the gathering of rigorous scientific data, followed by development of applications for management of animals based on those data. Likewise, the restoration profession is moving increasingly toward the gathering of sound data as a foundation for development of practical applications of those data. Thus, the *received wisdom* is growing evermore reliable, and restorationists, like wildlife biologists, are becoming increasingly versed in environmental science and ecology. As stated by Palmer et al. (2006, 2), the focus of their book, *Foundations of Restoration Ecology*, is "the mutual benefit of a stronger connection between ecological theory and the science of restoration ecology." The aim of this volume is to help interweave theoretical and practical aspects of wildlife biology with direct application to the restoration and conservation of animals.

Much of restoration involves, directly or indirectly, improving conditions for native species of wildlife. To be ultimately successful, then, restoration plans must be guided in large part by the needs of current or desired wildlife species in the project area. Such information includes data on species abundances and distribution, both current and historic; details on habitat requirements, including proper plant species composition and structure; food requirements; breeding locations; the role that succession will play in species turnovers; problems associated with exotic species of plants and animals; the problems of restoring small, isolated areas; and so forth. Thus, proper consideration of wildlife—their habitat needs and numbers—is a complicated process that requires careful consideration during all stages of restoration. Additionally, the success of a restoration project should be judged, in part, by how wildlife species respond to the project. Such monitoring will provide feedback for modifications of the specific project, as well as help refine future projects. The approach includes applications at all spatial scales, from broadscale (landscape) projects, down to small, site-specific projects. Throughout, however, I emphasize a holistic, integrated ecosystem approach.

In this volume, I provide ecologists, restorationists, administrators, and other professionals involved with restoration with a basic understanding of the fundamentals of wildlife populations and wildlife-habitat relationships. This knowledge will provide a good understanding of the types of information needed in planning, plus hands-on experience that will impart an understanding of the problems inherent in this type of work. It will provide the basic tools necessary to develop and implement a sound monitoring program. The two primary monitoring themes covered are experimental design and statistical analysis. Additionally, material is provided on the sampling of rare species and populations. With this knowledge, restorationists will be better equipped to discuss their needs with professional wildlife biologists. No special training or education is necessary, although knowledge of basic ecological concepts and basic statistics would be helpful. This book addresses wildlife-habitat restoration that uses the following techniques:

- Developing the concept of habitat, its historic development, components, spatial-temporal relationships, and role in land management
- Reviewing how wildlife populations are identified and counted
- Detailing techniques for measuring wildlife and wildlife habitat, including basic statistical techniques
- Discussing how wildlife and their habitat needs can be incorporated into restoration planning, especially concerning size of preserves, fragmentation, and corridors
- Developing a holistic approach to restoration of large landscapes (integrated, ecosystem approaches)
- Discussing the role that exotic species, competitors, predators, disease, and related factors influence restoration planning

- Developing a solid justification and reasoning for monitoring and good sampling design
- Allowing for the development and critique of individual monitoring projects
- Discussing and critiquing case histories of wildlife analysis in restoration projects
- Providing a firm understanding of sources available to the restorationists for further learning and implementation of wildlife-habitat relationships and monitoring in restoration planning.

Thus, this book provides a good thorough understanding of the conceptual and practical problems involved in sampling wildlife populations. It is critical that restorationists understand what wildlife biologists can and cannot provide within certain time and monetary limitations. Although I do not take a "cookbook" approach, I devote chapter 10 to case studies, tying the various concepts presented herein into a comprehensive package for your review and guidance. However, applying general prescriptions most often provides unpredictable results, some of which may cause more harm than good (e.g., attracting unwanted exotic species). This book provides the basic tools needed to understand ecological concepts that can be used to design restoration projects with specific goals for wildlife, as well as specific guidance and examples on how other projects have been designed and implemented.

Fundamentals of Habitat Restoration

In this book you will learn the fundamental principles of, and be exposed to, many of the most commonly used tools for evaluating the wildlife present in an area and determining their relationships with their habitat. Ecology is complicated, so there are many topics that must be thoroughly understood:

- species lists
- habitat uses
- ecological processes
- monitoring
- study design
- statistical analyses
- population processes
- exotic species, diseases, and parasites

Knowledge of these topics is necessary if you desire to develop the following:

- restoration plans
- endangered species recovery

- population monitoring
- impact assessments
- reserve designs
- habitat conservation plans
- basic ecological relationships

This book also provides guidance about where more advanced and detailed literature can be found.

Why a New Book?

This book builds on the material presented in *Wildlife Restoration* (Morrison 2002), because ecological principles remain the necessary foundation for discussing wildlife. I have, however, used the experience I have gained since its publication to reorganize and substantially expand this new edition. Additionally, I have incorporated recent topics in the broad field of wildlife-habitat relationships and wildlife-study design, topics that I and my coauthors have synthesized and discussed in new editions of other books (Morrison et al. 2006; 2008). This book is therefore more than a second edition, hence the new title. I also dropped the previous subtitle, *Techniques for habitat analysis and animal monitoring*, because those topics are now much more evenly balanced by an increasingly comprehensive coverage of the field of wildlife ecology.

Chapter 2 goes into detail on issues such as defining wildlife, single- and multiple-species approaches, and ecosystem management. Chapters 3 (Populations) and 4 (Habitat) have been revised and updated and are the core components of this book. Chapter 5, Assemblages, is new and explains how to understand wildlife restoration in the context of environmental stochasticity, recovery, succession, and related principles. Chapter 6, Desired Conditions, is also new, and it uses the practical work I have conducted as well as the literature that explains how to gather and then assemble the information needed to guide a restoration plan.

Chapter 7 (Design Concepts) develops and interrelates the core components needed to design and implement a conservation area. Topics covered include habitat heterogeneity and fragmentation, disturbance ecology and the dynamics of habitats in landscapes, corridors and buffers, and the landscape matrix as a planning area. Chapter 8 (Primer on Study Design) summarizes the basic principles of study design and monitoring that are required for the gaining of reliable information, in the context of the study of restoration and conservation of wildlife. Because most restoration projects cannot be replicated, I discuss how to apply the field of impact assessment to restoration. Chapter 9 focuses on monitoring, which is a centerpiece of how we gain information on the success of our projects, anticipate how we need to modify (i.e., fix) our projects postconstruction, and gain knowledge that can be used to plan future endeavors.

As mentioned earlier, chapter 10 introduces four case studies of wildlife restoration. Each case study reveals how the project was conceived, how goals were developed and criteria for success were determined, how monitoring was developed, and what changes (adaptive management) might be needed as the project develops following construction. These case studies take advantage of several recent and ongoing restoration projects that my colleagues and I have been conducting. They were developed by graduate students in my wildlife restoration course at Texas A&M University using information I provided on the specific projects as well as information they gathered from the literature. I am confident that the lessons we have learned in developing these case studies will be of benefit to readers of this book.

The book concludes with a synthesis discussion (chapter 11) on how to put knowledge gained in this book to work and a summary of the key points—the take home message—from each preceding chapter. I have also added a glossary that provides brief definitions of key terms used in the book and identifies the chapter where each term is initially defined and developed. Terms found in the glossary are also italicized in the text.

I am further confident that the field of restoration will continue to grow and become ever more important in the conservation and management of wildlife. I hope this book helps in some way to further promote the positive growth of restoration in general and, more specifically, the restoration of desired wildlife populations.

Chapter 2

Operating Concepts

There are many ecological concepts that must be considered when developing any study, including a restoration project. The foundation of the science of restoration is broad and includes such ecological concepts such as population ecology and genetics, ecophysiology, evolutionary biology, food webs, ecosystem dynamics, habitat and niche ecology, and other related topics (see Palmer et al. 2006 for a detailed discussion). Successful restoration obviously depends on the understanding of ecological principles, but restoring ecosystems is not limited to ecology, since it requires interdisciplinary approaches with other sciences, such as geography, chemistry, and physics (Halle and Fattorini 2004). In addition, economics, sociology, and politics must be considered for successful restoration projects but are beyond the scope of this book. In this chapter I discuss some of the key concepts and terms that underpin restoration in general; these concepts and terms thus serve as the foundation upon which I will build the remaining, more wildlife-specific chapters of this book.

Restoration Defined

The term *restore* is well understood to mean to bring back into existence or use, and *restoration* as the act of restoring. The difficulty in the terminology of restoration is determining the condition to which something will be restored. Automobile restoration is simple, given we know the year—indeed the specific day—that the car was produced. Drawings are usually available that detail every external and internal part of a car; photographs are likely available. Unfortunately, nature does not have serial numbers and parts catalogs, making the job of determining how to proceed with restoration of ecological systems fraught with uncertainties.

As reviewed by Halle and Fattorini (2004), there have been initiatives to establish a theoretical framework for restoration ecology (e.g., Hobbs and Norton 1996). The related fields of disturbance ecology and succession are fundamental to restoration ecology. Natural succession can be used and manipulated in ecological restoration to guide a system degraded by heavy disturbance back to its original state. Additionally, disturbances of smaller magnitude are often used to direct or speed up succession, such as to provide the required site conditions for establishment of desired species. Thus, developing the conceptual framework for a restoration project should include combining elements of disturbance and succession. It might be possible to restore the basic functions or ecosystem processes, but to achieve the former structure in full—including a particular assemblage of species—is usually impossible (Halle and Fattorini 2004). In chapter 5, I incorporate elements of disturbance and especially succession in the development of the assemblage of species. In this section I will lay out some of the key issues that must be considered when first venturing into the realm of wildlife restoration.

Time Frames and Historic Conditions

Anderson (1996) depicted the ways in which an ecosystem could be viewed with and without various human-induced impacts (figure 2.1). The dashed lines in figure 2.1 are a recognition that we cannot know the trajectory any

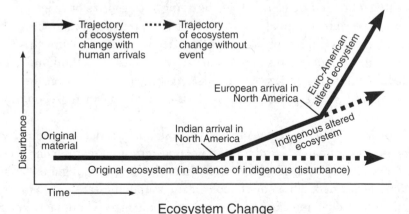

FIGURE 2.1. Humans represent an ecological factor that influences the trajectory of an ecosystem (from Anderson 1996).

ecosystem would take, with or without human influence. Similarly, we cannot know the trajectory that ecosystems will take into the future. The difficulty in determining historical conditions is, first, determining what time period is to serve as "historic," and then, determining what ecological conditions existed during that time period. For example, recent evidence suggests that controlled burning of vegetation to maintain a preferred ecosystem state was present as early as 55,000 years ago in southern Africa (Smith 2007).

Noss (1985) thoroughly reviewed the use of presettlement vegetation as a basis for restoration. He noted that presettlement vegetation systems were relatively ancient and stable and serve as a baseline against which we can measure humanized landscapes. When humans occupy areas in large numbers, we replace nonhumanized disturbance regimes with a new set of disturbances that impact native flora and fauna in different ways. I use the term *nonhumanized* rather than *natural*, because humans are, indeed, a natural part of the environment; humans evolved on this planet. Humans are increasingly making a choice to try and manage our activities such that we minimize our negative impacts on plant and other animal species.

Noss (1985) concluded that the question of whether Indians should be considered a natural and beneficial component of the environment cannot be answered conclusively. He noted that the potential impact that Indians had on the environment varied by time and location: in some regions Indian populations were sparse and probably contributed positive feedback to ecosystem functions, whereas in other regions their activities resulted in extensive alterations of vegetation and there is little justification to consider them any more natural than European settlers. I do not disagree with Noss per se, but rather offer that we gain nothing by trying to categorize or judge human impacts to the environment in a natural versus an unnatural manner; such a debate tends to put people into opposing factions (e.g., progrowth versus nogrowth, developer versus preservationist). As a scientist, I think my role is to help quantify the likely consequences of human activities on the environment so that the public and the decision makers can make informed decisions.

Human activities have, of course, had substantial impacts on assemblages of plants and animals. For example, most large, terrestrial mammalian predators have already been lost from more than 95% of the contiguous United States and Mexico. Thus, most ecological communities are either missing dominant selective forces or have new ones dominated by humans. For example, Berger et al. (2001) demonstrated a cascade of ecological events that were triggered by the local extinction of grizzly bears and wolves from the southern Greater Yellowstone Ecosystem. They found that the loss of grizzly

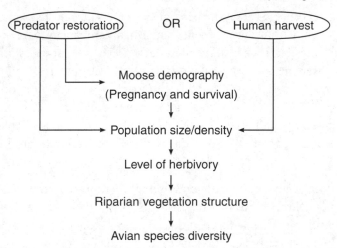

FIGURE 2.2. Overview of conservation options and the linkage among biological tiers of organization in a terrestrial ecosystem with large carnivores (from Berger et al. 2001, figure 4).

bears and wolves resulted in an eruption of moose, the subsequent alteration of riparian vegetation by the moose, and the coincident reduction of neotropical migrant birds in the impacted riparian zone. Berger et al. proposed a simple model that would restore some linkages among biological tiers of organization either through reintroduction of the large carnivores or through human harvest of overabundant herbivores (figure 2.2: the moose in the Yellowstone example). Thus, we see that even in areas that most would consider "pristine"—such as Yellowstone—humans have heavily influenced current ecosystem processes.

Historical ecology is the interface between ecology and historical geography that studies the relationship between human acts and acts of nature (and the response of nature to human acts). As summarized by Egan and Howell (2001), historical ecology centers on the following:

- Human influences, ranging from the subtle and benign to the overtly destructive, are pervasive throughout the earth's ecosystems.
- The interaction that takes place between the environment and human cultures is not deterministic, but rather a dynamic dialectic process that results in landscapes, which are, in effect, culturalized landscapes.
- Humans can produce and help maintain sustainable, productive ecosystems.
- Despite its emphasis on the past, the work of historical ecologists is future seeking.

Historic ecosystems are those ecological systems that existed some time in the past, literally from this moment to many millennia ago. As discussed by Egan and Howell (2001), historic ecosystems are important in restoration because they can be used as analogs or guides to current restoration plans. Thus, much discussion in restoration has focused on the condition that we wish a location to be restored to—the reference condition. *Reference conditions* are determined by analyzing the data obtained for each variable chosen for the historic range of variation study and assist the restorationist by the following:

- Defining what the original condition was compared to the present (composition, structure, processes, functions)
- Determining what factors caused the degradation
- Defining what needs to be done to restore the ecosystem
- Developing criteria for measuring success of the restoration

Reference conditions can be relatively easy to determine when the focus of restoration is repair of a location recently damaged by a catastrophic event, whether natural (e.g., fire, flood) or human caused (e.g., chemical spill). In contrast, reference conditions can be subject to great debate and uncertainty when the goal is to return a location to an ecological state that existed many centuries in the past.

There are basically four types of reference models, as depicted in figure 2.3. As defined by Egan and Howell (2001), these reference models can be referred to as (1) contemporary restoration sites, (2) historic models of restoration sites, (3) contemporary remnants, and (4) historic remnants (numbering refers to figure 2.3). In reality, all four types of models will provide valuable information in planning a restoration project. The combination of contemporary and historic information can help offset the deficiencies of each, because contemporary provides only a brief glimpse in time, while his-

		Location	
		Same	Different
Time	Same	1	3
	Different	2	4

FIGURE 2.3. Four types of reference models: (1) contemporary restoration sites, (2) historic models of restoration sites, (3) contemporary remnants, and (4) historic remnants.

toric conditions are fraught with uncertainties. In fact, Egan and Howell specifically discouraged the use of historic information from remnant sites (number 4, figure 2.3, different time and location) because they thought there was too much spatial and temporal uncertainty. Although I agree that spatiotemporal variation is of concern, this same concern applies to any location at a different time. I think, rather, that "different time, different location" is useful in gathering a complete picture of what was potentially occurring in and around a planned restoration site; I develop these issues and specific means of gathering historic data in chapter 6.

As noted by Palmer et al. (2006), although ecological restoration is an attempt to return a system to some historical state, the difficulty of reaching this goal is widely recognized. The authors suggested that a more realistic goal is to move a damaged system to an ecological state that is within some acceptable limits relative to a less disturbed system. The key here, then, is to define *acceptable limits*. Throughout this book I will incorporate discussion of the decision-making processes used to determine what the goals for restoration were and how they were developed. Here I provide two definitions that will serve as a foundation for the remaining discussions (adapted from Palmer et al. 2006).

> *Restoration ecology*: The scientific process of developing theory to guide restoration and using restoration to advance ecology.
> *Ecological restoration*: The practice of restoring degraded ecological systems.

As noted by Palmer et al. (2006), there is an obvious link between the practice of restoration (i.e., ecological restoration), and the concepts that were used to guide the restoration plan (i.e., restoration ecology). The conceptual foundation for restoration ecology comes, of course, from ecological concepts or theory. Recall my discussion of the link between theory and practice in my opening comments in chapter 1.

Natural Versus Desired Conditions

Although I will provide specific guidance on developing desired conditions for restoration projects (see chapter 6), it is not my purpose per se to judge the conditions that a restoration project should try and replicate. In many projects there has been debate over what the desired or *natural* time period should be—what should be the target of restoration. My purpose is to assist with understanding ecological processes as they relate to wildlife and their habitat.

Some restorationists feel that the baseline for natural conditions is one without human-induced impacts. Some take this further and include environmental impacts caused by the region's indigenous peoples under the rubric of natural. Willis and Birks (2006) reminded us that ecosystems change in response to many factors, principally including climate variability, invasions of species, and wildfires. Most records used to assess such changes on ecosystems are based on short-term data and remote sensing (aerial photography and satellite imagery) that span only a few to several decades. Thus, it becomes difficult to separate the effects that various types of forces have on ecosystem structure, function, and process. Willis and Birks emphasized the importance of examining paleoecological records—fossil pollen, seeds and fruit, animal remains, tree rings, charcoal—that span tens to millions of years to provide a long-term perspective on the dynamics of contemporary ecosystems.

Although I will develop the use of paleoecological records in chapter 6, the key point raised by Willis and Birks that is germane to our current discussion is the difficulty in defining what are natural features in ecosystems through time. Willis and Birks summarized several cases in which paleoecological records were used to clarify the origin of plant species—whether they were native or exotic to a region. For example, in a reexamination of British flora, several discrepancies between published records were found in which the same species were classified as *alien* or *native*, depending on personal interpretation. Part of the difficulty in determining the origin of a species is settling on how far back one takes human activity in deciding whether a species is native or exotic. Recall that in figure 2.1 there is a separation of indigenous people from European people, but both groups are clearly human. There are over 150 species of plants that have been introduced to the British flora by humans between 500 and 4,000 years ago, many of which were classified recently as threatened or nearly extinct. Thus, deciding which plants are native would have a substantial impact on efforts to either conserve, or potentially eradicate, many species. Is a plant introduced purposely 3,500 years ago, for example, native, whereas one introduced accidentally 500 years ago, exotic? I cannot answer this question; rather, the answer depends on the goal of a specific restoration project.

Noss (1985) suggested that we could enrich our concept of presettlement vegetation by evaluating individual areas not only in terms of their species and community content, but also in terms of their position in and potential contribution to broader patterns of diversity; that is, their landscape context. I agree that all restoration projects should be developed in a larger landscape context, a theme that I incorporate throughout this book. However, placing a project within a landscape (broad spatial extent) context does not imply that we should ignore or even deemphasize considerations of fauna and floral requirements on the local scale. As I develop in chapter 5, identifying the species that could potentially occur on a restoration project requires a stepped-down approach across spatial scales.

Those planning a restoration project must develop their own operating principles. However, achieving an ecological condition free from human-induced impacts is prevented by factors such as local plant and animal extinctions, introduction of exotic species, and requirements for dispersal of animal and plant species, all of which retard or prevent natural ecological processes.

Thus, developing a restoration plan requires prioritization of goals based on knowledge of (1) historic conditions, (2) current regional conditions, and (3) species-specific requirements. And, of course, one must evaluate legal requirements and consider the current political reality of the area.

Clearly, determining what is natural or native is open to interpretation. There are, however, some basic ecological principles that can help guide what is possible with regard to restoration. Just as there is no reason to attempt a research study that has no chance of success (because of time, funding, or logistical constraints), there is no reason to attempt to restore that which cannot be achieved. I strongly support attempts to place restoration plans in context of historic conditions (see also chapter 6) but caution that ecological reality must guide what we can and cannot achieve. Restorationists should clearly and openly state—and justify—the goals for a project.

Wildlife Defined

Wildlife is a term given to animals and plants that live on their own without taming or cultivation by people. However, in terms of traditional wildlife management, wildlife has been defined as mammals and birds that are hunted (game animals) or trapped (fur-bearing animals). For example, in 1933 Aldo Leopold named his classic textbook, which formed the foundation of wildlife management, *Game Management*. This definition was broadened gradually by wildlife biologists as increasing emphasis was placed on multiple species management to include nongame animals; songbirds, bats, amphibians, and reptiles are now frequently emphasized by wildlife biologists.

The definition of wildlife continues to be broadened, including a push toward more study of invertebrates. Fish are usually excluded from wildlife journals (e.g., *Journal of Wildlife Management*) because fish management issues differ greatly from those of traditional wildlife management, and

separate professional societies focusing on fisheries exist. Although the majority of work in wildlife ecology focuses on vertebrates, only about 15%
(each) of birds and mammals are classified as extinct, imperiled, or vulnerable, whereas about 50% of crayfish and 43% of stoneflies are included in
those risk categories (Stein et al. 2002). There is no reason why wildlife cannot be defined to include invertebrates.

Thus, the question for restoration becomes, how far do we go taxonomi-
cally to restore the animals in an ecosystem? As I develop throughout this
book, and as others have recently done (e.g., Palmer et al. 2006), there are
many key interrelationships within an ecological system that drive the functions and processes therein. We will discuss strategies for determining which
animals to incorporate into a wildlife restoration plan in chapter 6.

In part because of the historic focus by wildlife biologists on game species,
a general split developed in the 1960s and 1970s within the wildlife profession between *game* and *nongame* biologists. Concomitant with the professional split within the wildlife profession was a perception by many that scientific societies (e.g., ornithology, mammalogy, herpetology) tended to focus
too narrowly on basic biology while ignoring management and conservation
applications. To fill the need of many scientists who wanted to avoid traditional wildlife management but also emphasize conservation, the Society for
Conservation Biology (SCB) was formed, and journals such as *Conservation
Biology* (published by SCB), *Ecological Applications*, and *Restoration Ecology* (and the more applied *Ecological Restoration*, [formerly *Restoration and
Management Notes*]) were initiated.

Approaches to Ecological Restoration

Debates have also been ongoing over the emphasis on single- versus multiple-
species studies, and relative local-scale versus broadscale studies (Simberloff
2007). These debates over species and scale are linked to the desire of many
wildlife professionals to emphasize nongame animals. Because of the increasing loss of land area to residential and commercial developments, and
increasing impacts on other lands from recreational activities, many scientists
wish to emphasize multiple-species studies conducted over broad areas (i.e.,
landscapes) and deemphasize single-species and site-specific studies. As I explain in the following chapters, these species-scale debates fail to recognize
that each species requires unique combinations of resources that cannot be
identified by simply calling for multispecies studies. That is, one does not
count multiple species without counting individual animals. Additionally, although we must certainly develop restoration plans in a broadscale context

(see chapters 5–7), most land management occurs on small scales (i.e., from 1–100 ha).

As introduced earlier, the management of wildlife has recently been evolving along scientific and management, social, cultural, legal, and also ethical and aesthetic dimensions. An ecosystem approach is needed for effective conservation and restoration. By this I mean consideration of organisms from various taxonomic designations, along with their interactions with each other and with abiotic conditions and processes. Although recognizing an ecosystem on the ground is problematic, an *ecosystem* (i.e., an ecological system) can be generally defined as consisting of organisms of various taxonomic designations and levels of biological organization, along with their interactions among each other and among abiotic conditions and processes. Regardless of the specific definition of ecosystem, understanding wildlife in an ecosystem context requires understanding other concepts (from Morrison et al. 2006, 387):

- Population dynamics
- The evolutionary context of organisms, populations, and species
- Interactions between species that affect their persistence
- The influence of the abiotic environment on the vitality of organisms

An ecosystem context also necessitates understanding the role of humans in modifying environments, habitats, and wildlife populations (Grumbine 1994; Morrison et al. 2006, 387).

Synthesis

As I hope the preceding discussion has shown, the difficulty in the terminology of restoration is determining the condition to which something will be restored. Additionally, there is debate over what is considered natural or native: how do we treat plants and animals introduced to new areas by humans now considered indigenous relative to those introduced by more recent human populations? And, what exactly is considered wildlife, and over what geographic area must we evaluate in order to evaluate ecosystem processes? There are no established answers to these and related questions, but there is a process being established within the community of restorationists by which a structured methodology can be used to make informed decisions based on available data. The specific goals for each restoration project should be clearly stated and justified, including a thorough analysis of historic conditions and the role humans appeared to have had in shaping the direction of ecosystem development.

Thus, it is important that the restoration of wildlife and their habitats be holistically approached. In this manner we have a chance of understanding at least some of the factors responsible for determining the distribution, abundance, and success of animals. I further discuss and develop key ecological concepts as they apply to restoration in later chapters, including especially chapters 4 and 5, and develop creation of desired conditions for restoration planning in chapter 6.

Chapter 3

Populations

The ultimate goal of *wildlife-habitat restoration* is to provide for the survival and protection of individual organisms in sufficient numbers and locations to maximize the probability of long-term persistence. Management of habitat can provide for conditions in which organisms can maximize the number of viable offspring produced that in turn find mates and suitable environments and successfully reproduce. Fitness is influenced by the dynamics of interactions of individuals within a population, by interactions among populations and species, and by interactions between organisms and their habitats and environments. To properly restore habitat thus requires knowledge of population dynamics and behaviors. Successful restoration also requires that we understand the ecological processes that regulate population trends. Although habitat is essential to survival of all species (see chapter 4), by itself it does not guarantee long-term fitness and viability of populations (Morrison et al. 2006, 61).

In this chapter, I begin by discussing the spatial and geographic factors that influence habitats and environments, population structure, fitness of organisms, and, ultimately, the viability of populations. The restorationist must understand these population parameters because they relate directly to the size and location of the area that needs restoration to ensure survival of a species. It makes little sense, for example, to provide habitat for a species if its long-term survival depends on immigration of new individuals and no allowance has been made for such immigration in developing the restoration plan. I then explain the fundamentals of augmentation of animal populations through captive breeding and reintroduction. One or more species of interest might no longer exist in the restoration area such that they must be physically transported to the location(s). There are many examples of captive breeding and reintroduction in the field of animal ecology that have direct bearing on restoration.

Population Concepts and Habitat Restoration

The traditional definition of a *population* is a collection of organisms of the same species that interbreed. Other, broader, definitions are also in use, such as that of Mills (2007, 4), where a population is considered a collection of individuals of a species in a defined area that might or might not breed with other groups of that species elsewhere. As noted by Mills, his definition is by design rather vague to allow discussion of population-related issues in a broad sense. Development of a rigorous concept of a population is inhibited by many factors, including barriers to dispersal (e.g., water bodies, mountains, roads), and patchiness of resources prevents complete mixing of individuals and leads to heterogeneous distributions of individuals of a species. Therefore, the concept of the biological population is complicated by numerous factors, and these factors directly impact our ability to restore wildlife populations. A thorough understanding of how the distribution and abundance of a species of interest fluctuates across space and time is a core component of restoration.

More specific terminology has been developed to help describe how animals are distributed and how they interact through breeding. Individuals of a species that have a high likelihood of interbreeding are called a *deme*. The term *subpopulation* is used to refer variously to a deme or to a portion of a population in a specific geographic location or as delineated by nonbiological criteria (for example, administrative or political boundaries). Partial isolation of individuals and degrees of isolation among populations can result in *metapopulation* structures, which occur when a species whose range is composed of more or less geographically isolated patches, interconnected through patterns of gene flow, extinction, and recolonization and has been termed a population of populations (Levins 1970, 105; Lande and Barrowclough 1987, 106). These component populations have also been referred to as subpopulations. Metapopulations occur when environmental conditions and species characteristics provide for less than a complete interchange of reproductive individuals, and there is greater demographic and reproductive interaction among individuals within, rather than among, subpopulations. Metapopulation structure occurs frequently in wild animal populations. Metapopulations are linked by a multitude of factors, including dispersal and migration, habitat conditions, genetics, and behavior (Hanski 1996). Figure 3.1 depicts the multitude of interacting factors that hold metapopulations

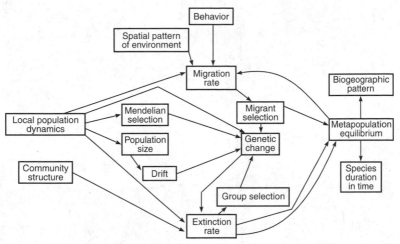

FIGURE 3.1. Factors associated with metapopulation structure and dynamics. (From Levins 1970, 75–107)

together, which obviously makes understanding of the likely population structure and dynamics of species of interest for restoration an essential core component of any plan.

As noted in the preceding paragraph, the structure of a population is a critical element of every restoration plan. If, for example, one or more of the species targeted for restoration are structured as metapopulations, your plan needs to consider how far apart ample areas of habitat can be located so that dispersal among subpopulations can occur. If areas of habitat are too far apart or cannot be located by the species, then extinction of the species within one location (subpopulation) might be permanent because there is no suitable area close enough to allow dispersal and recolonization. For example, Marsh and Trenham (2001) noted that ponds should be considered habitat patches—"ponds-as-patches"—when you are planning for regional conservation of amphibians. Consideration of population structure is a fundamental principle of animal ecology, which renders it a fundamental principle of restoration. Maschinski (2006) discussed how the spatial arrangement of patches will largely determine the success of a restoration project. Although clustering of patches can benefit species recovery due to increased dispersal opportunities, there is the possibility that clustered patches will enhance the likelihood of the spread of disease, predators, and other factors. Maschinski emphasized the need to carefully consider the biology of individual species when planning the spatial distribution of patches.

Each subpopulation within a metapopulation may vary in abundance of animals. See, for example, the diagrammatic but realistic representation of a

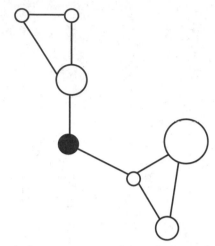

FIGURE 3.2. Hypothetical arrangement and size of subpopulations (circles) in a metapopulation. Large circle indicates a greater number of animals. Lines represent dispersal routes. The subpopulation indicated with a solid circle would be a key link between other subpopulations, and a top-priority site for restoration.

metapopulation in figure 3.2. Note the links between subpopulations: the loss of a subpopulation located between other subpopulations could lead to further extinctions because the linkage would be broken. Additionally, Figure 3.2 reminds us that habitat patches vary in both size and quality, which in turn drives the abundance of animals located therein. Thus, it is not only the identification of the metapopulation structure that is needed, but also an estimate of the abundance of individuals within each subpopulation. In restoration planning one goal would be to identify the metapopulation structure and then locate restoration sites to enhance this structure through, for example, promoting dispersal of individuals between subpopulations (i.e., sometimes called *stepping stone* habitat patches).

The rate at which animals disperse between subpopulations is related to the distance they must travel. The greater the distance between appropriate patches of habitat, the more difficulty an individual will have in locating the patch; it might die before finding suitable habitat. Information on dispersal abilities is thus another critical component you will need in planning for animal restoration. A map of the distribution of bighorn sheep in southeastern California, (figure 3.3) illustrates the complex metapopulation structure characteristic of many wildlife animals. Note that the loss of linkages between subpopulations caused the extirpation of some subpopulations. Here, identifying formerly occupied locations is a first step in prioritizing restoration

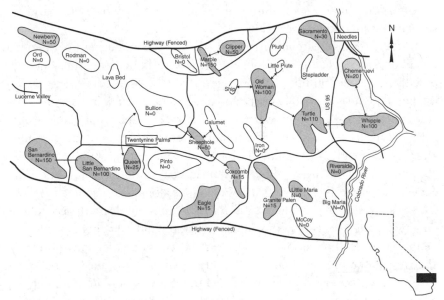

FIGURE 3.3. Map of a metapopulation of bighorn sheep in southeastern California. Stippled mountain ranges had resident populations of the size indicated. Mountain ranges with *N* = 0 are extripated populations; ranges with no *N* value are not known to have a resident population. Arrows indicate intermountain movements by sheep. (From Bleich et al. 1990, figure 1)

efforts, followed by determining the factors responsible for the loss of linkage (or direct loss of the subpopulation).

Population Dynamics and Viability

Population viability is the likelihood of persistence of well-distributed populations to a specified future time period, typically a century or longer. A well-distributed population refers to the need to ensure that individuals can freely interact, such as within a functioning metapopulation structure. The time span over which viability should be assessed should be scaled according to the species' life history, body size, and longevity, and especially population generation time. In population viability modeling, *thresholds* are conditions of the environment that, when changed slightly past particular values, cause populations to crash (Lande 1987; Soulé 1980). Such threshold conditions have given rise to the concept of *minimum viable populations* (MVP) (Lacava and Hughes 1984; Gilpin and Soulé 1986). An MVP is the smallest size population (typically measured in absolute number of organisms rather than density or distribution of organisms) that can sustain itself over time, and below which extinction is inevitable.

In the 1980s, researchers modeled MVPs by considering only genetic conditions of inbreeding depression and genetic drift. In theory, an MVP supposedly is of a size (number of breeding individuals, assumedly) below which the population is doomed to extinction and at above which it is secure. One guideline that was proposed was the *50–500 rule* whereby populations of at least 50 breeding individuals be maintained for ensuring short-term viability and 500 for long-term viability (Gilpin and Soulé 1986; Mills 2007, 250). However such guidelines seldom pertain to real-world situations and were devised based mostly on genetic considerations alone. Morrison et al. (2006, 69, 73) suggested that a rule of thumb may be to use at least 10 generations for gauging lag effects of demographic dynamics and 50 generations for genetic dynamics. Longer time spans can be considered if environmental changes can be predicted beyond that time period. Thus, for population of parrots with a generation time measured perhaps on the order of a decade, population viability should be projected over a century for demographic factors and five centuries for genetic factors, whereas for a population of voles—with more rapid reproduction, shorter life spans, and far shorter generation times—it may be projected only over a few years (Morrison et al. 2006, 73).

Population demography is the proximate expression of a host of factors that influence individual fitness and population viability (Morrison et al. 2006, 79). Vital rates may vary substantially in space and time as a function of food quality, weather, imbalance in sex ratios, and other factors. In some cases, populations respond to weather conditions, such as harsh winters, with lag effects measured in seasons or years. Falk et al. (2006), Morrison et al. (2006, 81–87), and Allendorf and Luikart (2007) review the influence of population genetics on conservation; we discuss the role of genetics in captive breeding later in this chapter).

Viable population management goals often specify providing for large interconnected wildlife populations and maintaining diverse gene pools. Rather, management goals should be ordered to first understand the natural conditions of a species in the wild. There are cases where populations are fully or partially isolated under natural conditions, and artificially inducing outbreeding among isolates, such as through captive breeding programs or manipulation of habitats, may contradict natural conditions.

As summarized by Smallwood (2001), demography has been deemphasized in conservation planning most likely because density—defined as the number of individuals per unit area—has emerged as the key term to quantify relative success of species across space and time. Density does not, however,

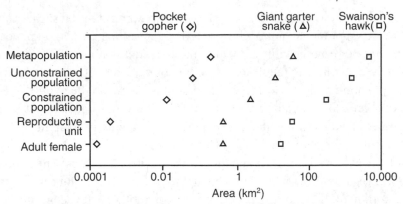

FIGURE 3.4. The spatial areas of habitat needed to support functionally significant demographic units can be plotted for each animal species. (From Smallwood 2001, figure 3)

provide information on the performance of animals in a particular area; it does not tell us if animals are successfully breeding or even surviving long-term in the area of interest. Smallwood (2001) showed that the areas needed to support viable demographic units—from adult female to metapopulation— change as we increase spatial area. In figure 3.4 we see that area determines the type of functionally significant unit we can expect to harbor in an area of given size; and, that different species require areas of substantially different size. In an area of a given size, we can expect to have a few breeding individuals of some species, constrained and unconstrained populations of other species, and full metapopulations for other species (here, constrained populations are those isolated in habitat fragments).

The distribution, abundance, and dynamics of a population in a landscape is thus influenced by species attributes, habitat attributes, and other factors. Species attributes include movement and dispersal patterns, habitat specialization, demography including density-dependence relations, and genetics of the populations. Habitat attributes include quality, size, spacing, connectivity, and fragmentation of habitat patches and the resulting availability and distribution of food, water, and cover. Other factors include a host of environmental conditions, such as weather, hunting pressure, and influences from other species (Morrison et al. 2006, 89–90).

Because of metapopulation structure, not all suitable habitats will be occupied at any one time. This in turn suggests that to conserve habitats, even if they seem unoccupied in any one year, monitoring wildlife use of habitats should proceed for more many years. Concluding absence of a species where it is actually present (see chapter 8) can be corrected with adequate sample size and monitoring duration to increase the power of the statistical evaluation. The appropriate approach in designing a restoration plan depends in part on the size and fragility of the population and its habitats, and on project objectives. Identifying the structure of the population(s) of interest is critical if a restoration project is to successfully establish or retain the population.

It is generally not advisable to combine observations of habitat-use patterns of individuals from different ecotypes, populations, geographic areas, or ecoregions (Ruggiero et al. 1988). Combining individuals in this way could mask the gamma diversity of habitat selection patterns that vary substantially across geographic areas. Providing habitat for one ecotype or for organisms in one portion of their range may be senseless for another ecotype or range portion if the organisms there select for different environmental cues (see the example of woodland caribou cited later in this chapter).

Distribution Patterns of Populations

The overall distribution and local abundance of many wildlife species are related in time and space. Many species have a *bull's-eye distribution* with their greatest areas of abundance toward the middle of their overall ranges and peripheral portions of the range in marginal conditions. Such distributions typically reflect several aspects of biophysical conditions and species ecology: (1) the geographic range of suitable biophysical conditions, (2) the range of tolerance of biophysical characteristics by the species, and (3) the occurrence of marginal-suitable conditions at the peripheries of the geographic range that often acts as sink habitat to hold nonreproductive individuals or individuals spread from higher-abundance areas during good reproductive years. Figure 3.5 depicts the situation where the abundance of a species declines toward the periphery of its range. Also note in figure 3.5, however, that our ability to identify how a species reacts to environmental conditions can improve as we move toward the edge of its range (i.e., the change in abundance along an environmental gradient). Brown et al. (1995) concluded that patterns of spatial and temporal variation in abundance should be factors considered when designing nature reserves and conserving biological diversity.

Identifying areas of high density of organisms or areas of high environmental suitability (remembering that these are not necessarily always synonymous) can be important for management purposes. Wolf et al. (1996) found that release of organisms into the core of historical range, and into habitat of high quality, were two key factors contributing to success. Other factors included use of native game species, greater number of released animals, and

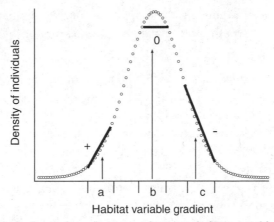

FIGURE 3.5. Limited sampling along a gradient in a habitat variable may produce a positive (a), nonexistent (b), or negative (c) correlation with a species response variable such as density. (From Van Horne 2002, figure 24.2)

an omnivorous diet. It also should not be forgotten that many factors other than those listed here also contribute to population viability.

Here again the restorationist can improve the probability of success by surveying the literature and seeking expert opinion concerning species' overall distribution relative to the restoration site(s). For example, is the restoration site near the center of the range or near the edge of range of the species? If the site is near the center, colonization of the site might be enhanced because immigration could occur from several directions and at relatively high frequency depending on the species and distances involved. But if the site is near the periphery of the range, colonization might be problematic, and overall environmental conditions might be harsh relative to locations near the center of the range (as discussed earlier).

Animal Movements and Habitat Management

Movements of wildlife through their habitats and environments impart particular dynamics to their populations. Movements particularly important for habitat management include the following:

- *Dispersal*: the one-way movement, typically of young away from natal areas
- *Migration*: a seasonal, cyclic movement typically across latitudes or elevations to track resources or to escape harsh conditions changing by season of year

- *Home range*: movement throughout a more or less definable and known space over the course of a day to weeks or months, to locate resources

The use of such a movement classification system can aid restoration planning and habitat management by determining (1) which species are likely to occur in an area in a given season, and thus the resources and habitats required during that season; (2) the number of species expected in an area over seasons, and thus the collective resources and habitats required; and (3) the need for considering habitat conservation in other regions beyond the area of immediate interest. It may also aid in identifying habitat corridors used during movement, and thus habitats and geographic areas needing potential conservation focus (see chapter 7). Morrison et al. (2006, 93–100) reviewed the influence of animal movements on habitat management.

Information on dispersal has direct applications to planning for restoration of animal habitat and ultimately the individual animals. For example, we are in the fourteenth year (in 2008) of a study of the willow flycatcher in the Sierra Nevada, with an overall objective of reversing the decline of this meadow-inhabiting songbird. Two initial objectives were to determine the direction and distance that fledglings dispersed from natal areas, and how frequently adults changed breeding locations. Data on dispersal distances and locations provided us with information on where to prioritize restoration of meadows. Figure 3.6 depicts part of the information on dispersing birds that we have been gathering. Although the average distance between natal site and initial breeding site was about 8 km, we did identify several birds that dispersed about 135 km. Meadows that were not occupied by the flycatcher and were well outside the usual dispersal distance would not be a priority for restoration. Referring back to figure 3.2 as an example of this strategy, we could prioritize meadows that helped link or hold together a metapopulation structure.

Migration has implications for restoration in several ways. Animals must find food, water, cover, and other conditions that allow them to survive during migration. Although the specific paths taken by animals during migration are in general poorly known, researchers are continually adding to our knowledge on the sites and resources needed by migrating animals. Ungulates and songbirds have been fairly well studied. Songbirds usually migrate at night and feed and rest during the day. Thus, songbirds require what are termed migration *stop-over sites*. Such stop-over sites thus become an obvious target for identification, protection, and perhaps restoration. In the desert Southwest, stop-over sites are conspicuous because they usually involve riparian areas and isolated springs, as depicted in figure 3.7.

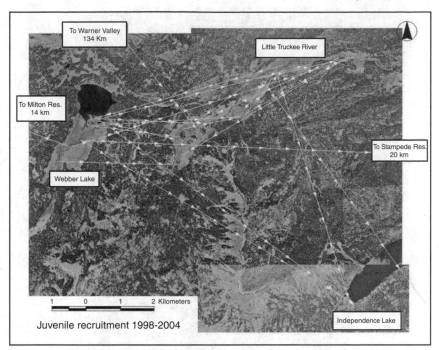

FIGURE 3.6. Example of dispersal in willow flycatchers in the Sierra Nevada, California. Lines represent straight line distances moved (dispersed) by birds born in one year and returning to breed in the following year (recruited into breeding population). (Unpublished data)

How organisms establish and use home ranges is important for managing habitat for populations. The term *home range* has been used in many ways in the literature (Powell 2000; Fuller et al. 2005). Researchers have typically followed animals for periods of time (say, through the use of marking or telemetry) and calculated the gross area used, by connecting outlying locations of the animal. However, it is unlikely that the animal used all of the area with equal frequency. Thus, determining the area actually utilized by the animal helps clarify the specific resources needed by an animal during specific periods of time; this utilized area is depicted in figure 3.8. The implications are clear for the restorationist: detailed information on what animals are actually using must be gathered to be able to specify the vegetation and special habitat features needed.

The area being used by an individual animal naturally varies through time. Thus, the home range of an animal must be placed in a temporal context. That is, are we discussing the area used within a day, week, month, season, or lifetime? All such definitions of home range might have meaning

FIGURE 3.7. Some vegetative types, such as riparian areas and this spring, can be of high value to animals relative to more ubiquitous vegetative types.

depending on the objectives of the study. Home range has ecological meaning when we are able to determine why the animal is using a particular area.

Stochastic Environments and Habitat Management

Populations can respond to a perturbation either functionally or numerically. A *functional response* of a population refers to changes in behavior of organisms, such as selecting different prey or using different substrates for resting or reproduction. Functional responses can also entail a temporary and localized increase in numbers resulting from immigration, or a decrease from emigration. A *numerical response* of a population refers to absolute changes in abundance of individuals through changes in recruitment. A disturbance of a habitat might elicit one or both kinds of responses (functional or numeric).

It is important to distinguish between such responses to understand if management activities, especially intentional habitat restoration or en-

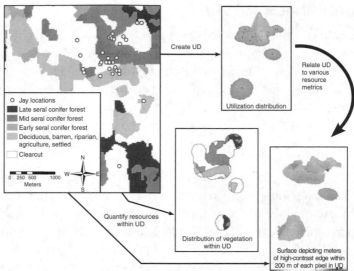

FIGURE 3.8. Calculation of a resource utilization function for a single Steller's jay. First, the jay's location estimates (upper left) are converted into a three-dimensional utilization distribution (UD; upper right) using a fixed-kernel home range estimator. The height of the UD indicates the relative probability of use within the home range. Greater heights indicate areas of greater use, as inferred from regions of concentrated location estimates. Second, resource attributes are derived from resource maps within the area covered by the UD: for example, the high-contrast edge at interfaces between late-seral forest and clear-cuts or urban areas (lower right) and a categorical resource measure of vegetative land cover (lower left) at each grid cell center within the area of the UD. The height of the UD (relative use × 100) is then related to these local (e.g., vegetation cover; lower left) and landscape (e.g., contrast-weighted edge density; lower right) attributes on a cell-by-cell basis. (From Marzluff et al. 2004, figure 1)

hancement activities, are truly serving to increase absolute population size or are simply redistributing organisms. In some cases, simple redistribution of organisms may be the goal, such as warding off foraging waterfowl from grain fields and agricultural lands and into nearby wetland refuges. In other cases, redistributions and local increases, such as displacement from disturbed or fragmented habitats, may obscure an overall population decline. The message for restoration is simple: monitoring of restoration activities is an essential component of determining *what* happened following restoration and, for both current and future plans, *why* the changes (or lack thereof) occurred. I cover the essentials of study design and monitoring in detail in chapters 8 and 9.

Linking Populations and Restoration Ecology

I will now discuss the interrelated concepts of multiple isolated populations, metapopulations, and source-sink dynamics as they link with restoration ecology. This section summarizes the concepts introduced earlier in a manner that simplifies the relationships and points us toward applications for restoration.

Multiple Isolated Populations

Populations of a species can be distributed in several ways across the landscape, including those arranged into isolated groups and those linked through movements between groups. For multiple isolated populations, each population experiences a susceptibility to extinction without the possibility of natural recolonization (Mills 2007, 211). As developed in chapter 4, individuals often aggregate—both conspecifically and heterospecifically—into locations that result in nonuse of suitable habitat. From a practical standpoint, we need to know if the separate populations are all likely to be impacted by the same environmental events, especially those that could cause a catastrophic decline in numbers, such as a hurricane, fire, or disease. If the fates of extinction of isolated populations are independent, then the probability of simultaneous extinction of all populations would be much less than that of any single population. In contrast, if the fates of the populations are completely correlated, the probability of total extinction would be the same as for one population (Mills 2007, 212). Thus, from a restoration perspective, understanding the distribution of populations of a species in space, the internal dynamics of each population, and if movement between populations is occurring, are all core aspects of designing a plan for conserving a species.

Metapopulations

The initial concept of a metapopulation structure has revised and expanded as researchers have conducted field investigations of animal movements and genetics of subpopulations. The key contribution of the metapopulation concept to applied wildlife population biology is a better understanding of the importance of animal movements—dispersal—in management and conservation. Studies of dispersal have shown that it is not only the area that animals live in for extended periods of time (i.e., a biological season such as winter or summer) that are critical for survival, but also the areas that animals must travel through to reach areas of seasonal use that must be identified, studied, and allowed for to ensure survival. Mills (2007) summarized several cases in which dispersal was identified and incorporated into management plans, including situations in which ensuring connectivity through corridor protection would enhance persistence; I discuss corridors and related topics in chapter 7.

Source-Sink

Related to the concept of metapopulations is the concept of source-sink dynamics. Pulliam (1988) was responsible for describing how some populations provide an excess of offspring (sources) into the environment relative to other populations who serve as a drain (sinks) on the system. As explained by Mills (2007, 214), the source-sink can be considered a metaphor in which some populations are strong contributors (sources) to the metapopulation while others are drains on the system (sinks). The crucial key to define sources and sinks for wildlife population biology is interpretation of population processes. As noted by Mills (2007, 214), you cannot use the relative abundance or density of a population to define a source because abundances vary across habitat in different seasons, and because poor (i.e., low-quality) habitat might have high numbers of animals because subordinate individuals are forced into them. Likewise, you cannot assume that a large area is a source whereas a smaller area is a sink. The source-sink concept is not restricted to metapopulation structures, however, because in reality all animals are distributed in a non-random fashion across space. Variations in habitat quality in a seemingly (to the human observer) homogeneous area can generate source-sink dynamics.

Restoration of a population determined to be a sink will depend on knowing what causes intrinsic vital rates to be low and why immigrants keep going into the area. The concept of the ecological trap applies here, where an *ecological trap* is a location that appears to be suitable to an animal because reliable environmental or behavioral cues are mismatched with the actual consequences for the individual's fitness. Ecological traps represent, in essence, an evolutionary lag between the quality of a location and the cues that lead to its preference, thus causing animals to make mistakes in choosing a habitat. For example, an individual could choose a location based on a composition of plant species that predicts future food abundance, but be unaware that nearby urbanization has allowed an exotic predator to now also inhabit the area. Or, an individual would be unaware that a new recreational site, which is now seemingly benign, will be swarming with people midway through its breeding cycle.

Thus, by definition an ecological trap actually contains conditions or resources favorable to an individual. Ameliorating the negative effects of a trap can thus be achieved by increasing the quality of the trap so that it no longer serves as a sink. Alternatively, decreasing the trap's attractiveness would ensure that an animal does not settle in the location (Mills 2007, 217). For example, if hawks are attracted to flooded fields because of enhanced rodent availability and use electrical distribution poles bordering the field for perching, but are in turn frequently electrocuted by the wires on the poles, modification of the cross arms of the poles could be done to prevent electrocution; or, a different agricultural practice could be used near the poles so that hawks would not be attracted to hunt there.

Van Andel (2006) summarized the concepts of remnant, source-sink, and metapopulation dynamics in relation to restoration ecology. As graphically depicted in figure 3.9, in remnant systems individuals survive long enough to bridge periods of unfavorable conditions (e.g., they can live but not reproduce or reproduce poorly). Individuals in sink habitats are maintained by continuous immigration from source habitats. The more complicated movement of individuals within a metapopulation dynamic is a reflection of frequent movements due to changing source-sink conditions among the individual populations.

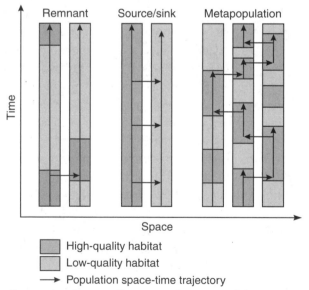

FIGURE 3.9. Space-time diagram illustrating the principal difference between three types of regional dynamics: remnant, source-sink, and metapopulation. (From Van Andel 2006, figure 6.1)

Exotic Species

Intrusion of natural environments by exotic species has become a major challenge for habitat conservation and restoration (Coblentz 1990; OTA 1993; Soulé 1990). Exotic species come in all taxonomic groups and all environments. Exotic plants can disturb native ungulate use of rangelands (Trammell and Butler 1995) and severely handicap management of natural conditions in parks (Tyser and Worley 1992). Exotic game bird and big-game introductions can affect distribution of native plants and animals (OTA 1993).

However, it is not always clear which species are exotic (nonindigenous) and which are native. A species is usually considered native if it is located in its presumed area of evolutionary origin and nonhuman-aided dispersal (Willis and Birks 2006). There are various cases of range expansions that may be natural, that may have been induced or enhanced by human alteration of environments, or that began as a minor introduction by humans. An example of a complicated invasion of North America is the cattle egret (*Bubulcus ibis*), which spread from Africa to South America about 1880, reached Florida and Texas in the 1940s and 1950s, and rapidly expanded north and west in North America (Ehrlich et al. 1988). The brown-headed cowbird (*Molothrus ater*), which apparently evolved in the Great Plains, has spread throughout most of North America during the 1900s, a movement likely a result of clearing of forests and agricultural development (e.g., see Morrison et al. 1999).

Exotic escapee species that have spread throughout the continent include the European starling (*Sturna vulgaris*); after two unsuccessful introductions, 60 birds were released into New York's Central Park in 1890, and within 60 years they spread to the Pacific and have since outcompeted and threatened many other bird species (Ehrlich et al. 1988). The house sparrow (*Passer domesticus*), also introduced from Europe, is another example of a species that has spread throughout North America by adapting to regional environmental conditions. Both the starling and house sparrow negatively impact native birds by usurping nest sites. Although confined primarily to urban areas, numerous species of parrots have now become established as breeding residents in many major cities throughout the world. And everyone is familiar with the nonnative house mouse and the several species of rats that have spread across North America. Morrison et al. (1994a) showed that the house mouse (*Mus musculus*), for example, was a dominant species of rodent in disturbed areas in southern California that were scheduled for restoration.

Restoration activities must assess the potential impact that exotic species, as well as native species whose range has expanded due to human-related activities (e.g., cowbird), will have on desired species. The presence of exotics can easily negate efforts to maintain or establish native species following restoration. For example, starlings and house sparrows can occupy the majority of nesting cavities that were created by restoration activities for native species (e.g., bluebirds, woodpeckers). Likewise, cowbirds are a major contributing factor to nesting failure in native songbirds, including in locations undergoing restoration activities focusing on endangered species (e.g., Kus 1999; Kostecke et al. 2005). Restoration plans that fail to thoroughly access existing wildlife and evaluate the likely response of wildlife to restoration could easily enhance populations of exotic and other nondesired species.

As summarized by Vander Zanden et al. (2006), the conventional wisdom suggesting that exotic species should be controlled during restoration does not always apply. There are situations in which an endangered, native species has come to rely on an exotic species. For example, the exotic plant salt cedar (*Tamarix* spp.) has become established in many riparian areas of the Southwest because of human-caused changes in hydrology due to channelization; native willow-cottonwood woodland has been substantially reduced in quality and coverage. The endangered southwestern willow flycatcher (*Empidonax traillii extimus*) now nests successfully in salt cedar. Removal of salt cedar would thus eliminate nesting and foraging locations for the flycatcher; restoration of native vegetation is extremely difficult because of inadequate hydrological conditions and would be unlikely to provide adequate nesting substrates. Thus, management of exotics must be considered in the overall context of the goals of restoration.

Roads to Recovery: Captive Breeding and Translocating Animals

Restoration of habitat and creation of other conditions that promote occupancy by desired species (e.g., reduction or elimination of exotic species) establish the foundation of restoration activities. Recovery of an animal population often requires, however, actions that directly involve breeding and movement of individual animals to one or more locations. As summarized by Falk et al. (2006), the focus of ecological restoration can be toward introduction, reintroduction, or augmentation of populations; table 3.1 presents a useful breakdown of the types of restoration and the sources for individuals that will be used in the restoration. I will follow their terminology herein.

Allendorf and Luikart (2007, 452) used the term *conservation breeding* to describe the general efforts to manage the breeding of plant and animal species that do not strictly involve captivity, such as moving (relocating) animals to a new location to avoid predators or disease. There is, however, some variation in the use of terminology associated with conservation breeding; for ex-

TABLE 3.1

Term	Definition
Type of restoration	
Introduction	Species or genotypes not presently at the project site, and not known to have existed there previously, are established at a site. Species may or may not be native to broader geographic area.
Reintroduction	Reestablishment of species or genotypes not presently at the project site, but that did occur there in the past (population was extirpated and reestablished).
Augmentation	Individuals of a species are added to a site where the species occurs presently. Also called *restocking*.
Type of restoration material	
Resident	Species, populations, or genotypes native to a local site. These can be extracted from a local site for onsite restoration or augmentation.
Translocated	Genotypes collected offsite for planting or release at a project site within the natural range of the species. Differs from usage in Gordon (1994)
Introduced	Species, populations, or genotypes collected offsite and introduced to a project site outside their historical range.

Note: Restorative materials can be native to a project site or brought in from elsewhere. If a species is not native to a project site, genetic appropriateness of the plant material can differ, compared to when a species is resident or connected to nearby resident populations by gene flow. Introduction, reintroduction, and augmentation may involve both rare and common species.
Source: Falk et al. 2006.

ample, the term *supplemental translocation* of Allendorf and Luikart (2007, 472) is associated with augmentation. In summary, *translocation* is the physical movement of individuals from one population to another population. Translocations can occur as an augmentation (or supplementation) into an existing population, as a reintroduction into an area where the species previously existed, or as an introduction to an area not previously occupied by the species (Mills 2007, 219).

Allendorf and Lukart (2007, 452–53) thought that there were three primary roles for conservation breeding as part of a recovery or management program:

- To provide demographic and genetic support for wild populations
- To establish sources for founding new populations in the wild
- To prevent extinction of species that have no immediate chance of survival in the wild

In this section I review the use of captive breeding and translocation in restoring animal populations. These techniques have been or are currently being used to assist with restoration of numerous rare and endangered species throughout the world. The goal of this section is to introduce readers to the major topics necessary to understanding restoration programs that incorporate captive breeding and translocation.

As developed earlier in this chapter, metapopulation biology has direct application to the design and management of restoration projects. Metapopulation dynamics include local population extinction, local population establishment or reestablishment, and movement or linkage among the various local populations. Metapopulation management, if properly applied, can decrease the probability of permanent extinction in local populations and help maintain genetic variability. Captive breeding facilities can maintain essentially fragmented populations. Thus, metapopulation dynamics has direct application to captive breeding and restoration planning (Bowles and Whelan 1994).

Captive breeding and translocation are not the ideal means of achieving recovery of rare populations. Captive breeding is expensive and translocation is problematic. Factors such as rates of gene flow among subpopulations, effective population size, mutation rates, and social structure must all be considered when planning restoration and translocation. Small populations might have been subject to past population *bottlenecks*, and genetic manipulation could be required to recover declining populations or to restore or maintain evolutionary potential. Maintenance of genetic variation and evolutionary potential are concerns for either rare or isolated populations and in captive populations. Captive breeding can lead to loss of genetic variation through random drift, genetic adaptation through selection to the captive environment, and thus inadequate genetic adaptation for reintroduction to a restored location (Bowles and Whelan 1994). Allendorf and Luikart (2007, 453–54) also discussed the difficulties and challenges associated with captive breeding, and noted that although captive breeding is often recommended as part of a recovery strategy, sufficient resources are seldom available to actually implement a viable program.

Ramey et al. (2000) provided five key issues that need to be addressed when considering the augmentation of populations:

• Are there two lines of evidence (e.g., genetic, demographic, and/or behavioral) that support the hypothesis that a severe population bottleneck has occurred?
• Would the introduction of additional animals degrade resource conditions, driving the wild animals to a more rapid extinction?
• Was the population bottleneck due to a disease outbreak (or other specific and known occurrence), and can the source of the problem be eliminated?

- Are there habitat patches nearby to establish a population (or metapopulation) of larger size, rather than a single, isolated population?
- How should the sex and age composition of an augmentation be structured?

Balmford et al. (1996; see also Allendorf and Luikart 2007, 454) also offered and discussed the economic and biological conditions to consider when selecting animal species for captive breeding, and Van Wieren (2006) presented a succinct description of issues and opportunities for reintroductions in ecological restoration (these references can be reviewed for details beyond the scope of this chapter). Careful consideration of all of these questions needs to precede initiation of an animal restoration project. Additionally, there is no reason to proceed with reintroduction if habitat and niche conditions are not appropriate.

An example of restoring a rare species is the program developed by The Peregrine Fund to establish a self-sustaining population of Aplomado falcons (*Falco femoralis*) in the southwestern United States and northern Mexico through captive breeding, release, and management, with the ultimate goal of removing the species from the Endangered Species List. Aplomado falcons were once widespread in the American Southwest, but their known northern range has become restricted to eastern coastal Mexico and a few other areas in that country. Changes in vegetation following the Spanish invasion and the grazing excesses of the late 1800s played important roles in the decline of the species. The last known pairs in the United States were near Brownsville, Texas, in 1946, and Deming, New Mexico, in 1952. Full-scale restoration began in the 1990s when The Peregrine Fund began breeding Aplomado falcons in captivity and releasing them in southern Texas. By the early 2000s, a wild-breeding population of about 40 territorial pairs had arisen from these efforts. On Matagorda Island, Texas, the falcons had to nest in low bushes because of the lack of suitable trees, which made them vulnerable to ground predators such as raccoons. The Peregrine Fund solved this problem by supplying each of 13 falcon territories with a nest box, set upon a pole. The falcons used these artificial nest structures, and their annual nest success increased. The use of such artificial structures is a good example of how restoration of populations often requires direct intervention.

As has been the case for numerous projects that seek to restore rare songbirds, waterfowl, and other species, direct actions also include the removal of native and exotic predators. The practice of removal of cowbirds to enhance songbird populations is controversial because, in the absence of changes in land use (e.g., livestock grazing) and substantial habitat improvement, cowbird removal likely has no ending date.

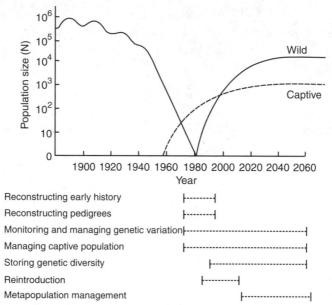

FIGURE 3.10. Chronology of a captive breeding and restoration program. (From Mace et al. 1992, figure 1)

Captive breeding and translocation has been successful, and is sometimes the only alternative to extinction in the wild. A restoration and conservation program that includes captive breeding and translocation involves numerous overlapping steps (figure 3.10). I will review many of the issues that should be considered, and the predominant techniques that can be used, when planning for captive breeding and translocation. Lacy (1994) provided a thorough review of genetic considerations, which I incorporate into my summary.

The goal of captive breeding programs is to support survival of the species, subspecies, or other defined unit in the wild. Meeting this goal requires (deBoer 1992) the following:

- Breeding and management of highly endangered taxa, with prescribed levels of genetic diversity and demographic stability, for defined periods of time, to prevent extinction.
- Using such captive programs as part of conservation strategies that manage captive and wild populations so as to ensure survival of these taxa in the wild. This requires using captive populations to reestablish, reinforce, or recreate wild populations.
- Developing self-sustaining captive populations of rare or endangered taxa for education programs that can be beneficial to survival of conspecifics in the wild.

Regardless of the species, the prerequisites to meet these goals are, as follows: At the level of the individual, sufficient longevity and physical, physiological, and psychological well-being should be assured in the captive situation. This involves species-specific zootechnical, medical, and biological knowledge and research. At the level of breeding pairs and groups, sufficient reproduction should be assured to guarantee continuity over generations. This entails species-specific knowledge and research on reproductive biology, ethology, and related topics. And, at the level of the population, the preservation of a genetic population structure that resembles the wild one as closely as possible should be assured (deBoer 1992).

In the following sections I will describe in detail many of the issues that are involved with restoring animal populations. The International Union for the Conservation of Nature (IUCN) (1998) outlined general guidelines for reintroductions of plants and animals, including an initial feasibility study and background research, evaluation of reintroduction sites, availability of suitable release stock, release of captive stock, and postrelease activities. Guidelines are also available from the Association of Zoos and Aquariums (AZA; http://www.aza.org/AboutAZA/reintroduction).

Genetic Considerations

Falk et al. (2006) summarized the central importance that consideration of genetics has in restoration ecology. Because genetic variation largely determines the ability of populations to persist through changing environments, the magnitude and pattern of adaptive variation across populations of a species becomes a consideration in restoration of animals. Severe genetic problems (e.g., inbreeding, bottlenecks) can occur as a result of conservation actions, including captive breeding and translocation. These problems might not be readily apparent in the timescale of management programs because few generations of vertebrates will have elapsed since the implementation of the program (Ramey et al. 2000).

There are two general types of genetic changes that can occur in a captive population having ramifications for restoration. First, selection can eliminate alleles that are maladaptive in the captive situation yet important for survival in the wild. Second, random *genetic drift* can cause the cumulative loss of both adaptive and maladaptive alleles. The primary problem in captive breeding is that each successive generation is a sample of the previous generation. Thus, the gene pool of the population that will eventually be released is invariably changed through captive generations. Rare alleles are especially susceptible to loss through genetic drift.

As noted by Falk et al. (2006), in a restoration context it is critical to distinguish between what they termed the *census population*, defined as the number of individuals that occur in the population, from the *effective population size* (N_e) (Wright 1931), defined as the number of individuals that contribute genes to succeeding generations. Because not all individuals in a population usually breed, N_e is usually smaller than the total population size and can vary through time based on both natural and human-induced factors. Thus, in all studies of animal ecology but especially those dealing with small population sizes, there is high value in determining the breeding status of individuals rather than just focusing on the number of animals present.

To preserve the genetics of animals that will be released, captive management must minimize adaptive and nonadaptive genetic changes. As discussed, the effective population size necessary to minimize genetic changes in captive populations has been the subject of much debate. Recall that the 50/500 rule called for >500 individuals for long-term programs. The rationale for the long-term criteria was to allow new mutations to restore heterozygosity and additive genetic variance as rapidly as possible as it is lost to random genetic drift. The concept of effective population size does, however, have several related meanings, including the number of individuals at which a genetically ideal population (one with random union of gametes) would drift at the rate of the observed population. The rate of genetic drift could be measured as the sampling variance of gene frequencies from parental to offspring generations. The primary goal of captive breeding is, however, to minimize all evolutionary (genetic) change, whether from random drift of selection to the captive environment. Another "rule" proposed that at least 90% of the genetic variation in the source (wild) population be maintained in the captive population (Lacy 1994). Ramey et al. (2000) suggested that it is justifiable to intervene when there has been a severe genetic bottleneck, which they defined as an effective population size of <10 individuals, and a lack of gene flow with other outbred populations. Some reintroduction efforts have been successful with as few as ~10 individuals.

Genetic variation includes many related concepts, including genetically determined variation in morphology or behavior, variation in chromosomal structure, molecular variation in genes, allelic diversity, heterozygosity of genes, and so forth. The phenotype of an organism determines its physical properties and fitness. Thus, quantitative genetic variation in phenotypes is of importance in designing genetic management for reintroduction. Much attention has focused, however, on theoretical models and prescriptions related to the management of underlying molecular genetic variation (i.e., the presence of multiple genetic variation within a population; the mean or expected diversity per individual).

Heterozygosity can refer to (1) diversity within individuals and (2) diversity among individuals in a population. The proportion of loci for which the average individual is heterozygous is usually termed *observed* heterozygosity. The probability that two homologous genes randomly drawn from a population are distinct alleles is termed the *expected* heterozygosity or *gene diversity*. The 90% retention of genetic diversity guideline mentioned earlier referred to expected heterozygosity, and many management programs have used expected heterozygosity as an index of genetic variability. The possibility of a population to adapt at all depends on the presence of sufficient variants, so allelic diversity might be critical to long-term persistence (Lacy 1994).

We cannot, of course, replicate past populations exactly, which means we must evaluate how similar the source population must be to the population we wish to augment or reintroduce. When augmenting a population we have genetic material available with which to evaluate heterogeneity in the genome, which allows some assessment of how much mixing of genetic material we wish to allow from perhaps multiple source populations. With reintroductions, little existing genetic material is likely available. Based on the assumption that populations near one another will be genetically similar because of similarities in biotic and abiotic conditions, the usual practice is to specify a geographic region within which the source animals should be gathered (Falk et al. 2006). In the case of remnant, isolated populations, an argument can be made for introducing additional genetic variability in an attempt to reduce potential problems associated with inbreeding and low genetic variability (where inbreeding is the mating between related individuals). In mixing genetic material from several populations, the idea is that biotic and abiotic conditions will determine what combinations of animals are most fit to survive into the future. Because restored populations usually represent a genetic subset of the available source population, some type of founder effect can be expected. A *founder effect* occurs when a few individuals drive the direction of genetic composition in a population. Thus, maximizing the genetic diversity of the source animals within an ecologically appropriate portion of the species range is an important consideration in planning animal restoration (see also Falk et al. 2006).

There are several reproductive technologies originally developed for agricultural applications that can be applied to conservation and restoration, including genome banking, cryopreservation, artificial insemination, and cloning (as reviewed by Allendorf and Luikart 2007, 457–8). *Genome banking,* which is the storage of sperm, ova, embryos, tissues, or DNA, can help improve genetic material without actually moving individual animals. For example, genome banking could allow you to manage gene flow into a population without actually relocating individuals, and also serve as insurance against population or even species extinction. As reported by Allendorf and Luikart (2007, 458), for example, the San Diego Zoo maintains samples from more than 7,000 species of endangered reptiles, birds, and mammals.

Cryopreservation is the freezing and storage of sperm, ova, embryos, or tissues to manage and safeguard against loss of genetic variation, and is the principal storage method for animal material. *Artificial insemination* is frequently used in captive breeding programs because it allows researchers to guide the mating of individual animals, and also reduces potentially unwanted behavioral interactions between mating animals. Typically, sperm is gathered from a desired individual, cryopreserved and moved to a captive breeding facility, and then used to artificially inseminate individuals.

Cloning is usually conducted by removing the nucleus from a donor egg cell of the animal that will carry the cloned embryo, and then injecting into the carrier's egg cell the nucleus from a cell of the animal to be cloned. Although cloning is controversial, the technology provides opportunities for protecting species and increasing genetic variation within species close to extinction. A disadvantage of cloning is, of course, that cloned individuals and populations thereof are generally identical and thus would be highly susceptible to the same infectious diseases and have low adaptive potential to environmental change. Allendorf and Luikart (2007, 458–9) questioned the value of cloning relative to the value of traditional captive breeding programs. Clearly, cloning is a technique that should at least be considered when developing an entire package of conservation measures for a species nearing extinction.

Captive Breeding

Lacy (1994) outlined three distinct phases in captive breeding:

1. First, a captive program is established with wild-caught animals; this is the founder phase.
2. Second, the captive population grows from the founders to the maximum size that can be supported by the program; this is the growth phase.
3. Finally, the population is maintained at the captive equivalent of carrying capacity, and animals above carrying capacity are available for reintroduction; this is the capacity phase.

Below I summarize key points in each of these phases.

FOUNDER PHASE

The goal of a captive breeding program is to mimic the genetic composition of the wild population. Unfortunately, most captive programs are begun as an effort of last resort after other plans (e.g., habitat restoration, removal of constraining factors such as predators) have failed. In such cases, much of the natural genetic variability has probably already been lost. Regardless of the size of the remaining wild population, attention should be given to taking a random genetic sample, rather than focusing on capturing animals that are relatively easy to obtain. However, sampling based on genetics is not necessarily equivalent to randomly sampling individuals from across the remaining geographic range. Knowledge of the structure of genetic variation in the wild would be necessary to sample the genetic variability adequately. For example, determining if recognized subspecies represent genetically separate evolutionary significant units would help guide the selection of individuals (Allendorf and Luikart 2007, 460). If such information is not available—which will often be the case—then sampling can be based on observational data on population dynamics and social structure. For example, if individuals are individually marked (e.g., color bands for birds, neck colors for large mammals), then data should be available on lineages, and efforts can be made to not oversample from a specific blood line. If animals are not individually marked or otherwise recognizable, then attempts should be made to avoid oversampling from individuals in isolated locations (i.e., unless dispersal of young is adequate, it would be likely that isolated groups of animals are closely related).

To obtain most of the heterozygosity present in a wild population requires moderate numbers of founders. For example, obtaining at least 20 unrelated (randomly sampled) founders yields ~97% of the expected heterozygosity of the wild population because little additional heterozygosity is added with additional founders (figure 3.11). Allendorf and Luikart (2007, 461) recommended a slightly higher value of at least 30, but preferably 50, founding individuals.

A lack of genetic variation can prevent the establishment of a captive population capable of providing genetically diverse animals for reintroduction, and thus impede recovery of the wild population. The probability that a particular allele will be included among the founder population, however, is determined by the frequency of the allele in the population. Very large numbers of founders might be required to give a high probability that an allele will be included in the founder population. As depicted in figure 3.12, a large number of founders could be required to achieve a high probability that rare alleles would be included captive population (Lacy 1994). Likewise, if the rate of

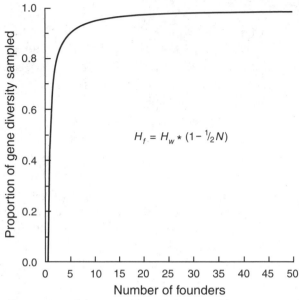

FIGURE 3.11. Proportion of the gene diversity (expected heterozygosity) present in a wild population (H_W) sampled in founders (H_f) drawn at random. (From Lacy 1994, figure 3.2)

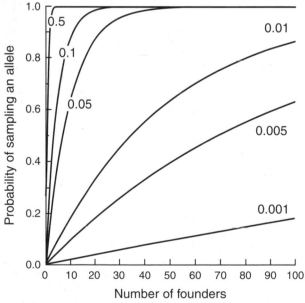

FIGURE 3.12. Probability that an allele will be sampled at least once among founders drawn at random for alleles of various frequencies (0.001, 0.005, 0.01, 0.1, and 0.5) in the wild (source) population. (From Lacy 1994, figure 3.1)

population growth is low, additional founders or subsequent supplementation with additional individuals would be advised. The most effective means of minimizing genetic drift and maximizing effective population size in a captive population is to equalize reproductive success among individuals, which is especially important in the founder phase (Allendorf and Luikart 2007, 461).

Pedigree analysis is generally defined as the genetic study of a multigenerational population with ancestral linkages that are known, knowable, or can be reasonably assumed or modeled. The emphasis is on examining the genetic structure imparted to the population by its pedigree relationships and evaluating the consequences of that structure on the long-term conservation of the population (Lacy et al. 1995). Available methods fall into three categories: (1) the analytical calculation of genotype probabilities in completely known pedigrees, (2) the simulation of possible pedigrees from known aspects of a population's structure, and (3) the determination of equations that broadly describe population genetic processes. A single gap in a pedigree can confound an analysis of genetically important individuals. Analyses can be restricted to those parts of the pedigrees that are completely known or to those populations with well-understood structure and dynamics. Additionally, assumptions can be made that allow an analysis of uncertain data. Often multiple methods can be applied to increase precision of results (Lacy et al. 1995).

Pedigree analysis is especially useful in captive breeding because it allows you to determine the level of inbreeding relative to some base population in which it is assumed that all individuals are unrelated to one another. The pedigree inbreeding coefficient is the expected increase in homozygosity for inbred individuals; it is also the expected decrease in heterozygosity throughout the genome. The inbreeding coefficient ranges from zero for noninbred individuals to one for totally inbred individuals (Allendorf and Luikart 2007, 307).

As developed by Allendorf and Luikart (2007, 308–9), an inbred individual may receive two copies of the same allele that was present in a common ancestor of its parents; such an individual is termed autozygous and is identical by descent at that locus. All autozygous individuals will be homozygous unless a mutation has occurred in one of the two copies descended from the ancestral allele in the base population. In contrast, allozygous individuals possess two alleles descended from different ancestral alleles in the base population (figure 3.13).

Growth Phase

Even if initial founder population is a random sample from the wild population, it is likely that some founders will have only a few descendents, and that

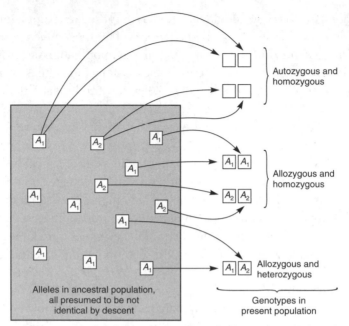

FIGURE 3.13. Patterns and definitions of genotypic relationships with pedigree in-breeding. Autozygous individuals in the present population contain two alleles that are identical by descent from a single gene in the ancestral population. In contrast, allozygous individuals contain two alleles from different genes in the ancestral population. (From Allendorf and Luikart 2007, figure 13.2)

specific alleles will be lost from the captive population. If founders produce disproportional numbers of offspring, then a conflict arises between the goal of balancing progeny and the goal of maximizing growth of the captive population. Ignoring a balance of contributors to the genetic diversity of the captive population will eventually lead to a decrease in heterozygosity.

Because of inevitable changes in the genetic composition of the captive population, there will often be pressure to begin releases into the wild before initial goals for the captive program have been met. Release of captive individuals while the population is in the growth phase is usually unwise. The argument that releasing the excess progeny from prolific founders would not harm the captive population is weak in at least two ways. First, the ultimate goal is to restore the wild population, which should include both specific numeric and genetic objectives. Releasing excess animals could further skew an already altered genetic composition in the wild animals. Second, a central conservation goal is to prevent extinction of the species of interest. Thus, it would be wise to ensure the viability of the captive animals by using excess individuals to establish another population whose survival is independent of the initial founder group (i.e., establish another captive population). This second point makes sense even if many of the animals are never released based on genetic priorities. Rather, they serve as a fail-safe in case the central founder population sustains a catastrophic loss. Sacrificing excess individuals, or allowing excess individuals to be used in various behavioral or ecological tests, is another option.

Maintaining a viable captive population requires minimizing genetic change caused by genetic drift and natural selection. As reviewed by Allendorf and Luikart (2007, 466–67), specific actions needed to minimize genetic change must be made on a species-specific basis. Genetic composition of the captive population can be tracked by maintaining individual pedigrees because it provides detailed information on the population and allows you to control the reproductive activity of individuals chosen for mating. Maximizing N_e is usually not the best approach for maintaining genetic variation in the pedigreed population. Genetic variation can be tracked by measuring either heterozygosity or allelic diversity. Maximizing N_e will minimize loss of heterozygosity but might not result in retaining adequate allelic diversity; a well-developed pedigree may allow both heterozygosity and allelic diversity to be managed. A simple example developed by Allendorf and Luikart (2007, 467–68) is shown in figure 3.14, where we see that one allele at each locus has been lost from founder F1 because he left only one descendent in the captive population. Thus, equalizing the contributions of these four founders in future generations would lead to an overrepresentation of genes from F1.

The pedigree of a captive population can be extremely complicated and takes much effort to maintain. Likewise, the kin relationships of founders of a

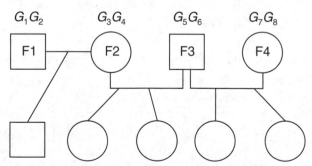

Figure 3.14. Hypothetical pedigree of a captive population founded by four individuals. We know that one allele at each locus has been lost from founder F1 because he left only one descendant in the captive population. Therefore, equalizing the contribution of these four founders in future generations would lead to an overrepresentation of genes from F1.

captive population might not be known. Techniques such as molecular genetics can help resolve the unknown relationships of founders. Potentially substantial errors will be introduced into a pedigree analysis if mistakes are made in assigning founder relatedness (Allendorf and Luikart (2007, 467–68).

Capacity Phase

The capacity phase is reached when the number of individuals in captivity of the ideal genetic composition is reached (i.e., carrying capacity). The goal now is to maintain the health and genetic composition of the captive population, while allowing for release of sufficient individuals (which is project specific) into the wild. Although alleles that are lost from the captive population cannot be replaced except by acquisition of more wild animals or by mutation, disparities in founder allele frequencies due to random drift or to selection (to the captive environment) can be partly reversed by giving priority breeding to animals most likely to have unique or rare genes. That is, heterozygosity will be maximized if animals with the lowest overlap of genes are bred.

Lacy (1994) listed the many strategies that have been used for maximizing retention of genetic variation in captive breeding programs and concluded that no strategy outperformed a prioritization based on mean kinship, although comparable results could be obtained by selecting breeders with the highest probability of carrying unique alleles or by using a circular mating scheme designed for maximum avoidance of inbreeding. Regardless of the technique used, very close inbreeding should be avoided because of the high probability of inviable or infertile offspring (e.g., Moore et al. 1992; Kalinowski et al. 2000; Ramey et al. 2000; Allendorf and Luikart 2007, 463–64). Loss of genetic variation by genetic drift does, however, have more serious and lasting effects than inbreeding. The harmful effects of inbreeding last for a single generation because mating between an inbred individual and an unrelated mate will produce non-inbred young. Genetic drift will, however, have long-lasting (across generational) effects on the captive population, and ultimately on animals released into the wild (Allendorf and Luikart 2007, 463–64).

Because the goal is to minimize losses of the founder population's genetic variation, breeding priority is given to those animals that descend from founders with the lowest representation in the descendant population. No completely satisfactory method exists for summarizing multiple contributions to an individual's genome into a single measure of founder value. Kinship, however, provides a genetic ranking that allows for optimal retention of heterozygosity (Lacy 1994). Accurate calculations of kin relationships, inbreeding coefficients, and retention of founder alleles may be tracked through accurate development of the pedigree Allendorf and Luikart (2007, 468).

As noted, maximizing retention of heterozygosity usually, but not always, optimizes allelic diversity. The ideal genetic management would be to immortalize the original wild-caught founders (Lacy 1994). Allendorf and Luikart (2007) develop related strategies for managing captive populations.

Metapopulation Structure

As reviewed earlier in this chapter, most wild populations are divided into many subpopulations, forming a metapopulation structure. Such a metapopulation structure can be incorporated into a captive breeding strategy. Dispersing a captive population over a diverse environment can avoid directional selection that depletes genetic variation and promote selection that enhances genetic diversity. Dispersal also helps protect the population from epidemic disease and other catastrophes. Isolated or partly isolated populations tend to diverge genetically and thus lose different genetic variants. The metapopulation (as a whole) will, therefore, retain greater gene diversity (expected heterozygosity) and greater allelic diversity than would a single, large (panmictic) population (Lacy 1994). Management of metapopulation structure is, of course, difficult. Small subpopulations are subject to extinction and will often require intensive management and repeated augmentations for extended periods of time (Mace et al. 1992).

Division of a population into several smaller populations does, however, come with potentially serious risks. Animals within each isolated population will become more inbred because of the fewer mate choices and genetic drift. However, if inbreeding depression was not severe, a larger population reconstructed later by mixing animals from the isolated populations would be expected to have greater genetic variation and perhaps higher individual fitness than would be the case if the population had never been subdivided. Moving approximately one animal between populations per generation will usually prevent excessive inbreeding within populations, but at a cost of reduced effectiveness of the subdivided population structure in retaining variation overall. If 5–10 animals are moved per generation, genetic divergence among subpopulations is largely prevented, and the metapopulation is therefore equivalent to a panmictic population. Lacy (1994) concluded that perhaps the prudent approach to managing population structure is to mimic the amount of isolation typical of the wild population (or the best estimate of the characteristics of the wild population) before human-induced decimation and fragmentation. Allendorf and Luikart (2007, 366–69) also developed the role of metapopulation structures in genetic management, and noted that the pattern of genetic variation depends on the spatial and temporal scales that are appropriate for each situation. They noted, however, that a small amount of gene flow between subpopulations can maintain the genetic integrity of an entire species.

Dividing or not dividing the population is not a mutually exclusive decision. In long-term breeding programs, occasional and temporary dividing is a potential tool for managing genetic diversity. And as noted, there is seldom any justification for keeping all of the captive animals in the same location.

Restoring a Population

As developed earlier, translocating animals is a viable option for speeding restoration. In fact, because of limited dispersal abilities, in many cases reintroduction might be the only method by which a desired species can reinhabit a site. There are, however, numerous components involved with successful reintroductions. In the following sections, I discuss issues involved with characterizing the source of animals for use in restoration, evaluating the release site, and determining the size of the released population.

Characteristics of the Source Population

The choice of specific individuals for reintroduction or augmentation is a key component of a successful animal restoration program. Animals for reintroduction may come from a captive population or an existing wild population. Given that mortality is high among reintroduced animals, only those animals that are surplus to the genetic needs of the source population should be released. This is why, as developed earlier, the genetic composition of the source population deserves so much attention. In practical terms, this usually requires that animals initially released are those descending from the most prolific lineages in the source population. Once reintroduced individuals survive at an acceptable level, the genetic composition of the restored (or augmented) population can be diversified by the descendants of other lineages (Lacy 1994). As summarized by Allendorf and Luikart (2007, 476), the source population should have high genetic diversity, genetic similarity, and environmental similarity when compared to the new population.

Evaluation of Introduction Sites

The initial success of the reintroduction program depends, of course, on survival of the released animals. Longer term, however, the released animals must not only survive, but produce viable offspring. Thus, the success of a reintroduction program is enhanced when animals are released into high-quality habitat. Later chapters, especially chapters 3 and 4, discuss those factors leading to successful survival and reproduction.

Resources central to survival and reproduction must be available in the release site. As such, a key step in a reintroduction program is to identify critical factors and summarize their status in the release location. An initial and simple site evaluation guide for woodland caribou (*Rangifer tarandus caribou*) is shown in table 3.2. Note that both key habitat components (habitat quality) and constraints on the use of those habitat components (predators [wolf density] and competitors [deer density]) will be included). Allendorf and Luikart (2007, 172–76) also provided a brief summary of habitat and other environmental considerations for reintroduction programs that can be reviewed for additional discussion.

Size of the Released Population

Reintroduction has been frequently used in attempting to conserve ungulate populations, and has met with varying success. For native game species, 20–40 founding animals have been sufficient for success (Gogan and Cochrane 1994). Allendorf and Luikart (2007, 474) recommended that at least 30–50 individuals be released if possible. Rock iguanas (*Iguana pinguis*) were translocated between islands in the West Indies to establish a second population to serve as a reservoir for this endangered species. The translocation involved only eight individuals, but resulted in a successful new population (Goodyear and Lazell 1994). The probability of success is based, of

TABLE 3.2

Factors considered critical to success of woodland caribou restoration at three sites in the western Lake Superior region

Site	Predators		White-Tailed Deer	
	Wolves per 1000 km^2	Black bears	Density	Incidence of brainworm
A	15	Pending	Low	44%–60%
B	30	Common	High	>90%
C	20	Absent	Absent	-

Source: Gogan and Cochrane (1994, table 9.3).

course, on the specific condition of the habitat at the new site (see chapters 3 and 4 for details). Not only must the physical environment and general habitat conditions are appropriate, but also resources must be of adequate quantity and quality, and be available to the relocated animals. Thus, careful analysis of the realized and potential constraints (e.g., predators, competitors, human-related disturbance) on access to necessary resources must be a prerequisite to any program. Most of the discussion below applies equally to the release of animals from a captive breeding program. Although beyond the scope of our discussion, the issue of introducing a species to an area not previously inhabited by it should be approached with caution (see review by Mills 2007, 233–39).

Because of the multitude of methods available (e.g., traps, snares, nets, immobilization drugs), I will not discuss capture techniques in this section. In chapter 8, I review many common capture methods (and many fine discussions of capture techniques for specific species are available, e.g., Bookhout 1994; Heyer et al. 1996; Wilson et al. 1996). Once captured, it is essential that the time in captivity be minimized. Different species, and different individuals within a species, react differently to capture, transport, and handling. It should be assumed, however, that captivity results in both behavioral and physiological stress on the individual. Here again, the specific techniques necessary for minimizing stress and maximizing survival are species specific. Most relocation programs use veterinarians who are trained in the management of stress for the species being used.

There are two basic techniques for releasing animals, termed soft and hard release. In *soft releases*, captured animals are held in captivity for an extended period of time (days to months) for a variety of behavioral and physiological reasons. This captivity can be in a laboratory or other holding location, or it can be in a confined (e.g., caged, fenced) location near or at the eventual release site; the latter is more common. The rationale for a soft release is that animals can become accustomed to the release-site environment, and observers can monitor animal condition prior to release. Naturally, food and other requirements must be provided, which increases animal contact with humans. Care must be taken to avoid habituation to humans during extended soft releases. Soft releases are also frequently used in captive breeding programs. For example, swift foxes (*Vulpes velox*) in Canada were paired and held in field release pens (3.7 m × 7.3 m) in prairie environments for one to eight months. They were placed in the pens in October or November and held during the mating season (January or February). If they had not produced young, they were released the following spring. If they produced

TABLE 3.3

Survival of swift foxes using soft and hard releases

Release method	Number radio-collared	Survival to 6 mo.	to 12 mo.	to 24 mo.
Soft	45	55%	31%	13%
Hard	155	34%	17%	12%

Source: Carbyn et al. (1994, table 10.2).

young, they and their young were released in summer to early fall (Carbyn et al. 1994).

In *hard releases*, animals are transported from the capture site (or captive rearing site) and released into the wild without any conditioning to the release site environment. The rationale for a hard release is to reduce additional stress that might accompany captivity (for animals from a captive breeding program, this would entail further captivity in a new location). Returning to the swift fox example, animals were also released directly into the field without being placed in the release pens. In this program, although soft-released animals initially had higher survival, no difference in survival was noted after twenty-four months (table 3.3). Therefore, because hard releases were successful and cost-effective (i.e., no expense for conditioning in the field), the Canadian swift fox program continued using only hard releases. Carbyn et al. (1994) emphasized that, to reintroduce the swift fox, it was necessary to determine if its ecological niche was still present and if a minimum viable population could be established. The success of any restoration program—whether or not reintroductions are involved—rests fundamentally on the condition of habitat and niche of the species in question.

Another example of the use of relocation to enhance the status of a declining species is that of the mountain sheep (*Ovis canadensis*) in California. Mountain sheep have a naturally fragmented distribution. Many mountain sheep populations have been extirpated from historical ranges as a result of activities associated with human presence, including disease contracted from livestock, habitat destruction, and possible overharvest (Thompson et al. 2001). Two primary techniques have been used to release mountain sheep: direct (hard) release on the periphery of a mountain range, following vehicular transport from the capture site; and vehicular transport, followed by helicopter transport and holding animals in a temporary enclosure in the interior of a mountain range for a short time (six to eight hours) before release. Thompson et al. (2001) found no statistically significant differences in an

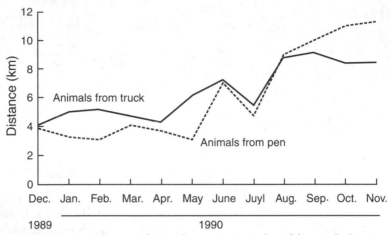

FIGURE 3.15. Average distance from release site (an index of dispersal), by month, for animals transported by helicopter and held in a pen prior to release (dashed line) and for animals released directly from a vehicle (solid line) in the Chuckwalla Mountains, southern California, 1989 to 1990. (From Thompson et al. 2001, figure 1)

index of dispersal (of animals from the release site) between the two methods employed (figure 3.15). However, 70% of the direct release animals survived, and only 30% of the penned animals survived approximately one year following release. They concluded that increased handling time required for penned animals negatively affected survival. This captivity plus exposure to helicopter noise were likely factors in the lower survival.

Methods for releasing animals from captivity, regardless of the amount of time in captivity, are in the development stages. No single method is guaranteed to perform better than another technique, even within the same or closely related species. This level of uncertainty is due largely to the unique environmental situations confronting each program. Research into release techniques for passerine birds has received little attention relative to large mammals. The captive breeding and release program for the endangered San Clemente loggerhead shrike (*Lanius ludovicianus mearnsi*), for example, is pioneering methods for passerines (Morrison et al. 1995). Numerous techniques are being tested, including releasing individual adults, pairing of adults prior to release, and releasing a group of captive-reared siblings simultaneously. The survival of released birds, however, appears to be largely dependent on control of an exotic predator (feral cats), indicating the critical need to evaluate the niche (constraints) available to the released birds, through personal observations. Likewise, release efforts for the endangered Hawaiian crow (*Corvus hawaiiensis*) are being inhibited by attacks on the

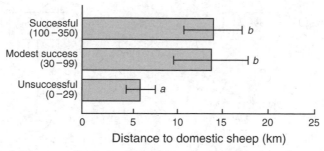

FIGURE 3.16. Success of bighorn sheep relocations in relation to distance to the nearest domestic sheep, western United States, 1923–1997. Different letters denote statistical significance ($P < 0.05$). (From Singer et al. 2000, figure 1)

crow by the Hawaiian hawk (*Buteo solitarius*), itself a federally listed endangered species!

Singer et al. (2000) evaluated the success of 100 relocation attempts of bighorn sheep within six western states between 1923 and 1997. They classified 30 attempts as unsuccessful, twenty-nine moderately successful, and 41 as successful. Relocations were less successful when domestic sheep were located within 6 km of the known bighorn sheep use area (figure 3.16). Projects in which wild omnivores were released into the core of their historical range using prolonged efforts up to 10 years have had the most success (Wolf et al. 1996).

With the exception of the Yellowstone island population, populations of grizzly bears (*Ursus arctos*) in North America are small and isolated. Because these populations are small, do not occupy a large protected area, and because grizzly bears have low reproductive rates, one potential management tool is population augmentation through relocating wild bears from healthy populations elsewhere. Kasworm et al. (2007) relocated four subadult female grizzly bears (two- to six-years old) from southeast British Columbia into the Cabinet Mountains in the summers of 1990 to 1994. Three of four transplanted bears remained in the target area for at least one year and satisfied the short-term goal for site fidelity. Ten years later, systematic hair-snag DNA surveys in the Cabinet Mountains determined that at least one of the original transplanted animals had reproduced thereby providing evidence of success for the long-term goals of survival and reproduction. An example of the most likely pedigree resulting from a relocated grizzly bear into the Cabinet Mountains is shown in figure 3.17.

Mortality due to predation is a primary cause of failure in reintroductions. Animals that have been isolated from predators, either throughout their lifetime (i.e., captive born and raised) or over evolutionary time, might no longer

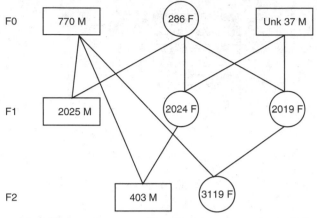

FIGURE 3.17. Most likely pedigree resulting from translocated grizzly bear 286F into the Cabinet Mountains, USA, 1993–2005. Pedigree contains family groups where both parents complementarily share one allele at all loci with the offspring and match as parents. Squares indicate males and circles represent females. Lines indicate a parent-offspring relationship. F0 is the initial generation, F1 is the first generation of offspring for translocated female 286, and F2 is the second generation. (From Kasworm et al. 2007, figure 2)

express appropriate antipredator behavior (Griffin et al. 2000). Griffin et al. reviewed attempts to train captive-bred or relocated animals to avoid predators. They concluded that prerelease training has the potential to enhance the expression of preexisting antipredator behavior. Potential training techniques involve classical conditioning procedures in which animals learn that models of predators are predictors of aversive events. Each technique is accompanied by potentially serious problems that must be avoided through careful study and planning (see Griffin et al. 2000: table 1). Saenz et al. (2002) conducted simulations of various strategies for reintroducing red-cockaded woodpeckers (*Picoides borealis*).

Synthesis

The goal of this chapter was to highlight the critical role that identification of the structure of populations has on the success of a restoration project for wildlife. No restoration project should be designed and implemented without some type of assessment of local and regional distribution of the species of interest. If time and or funding restrictions prevent a field assessment of the population(s), then a thorough literature review, combined with the opinions of several experts on the species and population ecology in general, might prevent the pursuit of a project whose goals are unattainable. At a minimum, such knowledge will allow all project stakeholders to be informed of the likelihood of colonization of the site by the target animals. Additionally, a clear understanding of the difference between simple occupancy and successful survival and reproduction within the project area can be imparted.

Reintroduction is gaining increasing attention throughout the world as a means of restoring and managing animals, as indicated by the increasing use of the technique. Reintroductions involve all vertebrate groups. As the literature on evaluations and improvements in reintroduction grows (e.g., Griffith et al. 1989; Trulio 1995; Hein 1997), these techniques will likely increase as a valuable restoration tool. Lastly, we need to continue to develop and evaluate techniques for genetic management (Lacy et al. 1995), especially those applied to metapopulations composed of small subpopulations.

Chapter 4

Habitat

The concept of habitat is a fundamental building block of contemporary ecology. This central role of habitat is based on numerous papers that relate the presence, abundance, distribution, and diversity of animals to aspects of their environments, and in which habitat was invoked to explain the factors and processes that contributed to the evolutionary history and fitness of animals (Block and Brennan 1993; Morrison et al. 2006). Other authors have likewise emphasized the importance of wildlife-habitat relationships. Anytime a land manager—including the restorationist—manipulates plants, he or she is consciously or unconsciously manipulating habitat and thus the animals associated with that habitat. However, as reviewed by Hall et al. (1997), there are several problems with current studies and discussions of habitat use that cause ambiguity and inaccuracy in our communication on the topic, which causes confusion for land managers and restorationists who are attempting to implement research findings to restore and ultimately manage land.

Issues of Scale

Many authors have developed the rationale for placing studies of wildlife-habitat relationships in an appropriate proper spatial and temporal scale (Wiens 1989ab; Morrison et al. 1992; Block and Brennan 1993; Litvaitis et al. 1994), and temporally and spatially explicit work is increasing rapidly (e.g., Scott et al. 2002; Morrison et al. 2006). Researchers are recognizing that their perceptions of wildlife-habitat relationships are scale dependent, due to the different scales at which different animals operate and at which we operate (Wiens 1989ab; Huxel and Hastings 1999). Johnson (1980) and Hutto

(1985), for example, proposed that animals select habitat through a hierarchical spatial-scaling process, with selection occurring first, at the scale of the geographic range; second, at the scale where animals conduct their activities (i.e., in their home ranges); third, at the scale of specific sites, or for specific components within their home ranges; and fourth, at the scale at which they will procure resources within these microsites. Hutto (1985) proposed that selection at the scale of the geographic range is probably genetically determined, and Wecker (1964) and Wiens (1972) demonstrated that selection at finer scales may be influenced by learning and experience, and so is more directly under the control of individual animals. As summarized by Askins (2000), restoration of animals demands an understanding of the requirements of specific species, which depend upon specific types of vegetation, breeding sites, and food.

Avoiding Pitfalls

There are many issues that restorationists and researchers must consider when setting out to quantify, evaluate, and ultimately restore habitat. Morrison et al. (2006, 182–85) stated that too many researchers ignore the fact that temporal variation in resource use occurs, or if they recognize that fact, they still only sample from narrow time periods where the resulting wildlife-habitat relationships only apply minimally to other situations. Also, researchers commonly sample from across broad time periods (i.e., years; summer or winter seasons) and then use averaged values for variables across the periods, which potentially masks differences in resource use.

The second issue that authors of habitat papers should consider is that if we want to advance wildlife ecology and ultimately restoration ecology, we must be sure that the fundamental concepts with which we work are well defined and understood. Peters (1991), for example, wrote about "operationalizing" ecological concepts if environmental scientists hope to further their science (76). By this, Peters meant that concepts such as habitat should have operational definitions, which are the practical, measurable specifications of the ranges of specific phenomena the terms represent. If the concepts are to be scientifically useful, then the original and subsequent definitions must be sufficiently measurable so that users can apply them in consistent ways.

The third problem in current discussions of habitat, and one that underlies all of the issues, is that the use of habitat terminology is imprecise and ambiguous. Block and Brennan (1993) stated that specific definitions of habitat are often vague, ranging from how species are associated with broad, landscape-scaled vegetation, to very detailed descriptions of the immediate physical environments used by species. It is easy to recognize a similar tendency among papers in wildlife science. This variability detracts from the ability to communicate effectively about habitat-related subjects.

A lack of explicit definitions leads ecologists to a variety of approaches for measuring the terms (e.g., habitat use, selection, preference; carrying capacity) (Wiens 1984, 398). I think that the abundance of the word habitat in the wildlife, restoration ecology, and conservation biology literature, and the prevalence of words that are related to habitat (e.g., community, ecosystem, and biodiversity) force the need for standard definitions at this time, or at least a clear knowledge of the different uses of the terminology.

When Models Fail: Conspecific Attraction

As reviewed in chapter 3, birds tend to aggregate their territories, even in continuous habitat. Conspecific attraction can be used to enhance the occupation of restoration sites by some species (Ahlering and Faaborg 2006). Although conspecific attraction can thus be used as a tool in restoration, it also has the potential to bias our evaluation of habitat requirements of species. Thus, although the presence of individuals provides information on habitat characteristics used by the species, the absence of individuals does not necessarily mean that a location is not appropriate for occupancy. As I review in this chapter, a number of factors other than physical habitat characteristics (e.g., predators, lack of mates) can result in lack of occupancy by animals. The lack of occupancy of locations that are appropriate for individuals of a species has several implications for determining habitat characteristics of a species (Campomizzi et al. 2008):

- We receive an incomplete assessment of habitat characteristics that can be used by species.
- We consistently underestimate the amount and distribution of habitat across the broad landscape for species.
- Our models of habitat use and selection are biased.
- Our conservation options are limited to protect existing habitat or to control limiting factors where birds choose to settle.

Experiments such as those conducted by Ward and Schlossberg (2004; see also review by Ahlering and Faaborg 2006) can be incorporated into studies of habitat use to assess if a behavioral mechanism such as conspecific attraction is biasing results of the study. Studies involving conspecific attraction can help determine the following:

- The cues that can be used to attract animals to a location
- The characteristics of locations into which animals are attracted
- The timing and intensity of cues that maximize attraction to a site
- If animals attracted to a site will return the following year
- The role that the presence of potential limiting factors (e.g., predators, competitors) play in attraction to a site

Results of studies of conspecific attraction and other behavioral studies of cues used to settle in an area can then be translated into recommendations for incorporation into restoration and management plans.

Definitions

Because the focus of this book is about transferring knowledge from the field of wildlife science to the field of restoration science, restorationists must be well versed in the way that wildlife scientists communicate. The definitions given below were developed based on Block and Brennan (1993), Morrison and Hall (2002), and Morrison et al. (2006), who relied on the original intents of ecologists such as Grinnell (1917), Leopold (1933), Hutchinson (1957), Daubenmire (1968), and Odum (1971). I first discuss the basic term habitat and the key ways in which wildlife scientists modify the term. I then place the overall study of wildlife-habitat into a framework that relates to different spatial extents (i.e., spatially explicit study of habitat). Lastly, I develop the concept of the niche and ways in which niche relationships relate to the development of restoration plans.

Habitat

Morrison et al. (2006) develop in detail the history and development of habitat and habitat-associated terms; I will summarize that material in relation to applications in restoration ecology. I define *habitat* as the resources and conditions present in an area that produce occupancy (for any reason) by an organism. Habitat is organism specific and relates the presence of a species, population, or individual (animal or plant) to an area's abiotic and biotic features. Habitat is thus the sum of the specific resources needed by organisms. Therefore, migration and dispersal corridors, and the land animals occupy during breeding and nonbreeding seasons, are habitat. An organism-based understanding of habitat is needed to determine appropriate restoration goals (Miller 2007).

Habitat is not equivalent to *habitat type*, which was coined by Daubenmire (1968,27–32) and refers only to the type of vegetation association in an area. When an author wants to refer only to the vegetation that an animal uses, he or she should use *vegetation association* or *vegetation type* instead (Hall et al. 1997; Morrison and Hall 2002).

The definition of habitat as organism specific is an absolutely critical concept. It means that restoring vegetation, regardless of how well it matches some desired condition, can easily fail to restore the desired assemblage of wildlife. Failure to plan simultaneously for plant and animal restoration results in a hit-or miss strategy for animals; restoring vegetation restores wildlife habitat for some species, but not necessarily the species desired.

I define the term *habitat use* as the way an animal uses (or consumes, in a generic sense) a collection of physical and/or biological components (i.e., resources) in a habitat. *Habitat selection* is a hierarchical process involving innate and learned behavioral decisions made by an animal about where it should be at different scales of the environment (Hutto 1985, 458). Johnson (1980) referred to selection as the process by which an animal chooses which habitat components to use. *Habitat preference* is restricted to the consequence of the habitat selection process, resulting in the disproportional use of some resources over others.

I have described how an animal perceives and occupies habitat. We also have an interest in determining how much habitat occurs in the environment. Thus, *habitat availability* refers to the accessibility and ability of an individual to obtain physical and biological components of the environment. This is in contrast to the *abundance* of these resources, which refers only to their quantity in the habitat, irrespective of the organisms present in the habitat (Wiens 1984, 402). In practice, however, it is difficult to assess resource availability from an animal's perspective (Litvaitis et al. 1994). For example, we can measure the abundance of food for a predator, but we cannot say that all of the prey present in the habitat are available to the predator because there are likely many factors that restrict their accessibility. Similarly, vegetation beyond the reach of an animal is unavailable for it to feed on. Although measuring actual resource availability is important for understanding wildlife-habitat relationships, in practice it is seldom measured because of the difficulty in determining exactly what is available and what is not (Wiens 1984, 406).

Habitat quality refers to the ability of the environment to provide conditions appropriate for individual and population persistence, and ranges from low- to medium- to high-quality habitats, based on their abilities to provide resources for survival, reproduction, and population persistence, respectively. High-quality habitat is usually equated with vegetative features that may contribute to the presence (or absence) of a species (e.g., Habitat Suitability Index models: Laymon and Barrett 1986; Morrison et al. 1991). However, that quality must be explicitly linked with demographic features if it is to be a useful measure. However, Van Horne (1983) demonstrated that density is a misleading indicator of habitat quality, and those confirming source and sink habitats in nature (Pulliam 1988; Wootton and Bell 1992) have caused ecologists to deemphasize this ranking. Thus, while the abundance of animals can be equated to some degree with habitat quality, the quality itself should be based on demographics of individuals and/or populations.

Spatial Scale

One of the major advantages of landscape ecology is the integration of ecological study across various scales of space and time. However, there is confusion between biological or ecological scale and level. The *level of biological organization* refers to the biological dimension of scale, and whether a study or plan pertains to ecosystems, communities, assemblages, species, individuals, or gene pools.. This dimension could also refer to classification levels of vegetation communities or ecosystems, such as plant associations, vegetation types, and ecoregions (e.g., Bailey 2005). Note that a landscape study or management plan might pertain to a fine-scale magnitude of biological organization such as an inventory of ecotypes, but across a broad geographic extent such a drainage basin. In this way, the various dimensions of scale may be applied at different magnitudes for a given purpose (Morrison et al. 2006, 155–58).

Applying a geographic information system (GIS) model at the wrong scales of geographic extent and resolution can lead to false conclusions. Likewise, understanding the dimensions of scale of a geographic-based analysis helps guide the scales at which such information should be used and not used. As summarized by Morrison et al. (2006, 155–58), I suggest that restoration and management plans clearly identify the magnitudes of geographic extent, map scale, spatial resolution, time period, organizational hierarchy (if appropriate), and levels of biological organization addressed and evaluated. In this way, much confusion over terms and methods can be avoided.

In summary, ecologists have learned that animals go through a series of increasingly refined selection decisions, beginning with selection of a geographic area; followed by selection of a specific combination of elevation, slope, and vegetation type; followed by selection of specific locations to forage, breed, or rest; followed by selection of specific items to use (e.g., food type) (figure 4.1). As we step down this hierarchy of habitat selection from

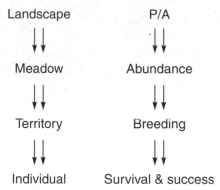

FIGURE 4.1. Relationship between the measurement of spatial extent and the appropriate measure of animal performance.

broad to specific we are able to understand an increasingly detailed amount of information about animals, and this information is matched with the spatial scale of study (figure 4.1). At broader spatial scales we can assess presence or absence of a species, for example, but we are usually unable to quantify much about survival or anything about reproductive success. Alternatively, information we can gather at or around the breeding site can tell us something about the number of young produced. Thus, we must take care to properly match the goal of a restoration plan with the appropriate scale of study.

Terms such as *macrohabitat* and *microhabitat* are frequently used in wildlife ecology (Johnson 1980). Generally, macrohabitat is used to refer to large spatial extent (i.e., landscape-scaled) features such as seral stages or zones of specific vegetation associations (Block and Brennan 1993), which equates to Johnson's (1980) first level (order) of habitat selection. Microhabitat refers to finer-scaled habitat features, such as would be important factors in levels 2–4 in Johnson's (1980) hierarchy. Macro- and microhabitat are such general categories that their use should be minimized; if used, the scales to which they apply should be stated explicitly.

Niche

As I discussed earlier, habitat is a valuable concept for developing general descriptors of the distribution of animals and lends insight into factors driving survival and fitness. However, habitat in general misses the underlying mechanisms that determine occupancy, survival, and fecundity. Other factors, including some often related to an animals' niche, must be studied to more fully to understand the mechanisms responsible for animal survival and fitness (Morrison et al. 2006, 56–57); such an understanding is critical to successful wildlife restoration. Morrison et al. (2006) provided additional details on how the niche can be used to enhance our understanding of wildlife ecology; I will now focus on how the niche concept can be used to improve our ability to restore habitat and the animals themselves.

O'Connor (2002) concluded that most species can be limited by a variety of environmental factors, and the influence of any single factor is not additive to the influence of any other factor. Thus, usually only one factor is limiting in any particular situation. It is unlikely that the same factor will always be limiting because of variation in nature results in a continual shifting of resources; this variation results in shifting among potentially limiting factors.

The observed abundance distribution of a species represents its *realized niche*, which usually will not include the full range of conditions under which the species could potentially be found, and which likely will not include the physical conditions where the species does best. This observed (realized) distribution of a species will differ from its potential distribution, the *physiological niche*, if the species is excluded from the conditions where it does best by factors such as competition (Morrison et al. 2006). Understanding the physiological optima of a species provides critical information for restoration of a species because these are the conditions in which, with appropriate management, the species actually does best. Actions such as control of predators or restoration of a different disturbance regime (see chapter 7) may allow some species to thrive in areas where they are rarely found under present conditions (Huston 2002).

Thus, habitat and niche are both critical concepts in ecological restoration. A difficulty with focusing on habitat alone is that measured features can stay the same while use of important resources by an animal within that habitat can change. For example, changes in the species or size of prey taken by a bird foraging on shrubs (e.g., Hejl and Verner 1990; Keane and Morrison 1999); the shrubs (habitat) do not need to change in physical dimensions or appearance. Crude differences we identify in habitat studies, such as changes in use of areas under study, often are caused by changes in use of specific resources that we fail to see. If habitat is described only as structural or floristic aspects of vegetation, we might fail to predict animal health because we do not identify constraints on exploitation of other resources that are critical limiting factors.

In summary, restoration planning must include factors not traditionally considered habitat because they influence whether an animal occupies a site and how it performs if it is able to have occupancy. Understanding how the niche is expressed across space and time will aid in understanding animal-habitat relationships, thus improving our ability to restore animals.

When to Measure

The behavior, location, and needs of animals change, often substantially, throughout the year. When to measure timing involves both between-season and within-season analysis. Many researchers, however, ignore temporal variations in habitat use, which can negatively impact habitat assessments. Without knowledge of an animal's total requirements, management recommendations have limited and perhaps faulty implications.

Many studies have shown that animals substantially change their use of resources between seasons (Schooley 1994; see review by Morrison et al. 2006, 182–85). Morrison et al. (1990) showed that birds wintering in mixed-conifer forests of the western Sierra Nevada fed heavily on an insect that overwintered under the bark of a specific tree species; birds seldom fed on that tree in the summer. Thus, a restoration project will likely fail if the proper mixture of plant species are not provided regardless of the development of the plants that are established. Information is usually available for most regions on the seasonal presence and activity of animals; such data should be considered when designing habitat resources in a restoration project. In chapter 6, I discuss development of desired conditions in more detail.

Naturally, the preferred study design is repetition of a study for every appropriate biological period (e.g., breeding, wintering). In figure 4.2 we see

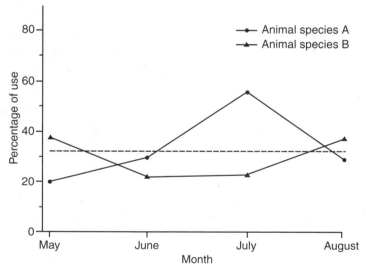

FIGURE 4.2. Use of a species of tree by two hypothetical animal species during summer. The dashed horizontal line represents the approximate average of use values for both hypothetical species calculated separately across time. (Reproduced from Morrison et al. 2006, figure 6.1)

that the "average use" over time by two species is not a close approximation of their actual behavior. The average indicates, however, that the animals use this tree species basically identically. Thus, the finer we stratify our sampling, the greater the number of levels of resolution we have available for subsequent analysis. Scenarios such as that depicted in figure 4.2 should concern researchers and should be evaluated early in a study by making sure that sampling is sufficiently intensive so that such relationships can be identified.

What to Measure

Determining the factors responsible for the occupancy, and eventually the survival and fitness, of an animal has obvious implications and applications to restoration. But, we must determine ways to understand how an animal perceives its environment; that is, we must determine what to measure so we can provide it for the animal through restoration. As noted by Morrison et al. (2006, 151–52), what we are doing is trying to identify patterns in the use of the environment by animals that can be translated into specific actions on the ground by humans.

Green (1979, 115) listed criteria for selection of variables to use in analysis of wildlife habitat:

- Spatial and temporal variability in biotic and environmental variables that would be used to describe or predict impact effects
- Feasibility of sampling with precision at a reasonable cost
- Relevance to the impact effects and a sensitivity of response to them
- Some economic or aesthetic value, if possible.

Understanding the variability inherent in the system of interest is critical in designing a study; the variability must be generally understood and the sampling designed to capture it. This variability includes natural, stochastic, or systematic change, and measurement and sampling error. We must explicitly identify the precision necessary to reach project goals, and then match all sampling to this needed precision (see chapter 8). Thus, does the project goal require determination of simple presence or absence of a species or actual density?

A typical problem inherent in habitat studies is to measure everything possible for the duration of the study, and then run correlations to try and identify variables important in describing the distribution of the animal of interest. A stronger and more efficient approach is to gather preliminary data and conduct analyses that can be used to reduce the variables being collected in the field (see chapter 8). Such a process provides more time overall to increase the precision of the remaining variables and, as necessary, to increase the sample size.

Spatial Scale

As developed earlier in this chapter, it is important to match the scale of analysis with the scale that we wish to apply our results for management purposes. In restoration, these decisions will be driven by the size of the project area and the goals regarding wildlife. In general, the smaller the area, the more attention that will have to be given to fine-scale habitat and niche parameters of the species of interest. This is because the probability that a specific habitat component will occur naturally increases as area size increases.

The definition of *small* and *large* is project specific. Many salamanders have home ranges of under 15 m^2 and are unlikely to move over 25 m (Grover 1998). In contrast, the home ranges for other small- to medium-sized terrestrial vertebrates can be upwards of 5–10 ha or more. Projects focused on one or a few specific species must be guided by the natural history of those animals. In contrast, projects of larger scale that have more general goals (e.g., enhance vertebrate diversity) will be guided by principles that relate general measures of the wildlife assemblage (e.g., species richness) to general measures of the environment (e.g., vegetation structure).

Wildlife managers are often frustrated at the failure of most models to apply in their specific location. This frustration comes primarily from trying to apply a relationship based on broad measurements of vegetation to local situations. And likewise, models developed at a fine scale can seldom be adequately applied (generalized) to other locations (Block et al. 1994). Morrison et al. (2006, 157) asked "How can this dilemma be solved?" They answered by noting that it is possible to adapt information developed for use in one location to the conditions that exist at another location. Studies that include information on the fundamental, mechanistic explanations of the activities and responses of animals to environmental conditions are more powerful than those that simply draw correlations between animal abundance and a list of habitat factors. For example, if we know that a bird species is switching use of plant species for foraging because of the prey available, we can in turn apply that knowledge to selection of plants to emphasize during restoration.

Measurements of the Animal

Earlier in this chapter I developed the relationship between the spatial scale of study and restoration and what you can expect to understand about animals at that scale. The type of information you want to know about animals in a project area depends on the goals you have for each species present (or all species collectively). That is, are you interested in simply knowing if a species is present or absent (i.e., presence/absence data), developing an index of relative abundance (e.g., numbers/count), or actually estimating absolute density (numbers/unit area)? Although it is beyond the scope of this book to review all of the methods available for estimating animal abundance (see chapter 9 for a review of some methods), I will discuss evaluating wildlife and habitat that focuses on counting animals and then relating these numbers to environmental features. Restorationists must be confident that they understand the relationships between animals and their environment if meaningful plans for habitat manipulations, or manipulations of the actual animals, are to be developed. Additionally, monitoring the response of animals to restoration activities often requires that valid estimates of presence-absence, abundance, or density are developed (chapters 8 and 9).

Many questions arise when discussing estimation of animal numbers:

- Are our estimates of absolute or even relative abundance a fair reflection of the number of animals actually present?
- Are we confident that we were able to detect the presence of a species of interest?
- If we use multiple observers, are their estimates comparable?
- Can estimates derived by one observer or group of observers in one study area be used to validate a model developed by other people at a different location?

There are many problems associated with counting animals, and unreliable estimates of animal numbers will negate conclusions drawn on habitat relationships based on even the most carefully collected environmental variables, and render monitoring of the response of animals to restoration difficult, if not meaningless. Much of the theory in ecology, as well as applications in resource management and conservation, depend on reliable expression and comparison of numerical abundance. Additionally, the development of many wildlife-habitat models is based, in part, on estimations of abundance (Morrison et al. 2006, 159).

The expression of density (or an index thereof) can vary widely depending upon the spatial scale of the study. Simply converting numbers of animals to some standard area—for example, extrapolating birds in a 1.5 ha study site to birds per 40 ha (100 acres)—does not standardize estimates for comparisons with other studies. Density for any one population is not likely to remain constant across spatial scales.

Thus, a conundrum awaits restorationists as you develop plans for maintaining or adjusting animal numbers, or restoring a species, to a project area. The size of a restoration area is usually based on some preexisting condition, such as landownership or economic constraint, and not on the species' relationship with the landscape. As developed in chapter 3, however, issues such as the structure of an animal population (e.g., metapopulation) must be considered if your goal is focused on one or multiple animal species. A restorationist usually cannot, of course, control the size or location of the project area. You can, however, place the goals that can be achieved for a project within the proper context of animal population ecology and habitat use. Variations in abundance are influenced not only by study area size, but also the year of study, site selection, sampling method, trap type, and various other factors. Determination of the effect of study area size on abundance values should be incorporated into the preliminary sampling phase of all studies.

Most restoration planning for wildlife is based on a review and summary of the available literature; I discuss development of desired conditions in chapter 6. The restorationist must be aware of the weaknesses inherent in most published studies of wildlife and wildlife-habitat use. For example, most studies of habitat use occur where the researcher has previous knowledge that the species of interest is in adequate abundance, thus ensuring accumulation of an adequate sample size. The home range or a finer-scale activity location (e.g., foraging or nest site) within the home range is usually the focus of study. The restorationist must then extrapolate the results of these studies to the project area, which is unlikely to be in close proximity to the location(s) from which the original data were generated. Such extrapolations are tenuous at best, however, because animals and their habitat are not uniformly distributed across the landscape.

Researchers tend to make decisions on study area size based on either convenience or on the location of small, relatively high animal density areas. Do not blindly follow the conclusions of a published study, but rather carefully evaluate the rationale given for the location and size of the study area(s), sampling methods used, and how the data were analyzed. All of these factors and more will influence what is concluded about animal numbers and the characteristics of the habitat used.

Measurements of the Environment

Two basic aspects of vegetation must be distinguished: the structure or physiognomy, and the taxa of the plants, or floristics. Many ecologists initially concluded that vegetation structure and configuration (size, shape, and distribution of vegetation in an area), rather than particular plant taxonomic composition, was most important in determining patterns of habitat occupancy by animals, especially birds (see review by Morrison et al. 2006, 160–61). Most recent studies have shown that plant species composition plays a much greater role in determining patterns of habitat occupancy than previously thought. The relative usefulness of structural versus floristic measures is again primarily a function of the spatial scale of analysis. That is, floristics become increasingly relevant as you seek to determine the mechanisms responsible for the performance (breeding success, body condition) of animals. Thus, relatively simple presence-absence studies of animals at regional or broader scales likely do not require a floristics analysis of vegetation. Broad categorization by physiognomy, probably including differentiation no more specific than life form such as deciduous and evergreen, or by general ecological classes or vegetative types, is probably adequate. As noted above, the best approach is to begin a study with a preliminary evaluation of the variables and sampling methods necessary to achieve the desired level of refinement; necessary sample sizes can also be determined from such preliminary work.

Broad Spatial Extents

Researchers have attempted to relate the numbers and kinds of animals to some measure of the gross structure of the vegetation. Most famous is the relationship between foliage height diversity (FHD) and bird species diversity (BSD): as foliage layers are added, the number of bird species tends to increase (see figure. 4.3). In vertically simple vegetation, such as brushlands and grasslands, FHD would not be expected to provide a good indicator of animal diversity (at least for most vertebrates). Recognizing this problem, Roth (1976) developed a method by which the dispersion of clumps of vegetation such as shrubs forms the basis for a measure of habitat *heterogeneity* or *patchiness* that could be related to BSD.

Note in figure 4.3 that there is considerable scatter around the regression line. Thus, the usefulness of this general principle as a site-specific predictor decreases as the scale of application becomes increasingly fine (i.e., as you go from relatively broad scale to relatively fine scale). Measures of diversity sacrifice complexity for simplicity; this is why they are useful primarily at larger spatial scales. These indices collapse detailed information on plants, such as species composition, foliage condition (vigor), and arthropod abundance, into a single number.

Many of the currently used habitat models operate at the broad habitat scale, including most statewide *wildlife-habitat relationships* (WHR)

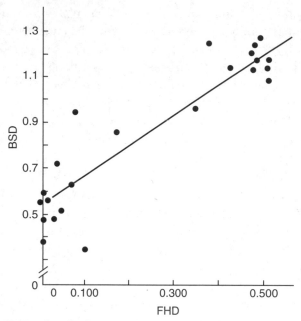

FIGURE 4.3. Foliage height diversity (FHD) versus bird species diversity (BSD). Bullets (respresent the study sites. (Reproduced from Willson 1974, figure 1)

constructs (Block et al. 1994), GAP models (Scott et al. 1993), and habitat suitability index (HSI) models (U.S. Fish and Wildlife Service 1981). Most of these models use broadscale categorizations of vegetation types as a predictor of animal presence.

Fine Spatial Extents

As we attempt to understand factors determining survival and fecundity, we increase the need to quantify detailed habitat and niche features of the environment. Dueser and Shugart (1978) listed four criteria that can help guide selection of fine-scale habitat variables for measurement:

- Each variable should provide a measure of the structure of the environment that is either known or reasonably suspected to influence the distribution and local abundance of the species.
- Each variable should be quickly and precisely measurable with nondestructive sampling procedures.
- Each variable should have intraseasonal variation that is small relative to interseasonal variation.

- Each variable should describe the environment in the immediate vicinity of the animal.

The first criterion of Dueser and Shugart (1978) indicates that natural history information will narrow the choice of variables. It is often worthwhile to let a biologist familiar with the area but not the species review your list of variables: a plant ecologist likely can offer valuable advice to a wildlife biologist in planning a study. Their concern regarding sampling variation indicates that measurements should be sufficiently precise so that variation within relatively short periods of time (within seasons or intraseasonal) are not obscured by the likely much larger variations that occur between seasons. High precision is required in studies of fine-scale habitat use. Their final point is intuitive in that habitat should be measured in close proximity to the animal. It is not incorrect to measure variables from many scales; it is often a mistake to include them in the same analysis. Morrison et al. (2006, 424–26) reviewed examples of the limitations of habitat features alone in predicting habitat quality, and Gawlik (2006) also recognized the importance and use of various performance measures in evaluation of restoration actions for wildlife.

Focal Animal Approach

Most studies of fine-scale habitat selection are variations of the *focal-animal* approach, which use the presence of an animal as an indication of the habitat being used by the species. No correlation between abundance and the environment is involved. Rather, the location of individual animals is used to demark an area from which environmental variables are measured. As detailed in the following section, an animal's specific location might serve as the center of a sampling plot, or a series of observations of an individual might be used to delineate an area from which samples are then made (e.g., see Wenny et al. 1993). In either case, the major assumption of this approach is that measurements indicate habitat preferences of the animal. For example, many studies have used the location of a singing male bird or a foraging individual as the center of plots describing the habitat of the species (e.g., James 1971; Holmes 1981; Morrison 1984a, b; VanderWerf 1993).

Morrison et al. (2006) presented a detailed review of the development of the focal-animal approach and the variables and methods used to gather data. Birds initially received the most attention with regard to the analysis of habitat-use patterns, which was likely a reflection of the conspicuousness of birds. James (1971) conducted one of the first and most-cited studies quantifying bird-habitat relationships in which she used 15 measures of vegetation structure to describe the multidimensional *habitat space* of a bird community in Arkansas; these methods are described in the next section (How to Measure). The conceptual framework James based her work on and general analytical techniques (multivariate analysis of focal-bird observations) used have led to a plethora of studies that expanded upon her basic ideas.

Dueser and Shugart (1978) had as their goal the description of microhabitat differences among the small mammal species of an upland forest in eastern Tennessee. They gathered information for vertical strata at each capture site of a small mammal: overstory, understory, shrub level, forest floor, and litter-soil level (table 4.1). Note that they did not collect species-specific information on plants beyond designations of *woodiness* and *evergreenness*, which can be considered an unfortunate omission for an analysis of fine-scale habitat use. They did, however, quantify features of the forest floor, such as litter-soil compactability, fallen log density, and short, herbaceous stem density. They found that certain of these soil variables played a significant role in describing the differences in microhabitats of the species studied.

The work of Welsh and Lind (1995) is a good example of a spatially-explicit analysis of habitat. Working with the Del Norte salamander (*Plethodon elongatus*), they presented detailed rationale for the selection of methods, including choice of analytical techniques, data screening, and interpretation of output. The variables they measured, separated by spatial scale, are shown in table 4.2. A similar example for multiple species of amphibians was given by Welsh and Lind (2002). The advancement in the way we sample habitat features is exemplified by the Welsh and Lind studies in that they carefully conceptualized the relationship between habitat use and spatial scale for their study species. As discussed throughout this book, analyses that are relevant to specific spatial scales are critical in restoration planning because we are able to relate the way animals use their environment to the size of the actual area being restored. Additionally, studies placed in a spatial context allow us to identify what animals appear to require to occupy an area, then to survive and mate in the area, and finally to produce offspring in the area.

How to Measure

In this section I briefly review some of the common methods used to measure wildlife habitat to give the restorationist a feeling for the available methodology. I do not present a survey of all literature available for all taxa; Cooperrider et al. (1986) provided a thorough review of basic sampling techniques for all major taxa of wildlife (see also Braun 2005; Morrison et al. 2006).

TABLE 4.1

Designation, descriptions, and sampling methods for variables measuring forest habitat structure

Variable	Method
1. Percentage of canopy closure	Percentage of points with overstory vegetation, from 21 vertical ocular tube sightings along the center lines of two perpendicular 20 m² transects centered on trap
2. Thickness of woody vegetation	Average number of shoulder-height contacts (trees and shrubs), from two perpendicular 20 m² transects centered on trap
3. Shrub cover	Same as (1), for presence of shrub-level vegetation
4. Overstory tree size	Average diameter (in cm) of nearest overstory tree, in quarters around trap
5. Overstory tree dispersion	Average distance (m) from trap to nearest understory tree, in quarters
6. Understory tree size	Average diameter (cm) of nearest understory tree, in quarters around trap
7. Understory tree dispersion	Average distance (m) from trap to nearest understory tree, in quarters
8. Woody stem density	Live woody stem count at ground level within a 1.00 m² ring centered on trap
9. Short woody stem density	Live woody stem count within a 1.00 m² ring centered on trap (stems ≤ 0.40 m in height)
10. Woody foliage profile density	Average number of live woody stem contacts with an 0.80 cm diameter metal rod rotated 360°, describing a 1.00 m² ring centered on the trap and parallel to the ground at heights of 0.05, 0.10, 0.20, 0.40, 0.60, . . . , 2.00 m above ground level
11. Number of woody species	Woody species count within a 1.00 m² ring centered on trap
12. Herbaceous stem density	Live herbaceous stem count at ground level within a 1.00 m² ring centered on trap
13. Short herbaceous stem density	Live herbaceous stem count within a 1.00 m² ring centered on trap (stems ≤ 0.40 m in height)
14. Herbaceous foliage profile density	Same as (10), for live herbaceous stem contacts
15. Number of herbaceous species	Herbaceous species count within a 1.00 m² ring centered on trap
16. Evergreenness of overstory	Same as (1), for presence of evergreen canopy vegetation
17. Evergreenness of shrubs	Same as (1), for presence of evergreen shrub-level vegetation
18. Evergreenness of herb stratum	Percentage of points with evergreen herbaceous vegetation, from 21 step-point samples along the center lines of two perpendicular 20 m² transects centered on trap
19. Tree stump density	Average number of tree stumps ≥7.50 cm in diameter, per quarter
20. Tree stump size	Average diameter (cm) of nearest tree stump ≥7.50

TABLE 4.1

Continued

Variable	Method
21. Tree stump dispersion	Average distance (m) to nearest tree stump ≥7.50 cm in diameter, in quarters around trap
22. Fallen log density	Average number of fallen logs ≥7.50 cm in diameter, per quarter
23. Fallen log size	Average diameter (cm) of nearest fallen log ≥7.50 cm in diameter, in quarters around trap
24. Fallen log dispersion	Average distance (m) from trap to nearest fallen log ≥7.50 cm in diameter, in quarters around trap
25. Fallen log abundance	Average total length (>0.50 m) of fallen logs (7.50 cm in diameter, per quarters
26. Litter-soil depth	Depth of penetration (<10.00 cm) into litter-soil material of a hand-held core sampler with 2.00 cm diameter barrel
27. Litter-soil compactability	Percentage of compaction of litter-soil core sample (26)
28. Litter-soil density	Dry weight density (g/cm^2) of litter-soil core sample (26), after oven drying at 45°C for 48 hr.
29. Soil surface exposure	Same as (18), for percentage of points with bare soil or rock

Source: Dueser and Shugart 1978: appendix. Reproduced by permission of the Ecological Society of America.

Even a cursory review of the methods sections in wildlife publications shows a reliance on standard, classical methods of quantifying the structure and floristics of vegetation: point quarter, circular plots and nested circular plots, sampling squares, and line intercepts. These methods are used because they have been developed and extensively tested by plant ecologists in a multitude of environmental situations. Standard methods provide an established starting point from which biologists can adapt specific methods as needed. Standard methods also provide comparability between studies. There are many fine books available that review sampling methods in vegetation ecology (e.g., Daubenmire 1968; Mueller-Dombois and Ellenberg 1974; Greig-Smith 1983; Cook and Stubbendieck 1986; Bonham 1989, Schreuder et al. 1993).

The most popular methods of measuring habitat originated with a protocol developed by James and Shugart (1970). They established 0.1 acre (0.04 ha) plots to estimate tree density and frequency. To estimate shrub density they made two transects at right angles to one another across the 0.1 acre plots, counting the number of woody stems intercepted by their outstretched arms. An ocular (sighting) tube was used to estimate vegetation cover. The methods used by James have had a positive and pronounced influence on most analyses of wildlife habitat that followed.

TABLE 4.2

Hierarchic arrangement[a] of ecological components represented by 43 measurements of the forest environment taken in conjunction with sampling for the Del Norte salamander (Plethodon elongatus)

Hierarchic scale
  Variable category
    Variables[b]

II. Landscape scale
  A. Geographic relationships
    Latitude (degrees)
    Longitude (degrees)
    Elevation (m)
    Slope (%)
    Aspect (degrees)

III. Macrohabitat or stand scale
  A. Trees: density by size[c]
    Small conifers (C)
    Small hardwoods (C)
    Large conifers (C)
    Large hardwoods (C)
    Forest age (in years)
  B. Dead and down wood: surface area and counts
    Stumps (B)
    All logs-decayed (C)
    Small logs-sounds (C)
    Sound log area (L)
    Conifer log-decay area
    Hardwood log-decay area (L)
  C. Shrub and understory composition (> 0.5 m)
    Understory conifer (L)
    Understory hardwoods (L)
    Large shrub (L)
    Small shrub (L)
    Bole (L)
    Height II—ground vegetation (B) (0–0.5 m)

III. Macrohabitat or stand scale (continued)
  A. Ground-level vegetation (< 0.5 m)
    Fern (L)
    Herb (L)
    Grass (B)
    Height I—ground vegetation (B) (0–0.5 m)
  B. Ground cover
    Moss (L)
    Lichen (B)
    Leaf (B)
    Exposed soil (B)
    Litter depth (cm)
    Dominant rock (B)
    Codominant rock (B)
  C. Forest climate
    Air temperature (°C)
    Soil temperature (°C)
    Solar index
    % canopy closed
    Soil pH
    Soil relative humitidity
    Relative humidity (%)

IV. Microhabitat scale
  A. Substrate composition
    Pebble (P) (% of 32–64 mm diameter rock)
    Cobble (P) (% of 64–256 mm diameter rock)
    Cemented (P) (% of rock cover embedded in soil/litter matrix)

Source: Reproduced from Welsh and Lind 1995 (table 1), by permision of the Department of Zoology, Ohio State University.
Note: Level I relationships (the biogeographic scale) were not analyzed because all sampling occurred within the range.
[a]Spatial scales arranged in descending order from coarse to fine resolution (Weins 1989a).
[b]Abbreviations used for the variables are as follows:
  C = count variables (number per hectare)
  B = Braun-Blanquet variables (the percentage of cover in 1/10 ha circle)
  L = line transect variables (the percentage of 50 m line transects)
  P = percentage within 49 m^2 salamander search area
[c]Small trees = 12–53 cm dbh; large trees = >53 cm dbh

Circular plots are easy to establish, mark, measure, and relocate, and estimates of animal numbers within such plots can be statistically related to vegetation data in a straightforward manner. Plots provide for the sampling of vegetation and animals at specific locations in space and time. Noon (1981) presented a useful description and example of both the transect and areal plot sampling systems. The problem with transects is that they cover relatively large areas, and thus make it difficult to relate specific animal observations (or abundances) to specific sections of the transect. Transects are, however, widely used to provide an overall description of the vegetation of entire study areas.

In summary to this point, fixed-area plots and transects can be used to provide site-specific, detailed analysis of wildlife-habitat relationships. The majority of sampling methods used since the 1970s to develop wildlife-habitat relationships—for subsequent multivariate analyses—have used fixed-area plots (usually circular) as the basis for development of a sampling scheme that may then incorporate subplots, sampling squares, and transects. I will present examples of some of the more widely used methods.

Dueser and Shugart (1978) developed a detailed sampling scheme that combined plots of various sizes and shapes, as well as short transects (figure 4.4). They used three independent sampling units centered on each trap: a

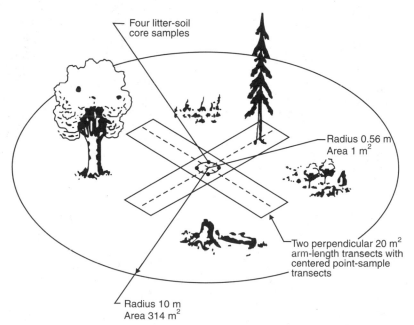

FIGURE 4.4. Habitat variable sampling configuration used by Dueser and Shugart in their study of small mammal habitat use. (Reproduced from Dueser and Shugart 1978, figure 1)

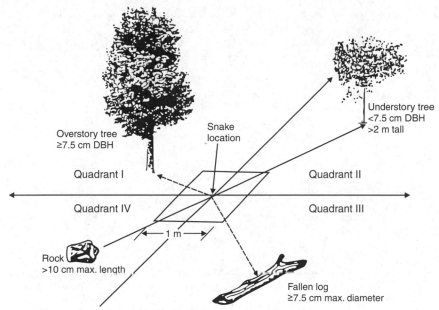

FIGURE 4.5. Sampling arrangement for snake locations. (Reproduced from Reinert 1984, figure 1)

1.0 m² ring, two perpendicular 20 m² arm-length transects, and a 10 m-radius circular plot. Reinert (1984) adopted techniques similar to those used in the bird study by James (1971) and the small mammal study by Dueser and Shugart (1978) in his study of snake habitat use (figure 4.5). Reinert used a 35 mm camera equipped with a 28 mm wide-angle lens to photograph 1 m² plots from directly above the location of a snake, and then determined the various surface cover percentages by superimposing each slide onto a 10 × 10 square grid. His variable list was previously presented in table 4.3. Reinert also sampled several environmental variables that measured air, surface, and soil temperature and humidity. Several of the more popular techniques for quantifying foliage cover are provided in figure 4.6.

Bibby et al. (2000) provided a basic summary of habitat assessment techniques, including a description mapping techniques for studies of avian ecology and how to relate bird counts to environmental characteristics. Figure 4.7 summarizes how these techniques can be applied in the field, ranging from mapping of general bird locations, to specific assessment of individual habitat use.

The use of radio telemetry usually results in a more complete description of the activities and areas used by animals than is available with visual

TABLE 4.3

Structural and climatic variables

Mnemonic	Variable	Sampling Method
ROCK	Rock cover	Coverage (%) within 1 m^2 quadrant centered on snake location
LEAF	Leaf litter cover	Same as ROCK
VEG	Vegetation cover	Same as ROCK
LOG	Fallen log cover	Same as ROCK
WSD	Woody stem density	Total number of woody stems within a 1 m^2 quadrant
WSH	Woody stem height	Height (cm) of tallest woody stem within 1 m^2 quadrant
MDR	Distance to rocks	Mean distance (m) to nearest rocks ((10 cm max. length) in each quarter
MLR	Length of rocks	Mean max. length (cm) of rocks used to calculate MDR
DNL	Distance to log	Distance (m) to nearest log ($\geq$ 7.5 cm max. diameter)
DINL	Diameter of log	Max. diameter (cm) of nearest log
DNOV	Distance to overstory tree	Distance (m) to nearest tree ($\geq$ 7.5 cm dbh [diameter at breast height])
DBHOV	Dbh of overstory tree	Mean dbh (cm) of nearest overstory tree within each quarter
DNUN	Distance to understory tree	Same as DNOV (trees < 7.5 cm dbh > 2.0 m height)
CAN	Canopy closure	Canopy closure (%) within 45° cone with ocular tube
SOILT	Soil temperature	Temp (°C) at 5 cm depth within 10 cm of snake
SURFT	Surface temperature	Temp (°C) of substrate within 10 cm of snake
IMT	Ambient temperature	Temp (°C) of air at 1 m above snake
SURFRH	Surface relative humidity	Relative humidity (%) at substrate within 10 cm of snake
IMRH	Ambient relative humidity	Relative humidity (%) 1 m above snake

Source: Reinert 1984 (table 1), reproduced by permission of the Ecological Society of America.

observations and traps. With the availability of increasingly small transmitters and batteries, researchers increasingly have been able to study smaller animals. Radio transmitters are extremely useful in quantifying habitat use because they allow the researcher to locate animals that might otherwise be unobservable visually, thus reducing one of the biases associated with habitat assessment. However, careful study designs are needed to avoid introducing other biases into the research. Many books and articles are available on radio tagging, including White and Garrott (1990), Kenward (2000), and Millspaugh and Marzluff (2001); see also Turchin (1998) for analyses of animal movements.

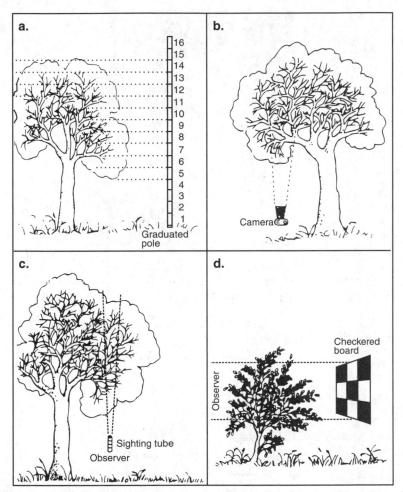

FIGURE 4.6. Some commonly used devices to measure habitat variables in wood-lands.

(a) Graduated pole held upright—most useful to measure the features of the foliage in the shrub layer, and low forests. (b) 35 mm camera with 135 mm or zoom lens—can be focused down through the forest profile (heights read off range-finder) and used to assess foliage density through a vertical section of the forest. (c) Sighting tube—observer looks directly up and assesses the canopy or shrub layer foliage density, or attempts to divide the profile into height bands and assesses vegetation cover within each. (d) Checkered board—used to assess vertical density of shrub layer. Observer walks away from the board until 50% of the board is assessed to be obscured by vegetation; this produces an index of the shrub density that can be repeated at a variety of heights. It is important that the same observer assesses when 50% of the board has become covered as observers may vary in this ability. (From Bibby et al. 2000, figure 11.8)

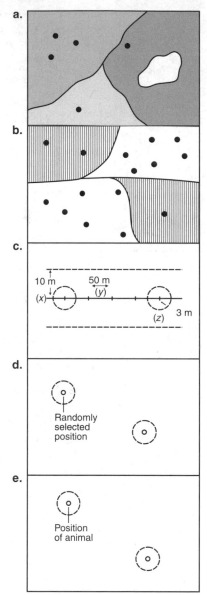

FIGURE 4.7. Scale of habitat recording for wildlife studies.
(a) All vegetation types are mapped without any habitat measurements, and the locations of animals are marked on the map (●). This method produces a broad understanding of habitat use, but it is difficult to test any relationships statistically.
(b) Habitat is subdivided into parcels on the basis of criteria such as vegetation age or plant species composition (□ = recent clear-cutting; ■ = old clear-cutting). Animal registrations (l), derived from a mapping census, are allocated to each parcel and compared with quantitatively measured habitat variables. The habitat data from

Synthesis

Advancing our understanding of habitat relationships will require increased cooperation among wildlife scientists, conservation biologists, and restorationists. Standardization of terminology will assist with such cooperation by promoting use of a common language. Wildlife scientists have also expended considerable effort studying efficient means of quantifying animal habitat. Restorationists will be able to accelerate achievement of project goals (for wildlife) by closely studying the strengths and weaknesses of previous wildlife-habitat studies. That is, there is no reason to reinvent the wheel, or repeat mistakes. Analyses that are relevant to specific spatial scales are critical in restoration planning because we are able to relate the way animals use their environment to the size of the actual area being restored. By placing our studies and subsequent restoration plans into a spatial context allows us to identify what animals appear to require to occupy an area, then to survive and mate in the area, and finally to produce offspring in the area.

---

the parcels are produced independently of the mapping and a statistical comparison between the two to test any significant relationships is possible.

(c) Habitat variables are recorded in standard sample plots at measured distances along the route of a transect count. This produces data on habitat variables in the same position as the transect count and allows the use of multivariate statistical methods to test relationships between animals and habitat variables. (x) = transect band width, (y) = measured transect segments, (z) = example radius of habitat recording circle.

(d) Habitat variables recorded in sample plots around the position of randomly located point counts. This produces detailed habitat data in the same position as the point count. Again this method allows the use of multivariate statistical methods to test relationships between animals and habitat variables.

(e) Habitat variables are recorded at the position of a territorial, feeding, or radio-located animal. This produces precise habitat data in an area selected by the animal. By also recording habitat variables at a random selection of plots within the study area it is possible to quantify habitat selection by the animals in terms of measured differences in habitat variables that were used and avoided. (From Bibby et al. 2000, figure 11.1)

Chapter 5

Assemblages

Based on results of over a decade of research on birds of New Guinea and its satellite islands, Diamond (1975) developed *assembly rules,* which described broad patterns of species co-occurrence. For example, Diamond found species with similar food habits would seldom co-occur: an island might harbor species A or species B, but never A and B together. Diamond called this pattern a *checkerboard* distribution and attributed it to competition between species for limited resources.

Assembly rules have been an important research focus of ecologists ever since, and were more recently reviewed in edited volumes by Weiher and Keddy (1999) and Temperton et al. (2004). Assembly rules became controversial because Diamond and others had difficulty showing that competition was, indeed, responsible for the patterns they found. For example, Connor and Simberloff (1979) showed that patterns similar to those reported by Diamond could be created if the individual species were randomly distributed across the extent of the study area (i.e., a null model) and did not require the driving force of competition. The Connor and Simberloff study represented a fundamental challenge to means of interpreting patterns and the speculation of assembly rules that define observed patterns in communities. As summarized by Gotelli (1999), the controversy in ecology over null models and competition continues to date.

In this chapter I review the concept of assembly rules and how it can be used in restoration planning. Although the concepts of competition and assembly rules are, indeed, controversial, these concepts do provide a means of framing some key aspects of how to plan for restoring wildlife. The concepts developed in this chapter help form the foundation for identifying the desired condition for wildlife in a restoration plan that I discuss in chapters 6 and 7.

Assembly Rules

As reviewed by Temperton and Hobbs (2004), the study of how to assemble species encompasses various approaches to finding rules that govern how ecological communities develop. Various schools of thought have emphasized biotic interactions (e.g., competition), whereas others have emphasized a more holistic approach in which the interaction of the environment with the organisms of a community, and the interactions among organisms, restricts how a community is structured and develops. Temperton and Hobbs concluded that no real consensus existed about what assembly rules constitute, and that as long as each school of thought clearly defined how it sees assembly rules, the debate will be able to progress accordingly and forge useful links to restoration practice. They went on to note, however, that it would be better if we could at least agree on the level of abstraction at which we look for assembly rules.

As implied in my review of community ecology in relation to assembly rules, I adhere to a comprehensive view in which biotic and abiotic factors, in combination with other constraints on species, determine the species and their abundance in a specific location. As developed in chapters 3 and 4, it is critical to identify the primary factors that can limit the occurrence, abundance, and performance of a species. For the restorationist, how these limiting factors function to put boundaries on the species that can potentially occur in a project area can be captured in the assembly rule process.

Temperton and Hobbs (2004) reviewed the three major models of community assembly: the deterministic, the stochastic, and the alternative stable states. In the *deterministic model*, a community's development is seen as the inevitable consequence of physical and biotic factors (Clements 1916). If natural and degraded communities follow the deterministic model of assembly, then reassembly of a community after disturbance should occur along predictable lines of development. In the *stochastic model*, community composition and structure is essentially a random process, depending only on the availability of vacant niches and the order of arrival of organisms (Gleason 1917, 1926). If communities follow the stochastic model, then there is no reason to expect exact reassembly of the community after a major disturbance. Any number of new states could develop, depending on the availability of organisms for invasion, on environmental conditions, and on historical events.

Developed more recently, the *alternative stable states* model is intermediate between the first two and asserts that communities are structured and restricted to a certain extent but can develop into numerous stable states because of an element of randomness inherent in all ecosystems (Sutherland 1974). If communities follow the alternative stable states model, then one would expect the recovery of a degraded system to follow one of several possible trajectories, depending on historical events, the availability and order of arrival of organisms, and the element of randomness inherent in all systems. Successful restoration in this case would entail knowing the history of the degradation and how best to restore the functions of the system, before or concomitant with the introduction of target species. Temperton and Hobbs (2004) concluded that most ecologists would tend to agree that the alternative stable states model of assembly is likely to be the one that most closely reflects reality in nature.

Keddy and Weiher (1999) described the procedure for finding assembly rules as a four step process:

1. Define and measure a property of assemblages.
2. Describe patterns in this property.
3. Explicitly state the rules that govern the expression of the property.
4. Determine the mechanism that caused the patterns.

They also noted that, contrary to common practice, documenting a pattern is not the study of community assembly. Additionally, demonstrating that a pattern exists relative to a null model is insufficient. Although null models are a valuable tool in identifying patterns, they do not identify the mechanisms causing the pattern to exist. Rules must be explicit and quantitative in nature (Keddy and Weiher 1999). The authors concluded that the search for assembly rules will benefit from a multitude of perspectives and approaches, and that we should remain open-minded to avoid pointless debates that really hinge on differences in style. I concur with Keddy and Weiher and add that we would be wise to focus on quantifying the relative contributions of biotic and abiotic factors and other constraints, rather than simply debating the role of competition in structuring assemblages and if an ecological "community" actually exists.

Echoing Keddy and Weiher (1999), Temperton and Hobbs (2004) noted that patterns are found more often in ecology than the mechanisms that caused them. The interactions between various abiotic factors makes it difficult to determine whether alternative stable states are due to species interactions, abiotic factors, or chance alone. As Temperton and Hobbs stated, restoration practitioners must try to produce a certain type of ecosystem and need guidelines to follow, and if possible, knowledge of mechanisms and the patterns involved in community development. We can develop general guidelines for application to large spatial scales and for predicting gross measures of animal performance (e.g., presence-absence). But, as I detailed in chapter 4, it becomes necessary to identify the mechanisms underlying animal performance as we work in smaller and smaller spatial areas and wish to restore a certain abundance and productivity of target animal species. I generally agree with Temperton and Hobbs (2004) that the more general guidelines or rules are, the more useful they will be to restoration ecologists working in different ecosystems. But, the rules must become more detailed (and likely complicated) as we wish to restore more than simple animal presence. Lockwood and Pimm (1999) presented a review of the success of restoration ecology projects and found that, although ecosystem function and partial structure could be restored to human-damaged systems relatively easily, restoring species composition and diversity (complete structure) was far more difficult. Hobbs and Norton (2004) concluded there is now evidence to suggest that different assemblages result from different starting conditions, order of species arrival or introduction, and type and timing of disturbance or management.

Terminology

As reviewed by Morrison and Hall (2002) and discussed throughout this book, ecological terminology is not well standardized and often poorly used in the literature. As described above, one term that is directly involved with assemblage rules is *community*. I will not venture too far into the debate over whether or not a community actually exists in nature. As summarized by Morrison and Hall (2002), a survey of the definitions of community reveals that the co-occurrence of individuals of several species in time and space is common to most. Also, most definitions stress the role of interdependences among the species (populations) under study. Wiens (1989b, 257–58) concluded that multispecies assemblages do occur in nature, and we should concentrate on identifying interactions among these species. Wiens also noted that we cannot hope to understand such groupings of species and interactions among them if we choose study areas arbitrarily. Such cautions on assuming that a community of species exists are relevant to assemblage rules, because it is clear that what we recognize as a community is driven by what we place as boundaries around our study areas. What we quantify as interactions among species will change as we change the spatial extent of our study areas; interactions also change as we change the temporal component of our investigations.

Thus, I prefer to follow Morrison and Hall (2002) and use the term *species assemblage* to denote the group of species that are present and potentially interacting within a study (or restoration) area. Such an assemblage could be part of a larger community, but there is no need to even invoke the community concept per se with regard to my treatment of assemblage rules for restoration in this book, or in ecological restoration. Rather, I focus on identifying the filters and constraints that will modify the species present in an area throughout a successional pathway. By doing so I do not discount the need to know if interactions among species are occurring on a larger spatial scale than the restoration site under study, but I only need to know if these interactions are serving as constraints to the species that could occupy the site of interest. Thus, I at least avoid the controversy surrounding the *do communities really exist* question. And, speaking in terms of assemblages seems appropriate when discussing potential rules that guide the assemblage of species. To avoid confusion I do, however, retain the original terminology used in literature that I cite.

Species Pool

Van Andel and Grootjans (2006) summarized the *species pool* concept, first defining a regional species pool as occurring within a biogeographic region and extending over spatial scales of many orders of magnitude larger than those of a local species assemblage. We can envision the species pool becoming increasingly smaller as we go from, say, the watershed scale down to the stream-reach scale, as depicted in figure 5.1. The local assemblage from the larger pool is determined by passing through a series of filters. Van Andel and Grootjans (2006) defined the following pools:

- *Regional species pool:* the set of species occurring in a certain biogeographic or climatic region which are potential members of the target assemblage
- *Local species pool:* the set of species occurring in a subunit of the biogeographic region, such as a valley segment
- *Community species pool:* the set of species present in a site within the target community

Following the terminology I am using herein, community and assemblage would be synonymous.

As reviewed by Hobbs and Norton (2004), the concept of *ecological filters* forms one of the main approaches in assembly rules theory. Out of a total species pool of potential colonists, only those adapted to the abiotic and biotic

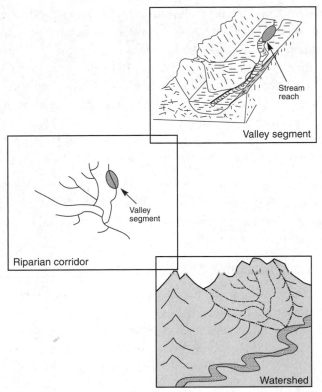

FIGURE 5.1. Schematic of the hierarchical scales used in watershed, riparian corridor, valley segment, and stream reach. (From Chambers and Miller 2004, figure 1)

conditions present at a site will be able to establish themselves successfully. A process of deletion takes place, analogous to a filtering out of those organisms not adapted to the habitat conditions. This approach focuses on the end product of numerous interactions between a colonist and the ecosystem components. Hobbs and Norton (2004) noted that the potential for assembly rules in restoration ecology is in its application at the beginning of restoration projects to ascertain what factors may be limiting membership in the community. Certain assembly rules are immediately obvious and seem trivial, such as the observation that predators without prey will starve; others go into more interesting aspects of interactions, such as what abundance of prey is necessary for a new predator to enter a community (Temperton and Hobbs 2004).

As noted by Hobbs and Norton (2004), in restoration ecology we start with a set of individual species as the raw ingredients and try to build them into a community. In some cases, we have an in situ assemblage, whereas in others we have little or no in situ assemblages. Hobbs and Norton suggested that there are a number of different filters that vary in their importance along relatively easily defined gradients, and that the effort needed to restore the system to a particular state will vary along these gradients. Thus, the approach taken to restoration in any given situation will be the product of a number of different filters reflecting the different factors that influence the course and success of the restoration. It is also critical to recognize that the action of filters will change over space and time. Hobbs and Norton (2004) recognized the following filters for restoration, which I have modified to apply more directly to animals and plants. Abiotic filters will include these factors:

- Climate: rainfall and temperature gradients
- Substrate: fertility, soil water availability, toxicity
- Landscape structure: landscape position, previous land use, patch size, and isolation

Biotic filters will include these factors:

- Competition: with preexisting and potentially invading species and between planted or introduced species
- Predation-trophic interactions: from preexisting and potentially invading species, and predation between reintroduced animal species
- Propagule availability (dispersal): bird perches, proximity to seed sources, presence of seed banks
- Mutualisms: mycorrhizae, rhizobia, pollination and dispersal, defense, and so forth
- Disturbance: presence of previous or new disturbance regimes
- Order of species arrival and successional model: facilitation, inhibition, and tolerance
- Current and past composition and structure (biological legacy): how much original biodiversity and original biotic and abiotic structure remains.

At least seven figures in Temperton et al. (2004) depict the generalized pathway from the potential pool of species, through various abiotic and biotic filters and other constraints, to the realized species pool in a location. In figure 5.2 I have synthesized these figures into a single diagram depicting pathways and filters, and how species fit into available niche space throughout the course of succession. Complicating such generalized diagrams as figure 5.2 is that the filters change in both influence on a species and in actual type as we proceed through succession and development of the species assemblage.

Restoration has two conceptual bases that relate to the development of restoration plans that emerged from different schools and from studies of dif-

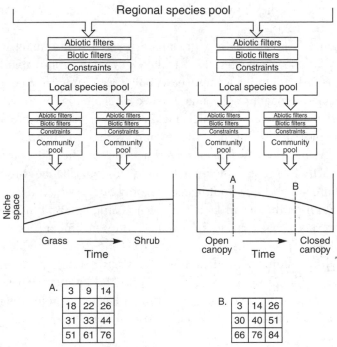

FIGURE 5.2. The species present at any area (community pool) are those remaining after the filtering process occurring at the local and regional levels. The figure represents two local species pools drawn from the same regional pool and co-occurring in time in two vegetation types. The community pool associated with a seral stage is drawn from the local pool that is specific to the more general vegetation type. The type and number of species present across the seral stages will also be a reflection of the size of the target area and niche space available. The cross-sectional cut depicted in the figure as A and B indicates how species (numbered squares) change in type and total number.

ferent taxonomic groups (White and Jentsch 2004). The succession concept emerged from plant ecology, whereas the assembly rules concept emerged from studies of animal communities in the context of island biogeography. These approaches are intertwined because filters, assembly rules, disturbance, and succession are part of a larger concept, namely community assembly. Studies of disturbance and succession contribute to our understanding of the processes that shape communities and contribute to restoration (White and Jentsch 2004). Nuttle et al. (2004) identified another relationship between assembly rules and succession: whereas succession describes the dynamics of changes in species composition, assembly rules deal with the interactions between organisms that determine the trajectory of those changes.

Thus, we witness the distinction between pattern (succession) and process (assembly rules).

In figure 5.2 we see that the regional species pool of Van Andel and Groot-jans (2006) that has been described is filtered through various abiotic and biotic factors to result in the local species pool. Species must first be physiologically adapted to occupy a given area based on general climatic conditions; additional abiotic factors (especially local weather conditions) will have additional filtering affects. But this filtering process has a temporal component that changes in at least two major ways: First, the specific factors that filter or constrain which species that will actually occur can change through time; that is, there are multiple limiting factors, only one of which need be acting at a time to limit the occurrence of a species. And, different limiting factors will be impacting each species or perhaps groups of similar species (e.g., severe weather). Second, the actual habitat or niche space available to species will vary through time as the area under study passes through various successional stages. Of course, the goals of the restoration project will help direct the type and speed of successional stages and, in fact, the project might be designed to hold an area in a rather constant sere (e.g., maintain a wet meadow).

But regardless of the restoration goal, figure 5.2 shows that a multitude of factors must be considered when trying to predict and guide the actual species that will be present on the target site—the community species pool-of Van Andel and Grootjans (2006). As I develop in chapter 6, organizing the objectives and specific target conditions and species within a restoration project must be accomplished in a spatially explicit (hierarchical) manner. You may set specific target conditions, such as the structure and floristics of vegetation, and specific lists of animal species to be achieved during different stages of restoration. For example, as depicted in figure 5.2 you might recognize several seral stages, describe the desired condition of the vegetation and other environmental features within each stage, and describe the potential list of species—the species assemblage—that such accompany each stage. Note in figure 5.2 how the various pools of species change as you proceed through filtering processes that accompany the conditions present within a seral stage. Also note that each subsequent community species pool need not be a subset of the pool at a previous seral stage. Rather, the community pool is under constant reshuffling as a result of the filtering that is occurring through time (succession) from the larger local and regional species pools.

Another advantage of developing such a comprehensive scenario for a restoration project is that it can be linked directly with monitoring and especially development of a valid adaptive management plan, topics that I discuss in chapter 9. That is, you cannot have effective monitoring unless you have specific targets. Even if the ultimate goal of a project is a later successional stage, knowing what should be occurring along the pathway to that desired stage (condition) allows you to make midcourse corrections and improve your opportunity for overall project success. Again, such planning is a core component of adaptive management.

Temperton and Hobbs (2004) argued that if the search for assembly rules is only able to elucidate rules of very specific applicability, then this could treat each project as a unique case study without a guiding conceptual framework. Thus, they advocated working on assembly theories that are as universal as possible and that include the dynamics of a system, because the higher-level characteristics of systems are those that are most likely to yield directly useful results for restoration, rather than rules pertaining to pairwise interactions of specific organisms in specific habitats. I disagree in part with the conclusion of Temperton and Hobbs (2004) because it treats assembly rules as an all-or-nothing approach. Rather, I view the conceptual framework for assembly as a hierarchy that moves from the broad rules to more specific rules as one desires to restore increasingly informative measures of the performance of a species (i.e., from presence-absence to abundance to productivity). I fully agree with Temperton and Hobbs (2004) that we need to include the role of the environment in the search for assembly rules. Thus, general rules are likely to exist for making predictions at broad spatial extents, such as for predicting presence-absence of animals and suitable habitat; rules for predicting breeding locations and ultimate success are much more problematic (and certainly more complicated).

Simberloff et al. (1999) concluded that the search for simple rules determining community structure, such as morphological size or functional group, has been fruitless to date. This failure to locate simple and general rules does not, however, mean that rules do not exist that govern assembly. Simberloff et al. (1999) suggested that assembly rules, if they exist, would be very local and not necessarily very simple, and they will not be very general. They suggested that we gather far more information than is usually done on the biology of individual species, including a focus on behavioral interactions. We need to be quantifying the often subtle responses of species to one another and to small changes in the physical environment. Assembly rules must contain a geographic, habitat, historic, and biotic (e.g., interspecific competition) component (Fox 1999; Simberloff et al. 1999).

Restoration Implications

Restoration efforts need not passively depend on the results of the filtering process but can instead try to modify the effects of filters to allow desired species in and prevent the establishment of undesired species (Hobbs and Norton 2004). Restoration can also try to speed up or direct the natural recolonization process where it is too slow or is impeded in some way (Hobbs and Norton 2004). Modification of abiotic filters for animals would include providing structures (natural and artificial) for use as shelter and water sources. Modification of biotic filters includes controlling exotic species, predator control, and introducing animals (as discussed in chapters 3 and 4). Overcoming dispersal barriers also may include increasing the overall species pool by transporting species from place to place and providing the opportunity for species from other regions to establish (barriers are discussed in chapter 7). Thus, restoration includes designed efforts to manipulate filters to arrive at a desired species composition (Hobbs and Norton 2004).

Belyea (2004) related that although ecological assembly is a branch of ecological theory that is highly relevant to ecosystem restoration, there are important differences in goals and approaches. That is, ecological assembly attempts to explain how the structure and function of a community and ecosystem develop, whereas restoration attempts to direct community development to a particular state. I add that we must, of course, understand ecological assembly if we hope to design and implement successful restoration projects.

Synthesis

The study of how to assemble species encompasses various approaches to finding rules that govern how ecological communities develop. Contrary to common practice, documenting a pattern is not the study of community assembly. Assembly rules must be explicit and quantitative in nature. My review indicates that if assembly rules do, indeed, exist, they would be very local and not necessarily very simple. As such, it is unlikely that we will be able to identify rules that will be generally applicable except at the larger spatial extents (which is a useful starting point for restoration planning). Restorationists need to pursue or otherwise promote studies that gather far more information than is usually done on the biology of individual species, including a focus on behavioral interactions. We need to be quantifying the often subtle responses of species to one another and to small changes in the physical environment.

Restorationists can design projects that try to modify the effects of filters to allow desired species in and prevent the establishment of undesired species, including managing recolonization, modifying abiotic and biotic filters, and overcoming dispersal barriers. Thus, restoration includes designed efforts to manipulate filters to arrive at a desired species composition. We might be able to develop general guidelines for application to large spatial scales and for predicting gross measures of animal performance (e.g., presence-absence). But, it becomes necessary to identify the mechanisms underlying animal performance as we work in smaller and smaller spatial areas and wish to restore a certain abundance and productivity of target animal species.

Chapter 6

Desired Conditions

The initial step in designing a restoration project is clearly establishing project goals and specific outcomes for plant, animals, and the overall environment. In some cases, establishing a goal involves establishing a time period for replicating all or part of a preexisting ecosystem. To be complete, this planning process should include evaluation of the historical animal communities (see Swetnam et al. 1999 for a review). As developed throughout this book, simply providing a general vegetation type or plant association is unlikely to provide the necessary habitat components to allow occupancy by many animal species.

Restoring an area to match some preexisting condition is difficult unless data on historical conditions are available. Thus, the goal of this chapter is to describe techniques useful in reconstructing the historical assemblage of animals in an area. This will be accomplished by (1) describing techniques for gathering historical data on animal occurrences, and (2) determining the uncertainty associated with historical data. Selected case studies of the use such techniques are also discussed. The material summarized herein is taken, in part, from Morrison (2001).

Most animal species currently occupying the earth are survivors of the abiotic and biotic influences of the Pleistocene. The Pleistocene epoch, which began about two to three million years ago, is thought to have ended about 10,000 years before the present. The Pleistocene was characterized by a series of advances and retreats of continental ice sheets and glaciers. Our Recent Epoch is, in fact, probably another interglacial period of the Pleistocene (Cox and Moore 1993). Thus, the distribution and abundance of animal species existing currently can be linked to the geological events of the Pleistocene. The retreat of the ice sheet allowed occupation of vast areas either by species preadapted to the newly developing vegetation or those able to adjust to the new environmental conditions. The present range of many taxa was probably reached in the early Holocene, but some survived in refugia until the late Holocene (Elias 1992). The influence of glacial-interglacial cycles on species geographic range has been substantial (e.g., review by Gutierrez 1997).

For example, in the Southwest both post-Pleistocene dispersal and subsequent colonization as well as vicariant events (the distribution that results from the replacement of one member of a species pair by the other) and subsequent extinction have influenced the current assemblages of mammals (Davis et al. 1988). The authors stressed that the degree to which each process influenced animal distribution should be considered in explaining current faunal composition. Johnson (1994) studied the range expansion of 24 species of birds in the contiguous western United States, and showed that climatic information from the region offered support for wetter and warmer summers in recent decades, which he related to the range expansions. The restorationist can use these phenomena in project design to help identify potential animal communities in the project area; these processes relate to the filters described in chapter 5.

Predicting the specific species composition of a locality is extremely difficult. We can understand this by reviewing the definition of habitat (see chapter 4). Habitat is a species-specific concept that includes more than vegetation. Identifying the vegetation type of a locality does not equate with identifying the many species-specific habitats of the locality. Without observations or specimens, we can never be positive that the species of interest ever occurred on the site. We can assemble lists of probable species occurrences based on the available evidence, such as distance to the nearest verified record(s) of the species. The more thoroughly we can describe the historical vegetative and environmental conditions of the site (e.g., presence of permanent water, soil conditions), the more complete can be our list of potential species. And again, the more thorough our understanding of the processes filtering species, the more likely we are to establish realistic goals.

Kessel and Gibson (1994) concluded that changes in a region's animal assemblages could be categorized as (1) those we believe are real changes, (2) those that merely reflect our increased knowledge of the animal community, (3) those that may be natural long-term fluctuations, and (4) those attributable to confused species identifications. Unfortunately, it is almost impossible to distinguish today whether some perceived changes are real directional changes or just fluctuations, and some species fall into more than one of these categories. For most species we have only either sporadic, often vague, comments on status in the historical literature or data too recent or too incomplete to be a basis for reconstructing past communities and changes through time. Here we explore some methods to try and determine the history of animal occupancy of specific areas.

Historical Assessments

The methods typically used to assess historical conditions are existing data sets, museum records, fossils, and field notes and literature. It would be best to look at the process of historical assessments as a puzzle: that is, you can still identify the picture if some pieces are missing; the number of missing pieces determines the uncertainty in your assessment.

Data Sources

The USGS-Biological Resources Division (formerly managed within the U.S. Fish and Wildlife Service) coordinates the nationwide Breeding Bird Surveys (BBS). Initiated in 1965, the BBS consists of >2000 randomly located, permanent survey routes established along secondary roads throughout the continental United States and southern Canada surveyed annually during the height of the breeding season, usually in June. Each route is 25 miles (40 km) long and consists of 50 stops spaced at 0.8 km intervals (Robbins et al. 1986).

The National Audubon Society (NAS) coordinates an annual bird-counting effort during December. Known as the Christmas Bird Count (CBC), this effort is a single-day count conducted by volunteers in a 15-mile radius (24 km) of a chosen location (usually a city, wildlife refuge, or any other area of interest). Begun in 1900, the effort has grown into a valuable database for long-term monitoring of population trends. CBC data have been summarized by the U.S. Fish and Wildlife Service (now summarized by the USGS-BRD) and by various independent researchers (e.g., Wing 1947). Similar count data were recorded in the *Canadian Field Naturalist* from 1924 through 1939. Counting areas were initially concentrated in the eastern United States. Since the 1950s, however, counting areas have become more common throughout the continental United States. The increasing coverage and density of counting areas is expanding knowledge of bird distribution, and this will allow more accurate assessments of trends in numbers in the future. The data are published annually by NAS (currently in *National Audubon Field-Notes*; formerly titled *American Birds* and *Audubon Field-Notes*).

The BBS and CBC data provide a relatively recent list of the distribution and abundance of birds. These data alone, however, do not provide historical information for many locations at this time. Nevertheless, they are a source of information that might be available for a project area.

There are many published papers and reports that can be used to reconstruct the fauna of most regions. These are the result of natural history surveys conducted during the late 1800s into the early 1900s. There are journals in every region of North America that emphasize natural history reports. A thorough literature survey will allow a reconstruction of fauna occurring in the location where the study occurred prior to the advent of massive human-induced development. Remember, of course, that natural catastrophies (e.g., floods, fires, tornados, drought) also substantially influence the local distribution of species. As I will explain, extending the results of these usually localized surveys to other locations reduces the confidence that can be placed on the reconstruction. Fleishman et al. (2004) discussed the use of historical data as an aid in understanding faunal distributions, with special reference to the Great Basin of western North America.

Museum Records

Natural history collections are housed at a variety of private and public universities, museums, and research organizations. These collections were usually accumulated to characterize the fauna of a region. Each specimen is accompanied by an original paper tag that lists the date and location of collection, the collector, the species identity, and perhaps a few natural history notes. Although museums naturally concentrate on species within their geographic region, many of the larger museums have gathered specimens from throughout North America and even the world. Thus, some museums contain specimens from outside their immediate area. Fortunately, museums are rapidly making their data available electronically. Funding from various entities, including grants from the National Science Foundation, has allowed museums to enter much of their holdings into online databases. Although it is easy to search for records, these online data should not be treated as primary data. While museum managers are constantly updating taxonomy and checking specimen identifications and associated data, issues such as outdated names, mistaken identifications, and erroneous localities do occur in the data sets. Readers should check with their local or regional natural history collection(s) for the most recent information on access to and use of these online databases.

Bird eggs are housed in some ornithological collections. The science of studying eggs, oology, was extremely popular in the early 1900s—several oological journals were published. Egg collections have been valuable in the analysis of natural history parameters (e.g., clutch size, breeding phenology), breeding distributions, and eggshell thinning. The data slip accompanying the egg sets usually included clutch size, nest location (height, plant substrate), and related data. Kiff and Hough (1985) provided a detailed summary of information on the location, specimen holdings, geographic coverage, and related information of egg collections in North America. The premier oological collection in North America is the Western Foundation of Vertebrate Zoology (WFVZ), Camarillo, California. The WFVZ can be contacted for additional information on accessing oological collections.

Limitations and Cautions

As discussed earlier, an increasing number of museums are entering their original data into computer databases. This is a favorable trend as it assists with managing the collection and rapidly answering questions regarding specimen holdings. However, users must be aware that errors occur during transcriptions of original data. In addition, few databases transcribe all of the information—including natural history notes—into the database. Thus, it is wise for the user to (1) first request a computer printout of specimen holdings (e.g., sorted by date and location) of the species of interest, and then (2) request photocopies of the original data slips. In addition, the identification of specimens in many collections has never been verified. Thus, it is often necessary to either visit the collection to confirm identification, request a loan of critical specimens, or ask the museum staff to verify a record.

The presence of a specimen in a collection only indicates that the species was present at the time of collection. The absence of a species in a collection cannot be used to conclude that the species did not occur at the time the collecting was under way. Unfortunately, certain people have misused museum collections by interpreting the lack of specimens of a species of interest (e.g., endangered species) as indicating a lack of occurrence. The novice users of museum data should consult with people experienced with such matters before using the information.

Persons contacting museums for specimen information should be aware of the poor economic climate in most institutions. Thus, even academic users should offer to pay (or expect to be asked to pay) at least for photocopies of data or even the labor involved in accessing and printing computer databases. It is critical that users of museum data clearly discuss the limitations and biases in any records used. This will help alleviate any reluctance of curators to make the data under their supervision available. Requesting that specimens be sent for personal use (e.g., to verify identification, take measurements) should be kept to a minimum: (1) requesting specimens places a large workload on museum staff; and (2) handling and shipping specimens causes wear that alters measurements of specimens and shortens their useful lifetime.

Fossils and Subfossils

The fossil record can sometimes be used to reconstruct the former range of species. For example, Harris (1993) reconstructed the succession of microtene rodents from the mid- to late-Wisconsin period of the Pleistocene in New Mexico, and Goodwin (1995) reconstructed the Pleistocene distribution of prairie dogs (*Cynomys* spp.). Hafner (1993) used the Nearctic pikas (*Ochotona principes* and *O. collaris*) as biogeographic indicators of cool, mesic, rocky areas. Fossil pikas have been found far from extant populations, especially in Nevada (figure 6.1). Such reconstructions not only help us understand why species change in distribution but also lend insight into factors limiting current distribution.

More recent subfossil remains have been used to reconstruct the environment. *Subfossils* are unmineralized remains that may be only a few hundred years old. They are often found in caves, mines, woodrat (*Neotoma* spp.;

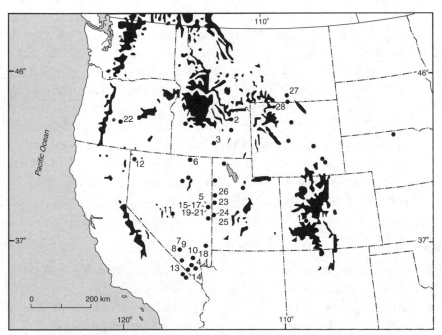

FIGURE 6.1. Distribution (shaded areas) of extant pikas and late Pleistocene-Holocene fossil records (solid dots) in western North America. (From Hafner 1993)

packrat) middens, and in numerous rocky crevasses. For example, Ashmole and Ashmole (1997; see also Olson 1977) used subfossils, in part, to reconstruct the prehistoric ecosystem of Ascension Island (equatorial Atlantic).

Identifying fossil or subfossil remains requires the availability of reference specimens for comparison with the unknown items. Because we are restricting our analyses to the Holocene and forward in time, it is likely that any major natural history museum will contain adequate specimens for comparison. Knowledge of vertebrate morphology is necessary to speed the identification process, although persons with good undergraduate training in wildlife science or zoology can usually perform the analyses.

Kay (1998) reviewed archaeological (and other) data to reconstruct the pre-Columbian ungulate and predator (including human) composition of North America. This work has implications for restoration because it focuses on the processes shaping the relative abundances of animals, and hence their impacts on vegetation and succession. Kay concluded that Native Americans were the ultimate predator that structured North American ecosystems from ca. 12,000 years before present to ca. 1870, especially in the western United States and Canada. Work such as Kay's have large implications for how restoration efforts are directed by establishing goals for the composition of animal assemblages, and thus vegetation types and successional patterns. For example, Kay showed that of nearly 60,000 ungulate bones unearthed at >400 archaeological sites in the United States and Canadian Rockies, <3% were elk and only about 10% were bison. Along with other evidence, he concluded that ungulates were being kept at low numbers (relative to present) by hunting by Native Americans, which—in combination with human-started fires—had large implications for what the vegetative landscape resembled. Smith (2007) reviewed evidence that humans were successfully domesticating animals and plants ~10,000 to 12,000 years ago. Restorationists can use this type of information to reconstruct not only the animal species present and their relative abundances, but also gain an understanding of the vegetation and successional processes occurring in the past.

Literature

Assessing changes in animal communities over time is a topic of particular interest to conservation biologists who are trying to separate human-induced changes from changes caused by other environmental factors (as developed in chapter 2, these are usually termed *natural* changes, although this is strictly a misnomer given that humans evolved on this planet). Thus, there is much literature on determining past animal communities.

Documentation of changes requires a temporal baseline against which subsequent records can be compared and evaluated. Prior to the early 1900s, such baselines were lacking for most regions. From the early 1900s through the 1950s, regional surveys accompanied by mass collecting established reasonably precise ranges for many species.

Power (1994) published a good example of assembling historical avifaunal records by using documents beginning in the 1850s (e.g., from the U.S. Pacific Railroad Survey) and continuing through the early 1900s to reconstruct the distribution and abundance of birds of the coastal islands of California. Other useful papers appeared in *Proceedings of the California Academy of Sciences, Pasadena Academy of Science Publication, Proceedings of the National Academy of Sciences, Pacific Coast Avifauna,* and the journals *Condor* and *Auk*. Note that both national and relatively regional outlets are a source of substantial information.

The National Audubon Society has published a compilation of bird observations submitted by the public throughout the 1900s to date. These observations are divided by geographic regions of North America and summarize seasonal (winter, summer, fall) observations. Each region has experts in the birds of the region who compile and verify the records submitted by field observers. The narrative accompanying many of the observations provide explanations on the status of the species in the region. These data are useful in helping to interpret the status of rarer species of a region, although they have little if any long-term data on site-specific changes in occurrence. Of course, summarizing such information requires many hours of tedious searching through thousands of records.

The field notes, journals, and other written records of scientists and other observers are contained in various publications (books, journal articles) or housed in their original form in museums. These written records often contain species-specific information on distribution, abundance, breeding status, and other natural history notes. They can help reconstruct the animals present in the region surveyed, along with the general environmental conditions.

Uncertainty

Without direct observations of the presence and activity of a species, we cannot assign a specific probability to our reconstructions of historical animal assemblage. Even if specimens exist for the specific locality of interest, we will usually not know the status of the species on the site. That is, even the presence of an adult of a nonmigratory species at the location does not mean it was actually breeding there. Naturally, we would assign a higher probability of residence on the site in this situation compared to the finding of a specimen for a migratory species. Thus, we can assess the quality of our historical reconstruction by assigning probabilities of certainty to each data source. Such assignment of uncertainty is primarily relative and qualitative. Factors to consider in assigning uncertainity include the following:

- Age of data source
- Distance from data source
- Quantity and quality of data sources:
  - one record versus numerous records
  - records from brief time period versus samples across time
  - actual specimens versus visual observations
  - completeness of data record
  - reputation of data source

The uncertainty of each conclusion needs to be discussed, and all assumptions must be stated.

Developing Desired Conditions

As developed in chapter 5, a conceptual model of ecosystem processes and functions, filters, and assemblages of species is a fundamental step in designing a restoration project. Development of this model involves a process of feedback between the desired ecological conditions and what is ultimately feasible to implement and maintain given current environmental conditions, budgetary constraints, and legal mandates. For any restoration project, such a model must consider the environmental conditions in areas surrounding the overall planning area. That is, it makes little sense to try and restore habitat for a specific suite of animal species if immigration and emigration are impossible, or if predators or competitors cannot be managed either within the project area or on surrounding lands.

Thus, simple conceptual models consider the main ecological processes that are necessary to achieve the desired ecological condition, and how restoration can be achieved within the context of these processes. Such visual models are also valuable in explaining the rationale for the overall project design. For example, if fire is an essential component in maintaining the condition of the vegetation, then the frequency and intensity of fires must either be incorporated into the restoration plan, or alternatives to fire must be developed (e.g., cutting trees and thinning shrubs). Regardless of the management tools ultimately used, the models provide a simple visual representation of the ecological processes necessary to maintain the desired condition (where the desired condition has been developed in partnership with stakeholders). For example, spring snowpack is ultimately an essential driver of summer meadow wetness; and meadow wetness is an essential component of the breeding success of meadow-inhabiting birds. In this scenario, project managers must determine if meadows can be maintained through spring snowpack over the long term, or if other management actions must be planned to compensate for overall drying of climatic conditions (e.g., modifying stream flows and channels to ensure meadows stay wet regardless of natural snowpack). It would be difficult at best to plan for this type of scenario without first developing a conceptual model of ecosystem processes and functions. One of the case studies presented in chapter 10 deals with this issue of managing meadow wetness.

An essential component of this framework is incorporation of the ecological requirements of key animal species. While developing and maintaining a functioning ecosystem in the desired condition will provide habitat and niche components for the majority of species, it is unlikely that all key requirements for many species can be achieved in this manner given the degraded environments we are working in and ongoing human impacts and disturbances (as detailed in chapter 4). For example, unnatural numbers of native predators and exotic species can prevent restoration of essential niche components for many species regardless of the restored condition of the physical habitat. Additionally, because of legal and administrative mandates for conservation of selected species (e.g., legally threatened or endangered species, species of special concern), individual species or small groups of similar species must usually be incorporated into this framework (see discussion of focal species later in this chapter). Thus, in addition to rehabilitation of vegetative structure and hydrologic functions, additional actions must be taken to account for the niche components of animal species in the restoration activities and subsequent management actions.

As developed in earlier chapters (especially chapter 4), it is critical to the success of planning for successful restoration of wildlife that the distinction between habitat and niche factors, and their relevance to the viability of animals, be recognized. Regardless of the apparent appropriateness of habitat, an animal may be absent because the niche factors are inappropriate. This is a primary reason why models of the presence or viability of animals based on habitat factors often result in poor predictions, and why different habitat models are needed for different locations and times.

Thus, failure to account for niche factors in restoration planning often results in poor success for wildlife diversity and viability. For example, regardless of the physical appropriateness of vegetation, the presence of cowbirds usually results in breeding failure for a host of songbirds, including such focal species as the willow flycatcher. Likewise, failing to account for changes in the size distribution and availability of arthropod prey—through provision of appropriate plant species—across the spring and summer will likely result in inadequate food resources for many species. In addition, the complexity of food webs also needs to be considered in restoration planning. For example, the patterns of insectivorous bird predation were altered by elk browsing on aspen trees, because the consumption of aspen shoots by elk reduced the quantities of galls produced by sawflies, the presence of which had significant and positive effects on the species richness and abundance of other arthropod species. This latter example again highlights the need for development of a sound conceptual ecological framework prior to initiation of restoration planning.

Although a thorough evaluation of niche relationships adds time and effort to the front-end planning of a restoration project, such work substantially improves the efficacy of the final restoration plan and drives anticipated postrestoration management activities. For example, if initial analyses indicate that a songbird species would experience high nest parasitism and thus poor breeding success, there would be little reason to attempt to restore habitat for the species unless intensive cowbird management was an integral part of the restoration plan. Likewise, in the previous example of the web involving birds, elk, aspen, and arthropods, failure to incorporate control of elk in the project area would likely result in poor aspen regeneration and negative responses (e.g., breeding failures) by the bird community.

Of course, it would be a Herculean task to try and resolve all ecological relationships in a restoration project. However, following the more practical yet extensive process of identification of key ecological attributes, in combination with more detailed efforts for focal species, will result in a comprehensive plan that has a high probability of achieving the desired ecological condition and substantial biological integrity. The successful establishment and breeding of willow flycatchers would, for example, be indicative of restoration of properly functioning wet meadow systems.

Focal Species

Wildlife management has been mostly defined in terms of economic impacts and legal mandates, often focusing on meeting regulatory edicts and litigative exigencies. Thus, the focus is often on the lists of threatened, endangered, and rare or sensitive species. While other species are not excluded from the realm of wildlife management, they often receive diminished or no formal management attention. Arguments have been raised that it is too compli-cated to address the full ecological community and that selected signposts or indicators, largely species of legal concern or consumptive interest, must be chosen. Management agencies and many researchers have thus attempted to devise ways of simplifying how we study and manage species, including a fo-cus on broadscale management of vegetation (i.e., landscape approach) and/or a focus on a selected list of animal species. This latter approach has been broadly termed a *focal species* approach (not to be confused with focal animal sampling in behavioral ecology; see chapter 4).

Thus, many management agencies have relied on a *coarse-filter* approach, the basic tenet of which is that managing generally defined or broadscale habitat conditions, such as managing forests within historical ranges of natu-ral disturbances such as wildfire, will provide for the needs of all associated native species (Hunter 1991). According to Morrison et al. (2006, 382), the coarse-filter approach usually fails to protect a large number of native species. In the coarse-filter approach, *wildlife* is operationally defined as the often un-specified wildlife community that is associated with some general macrohab-itat condition as defined for one or a few species (the indicator species ap-proach, e.g., Abate 1992), or defined as in some other general way such as by reconstructing historical conditions (the range of natural variations ap-proach, e.g., Morgan et al. 1994; Fule et al. 1997). However, the environ-mental requirements of many species are not necessarily met by this ap-proach (Landres et al. 1988; Niemi et al. 1997).

The focal species approach as a tool for defining the environmental attri-butes and, in turn, the management practices needed to meet the conserva-tion requirements of animals and plants, and orienting restoration efforts was synthesized by Lambeck (1997). Lambeck proposed that a group of species would be selected as a focus of management that were the most influenced by specific threatening processes, were area sensitive, dispersal limited, and re-source limited. Related concepts such as indicator, umbrella, and flagship spe-cies have generally been shown to fail to provide for effective management of multiple species. For example, Lindenmayer et al. (2002) reviewed and criti-cized the focal species approach, in part, on grounds that it incorporated such troublesome concepts such as indicator species. Lambeck (2002) countered Lindenmeyer et al. and noted that the focal species approach is multifaceted in that it can incorporate multiple species with many different resource re-quirements and responses to different threatening factors.

Fleishman et al. (2004) briefly reviewed the reasoning behind a concept re-lated to the focal species concept, namely the surrogate species concept. They noted that, in theory, certain species might serve as reliable and cost-effective

TABLE 6.1

Definitions of different categories of surrogate species

Category	Definition
Umbrella	Species whose conservation confers a protective umbrella to numerous co-occurring species
Indicator	Species whose distribution, abundance, or population dynamics can serve as substitute measures of the status of other species or environmental attributes
Keystone	Species that significantly affects one or more key ecological processes or elements to an extent that greatly exceeds what would be predicted from its abundance or biomass
Ecosystem engineer	Species that, via morphology or behavior, modifies, maintains, and creates habitat for itself and other organisms
Flagship	Charismatic species that serves as a symbol to generate conservation awareness and action
Focal species	Species used, for any reason, to help understand, manage, or conserve ecosystem composition, structure, or function

Source: Fleishman et al. (2004).

measures of other variables that are difficult and expensive to measure directly, including total species richness and various ecosystem functions. They classified under the rubric *surrogate species* concepts such as *umbrella species, indicator species, keystone species, ecosystem engineers, flagship species,* and *focal species* (table 6.1). They also noted that evidence to validate the utility of surrogate species, and guidance on how to select effective surrogates, is sparse.

A key assumption of the focal species, surrogate species, and related approaches is the assumption of the nested subset. In figure 6.2 we see (figure 6.2a) that nested niche requirements lead to a close similarity in responses of different species to perturbations. But if species are not nested (figure 6.2b), responses to perturbations could be largely dissimilar depending on the nature and intensity of the perturbation. Clearly, as more species are added to consideration, the more unlikely that nestedness will occur. Nested subsets have been used to manage based on *guilds,* although the guild concept is usually misused in ecology (e.g., Morrison and Hall 2002).

Studies of actual applications of the focal species approach generally conclude that the approach can be a useful starting point for conservation. Lindenmeyer et al. (2002) and Freudenberger and Brooker (2004) showed that weaknesses of the focal species approach included the following:

- Focal species do not serve as surrogates for other species.
- Selection of species often shows high social bias (i.e., species of interest to the public such as large predators).
- The assumption that threatening processes are independent is false.

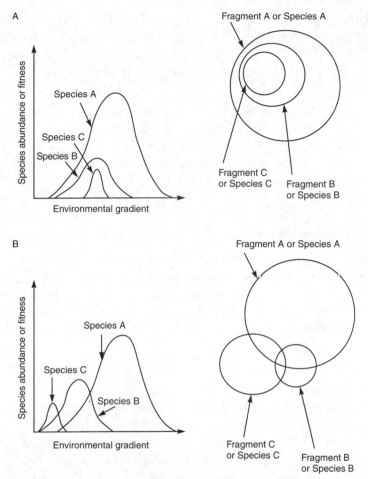

FIGURE 6.2. The principle of nested-subset theory in relation to niche requirements and habitat fragmentation. (A) Nested-niche requirements may lead to
nested-species assemblages. (B) Alternatively, species may not have overlapping
niche requirements, which would lead to nonnested species assemblages. (From
Lindenmayer et al. 2002, figure 1)

- We do not know enough about every species to correctly choose focal
  species.
- Empirical testing of the response of species to management actions is
  minimal.

Thus, there is no single approach to developing a list of species—the focal
species—for emphasis during project planning. There will be a multitude of
reasons for selecting species. Where we run into trouble is when we try to use

one or a few species as surrogates (indicators) of how other species will respond to a threatening process and management action. As summarized by Morrison et al. (2006, 382–83), species are selected as a focus for study and management based on economic, cultural, ethical, and even esthetic dimensions.

In the next section I outline a general procedure for identifying focal species for use in restoration planning that can accommodate a multitude of ecological and nonecological criteria. The key element of any focal species approach is to clearly document the rationale used for all selection criteria. Thus, although people might not agree with your choice of focal species, at a minimum they can at least understand why the selection was made.

Implementation Steps

Here is an outline of specific implementation steps for wildlife restoration that are structured in a spatially explicit context; that is, from broad spatial extent on down to local scale applications. Each step in selection of species should be clearly documented in all restoration planning processes.

STEP 1: PLANNING AREA (E.G., BASIN COMPOSED OF MULTIPLE WATERSHEDS)

Determining the desired ecological condition of the largest planning area, and the ecological process that must be present or provided, is the essential initial step in developing a restoration plan. This is because each management unit (e.g., watershed) within the overall planning area will be influenced by the surrounding areas, and further, the plan developed for each management unit must reflect the needs of the overall planning area. For example, it will be impossible to adequately plan for the distribution and condition of a specific ecosystem element, such as old-growth forest, if old growth cannot be allocated across the management units within the overall planning area. Within the framework of the desired ecological condition, the requirements of specific focal ecosystem elements and species must also be planned and allocated across the management units. Priority steps to conduct include the following:

- Determine desired ecological condition of the planning area.
- Develop conceptual ecosystem model of the planning area that identifies major factors that will lead to desired ecological condition.
- Identify key ecological attributes that drive the ecosystem model(s).

- Identify suite of species characteristic of the desired condition.
- Identify focal species
  - Legal requirements
  - Special status (state, federal, local)
  - Other
- Identify key constraints and stressors that inhibit proper functioning of the pathways in the ecological model (and thus prevent attainment of the desired condition)
- Identify constraints/stressors that can be alleviated
- Identify constraints/stressors that cannot be alleviated
  - Reevaluate desired condition and associated conceptual model(s) in light of constraints and stressors and modify as necessary.
  - Develop monitoring plan using the key ecological attributes at the overall project scale.
  - Determine response variables, establish quantitative goal, time frame for attainment, and threshold for additional action (see step 3).

STEP 2: PROJECT AREA (E.G., SPECIFIC WATERSHEDS)

As noted in step 1, the restoration plan for each project area (e.g., management unit, such as a watershed) must follow from the desired ecological condition for the overall planning area (e.g., basin). By assessing the current condition of each project area, planners can determine how to allocate target levels of each ecosystem component across the specific projects based on the desired condition for the entire planning area. For example, a restoration plan for a specific project area may include an increase or decrease in the amount of a specific ecosystem element (e.g., old growth, riparian vegetation, goshawk habitat) depending on the current conditions at the project area and what is available, or available for restoration, in other project areas. Additionally, because of the location of a project area, certain management actions (e.g., cowbird control) may be deemed infeasible (e.g., proximity to residential areas). Specific actions to pursue include the following:

- Describe current vegetation and other environmental features.
- Identify location and amount of special features (e.g., old growth, riparian, aspen).
- Identify constraints/stressors spatially and temporally.
- Compare special features to project-wide distribution, area, and

condition. Develop species list from literature, available data, and expert opinion.

— Identify focal species. Evaluate potential of site to maintain each
  species and the constraints on maintenance.
— Develop management plan for all special features and focal species.
• Develop preliminary restoration plan.
— Evaluate efficacy of plan in relation to neighboring (including
  outside project area) environmental conditions, constraints, and
  stressors.
— Revise plan.
• Develop monitoring plan at the project scale.
— Determine response variables, establish quantitative goal, time
  frame for attainment, and threshold for additional action (see
  step 3).

Step 3: Adaptive management implementation

Adaptive management requires the specification, during the development of
the restoration plan, of potential actions that could be taken if monitoring
thresholds are triggered (see chapter 9 for details on adaptive management;
each case study in chapter 10 includes discussion of adaptive management).
Failure to develop potential actions negates many of the core benefits of
adaptive management and results in a nonadaptive, trial and error approach
to resource management. Potential scenarios or pathways that the project
area is likely to follow are a consequence of the initial ecosystem model upon
which the plan was formulated. Varying degrees of uncertainty will be associated with each pathway: if thresholds are triggered then actions may include
management actions within the existing ecosystem, physically modifying the
pathway that the project area is following, and revision in the planned treatment schedules.

Determination of actions to take involves a comparison of the status of key
ecological attributes and focal species between the overall project area and
within project subareas. For example, a relative failure of a restoration action
in one project area may have been offset by a relative success of the same or a
similar restoration action at other project areas, resulting in an achievement
of target values for the project area as a whole. Here again we see the value in
viewing restoration in an overall context rather than as a series of individual,
project-level endeavors (recall the discussion of metapopulation management in chapter 3). This concept is especially relevant for wildlife that is often highly mobile and known to change centers of activity for reasons that are

largely speculative (e.g., changes in the distribution and abundance of competitors and predators, distribution of food quality, population responses outside of the planning area). Specific steps to follow for developing an adaptive management approach include the following:

- Develop specific management actions if a threshold is triggered at either the overall project or sub-project area scale.
  - Implement of additional management actions (e.g., additional plantings or weedings, predator control).
  - Modify vegetation structure (e.g., change tree density).
  - Revise treatment schedules (e.g., increase or decrease burning intervals or intensities).
- Revise conceptual model or models as necessary based on new information and changes in management actions and restoration activities.
  - Revise thresholds and triggers as indicated by revisions in actions
- Continue monitoring at appropriate spatial and temporal scales.
- Feedback to step 3, #1.

Chambers and Miller (2004) presented a succinct example of application of a hierarchical approach to restoration and management using Great Basin riparian systems as a backdrop. Their four-part spatial scheme (diagrammed in figure 5.1) is composed of a watershed, a riparian corridor, a valley segment, and a stream reach. Chambers and Miller noted that, historically, watersheds have been described as the most appropriate unit for ecosystem management and fluvial analyses. For example, upland watersheds in the central Great Basin range from under 5 to 100 km^2, are characterized by second- and third-order streams, are subject to localized climatic conditions that vary with elevation, and contain a variety of vegetation types. Analyses at this spatial scale are used to provide an understanding of the types of responses that result from environmental perturbations. The riparian corridor is embedded within watersheds and represents an integrated network of stream channels and adjacent floodplains and terraces. Analyses at this spatial scale focus on the interactions between hillsides and channel processes, and the types and patterns of vegetation. Then, nested within the riparian corridors are valley segments that have semi-uniform characteristics of slope, width, geologic materials, and climate. The final unit is stream reaches, which are sections of the valley segments with relatively uniform channel morphology, bed material composition, bank conditions, and woody debris. Studies conducted at this final scale are mechanistic in nature and are designed to increase understanding of disturbance processes and potential restoration treatments (Chambers and Miller 2004).

Case Studies from the Literature

In chapter 10 I present four case studies that developed restoration plans for wildlife, all of which used the basic stepwise procedure outlined above. Here I provide some additional background material on development of desired conditions from the literature.

Morrison et al. (1994a, b) analyzed the past and current distribution and abundance of vegetation and wildlife to develop a wildlife habitat restoration plan for the Sweetwater Regional Park, San Diego County, California. They compared new survey results for amphibians, reptiles, mammals, and birds with historical data obtained from specimens housed at the San Diego Natural History Museum. *Historical* was defined as pre-1975 collections because most intensive residential and commercial development began in the mid-1970s. Literature sources were used to supplement the museum records.

In summary, Morrison et al. indicated a substantial loss of native amphibians and reptiles, including four amphibians, three lizards, and eleven snake species. The small mammals community was depauperate and dominated by the exotic house mouse (*Mus musculus*) and the native western harvest mouse (*Reithrodontomys megalotis*). There was an apparent net loss of thirteen mammal species, including nine insectivores (e.g., shrews [*Sorex*]) and rodents, a rabbit, and three large mammals. There was an absolute loss of eighteen bird species and a gain of six species. They used these data, in part, to develop a plan for restoration of the plant and animal communities.

Johnson (1994) calculated distribution changes by comparing this mid-century baseline with subsequently published regional avifaunal compilations, including newer editions of the *Check-list*. For detailed information on pioneers and extralimital nesting over the last three decades, he tallied nesting season records cited in *Audubon Field Notes* and its successor, *American Birds* (which has again reverted to the name *Audubon Field Notes*). He emphasized extralimital late spring records because they often point to pioneering and imminent summer residence. He defined pioneering as the presence of a singing male or a pair in appropriate breeding habitat.

As an example, Johnson (1994) reconstructed the steadily expanding range of the Summer tanager (*Piranga rubra cooperi*) in the Southwest (figure 6.3). His analysis (1) identifies the species' range as of the midcentury, and (2) its steady northward and westward expansion. For this species and most of the other species he analyzed, Johnson concluded that recent range expansions were due to climatic changes (increased summer moisture and higher mean summer temperatures), and not responses to anthropogenic causes. This analysis allows restoration planners to understand the history of

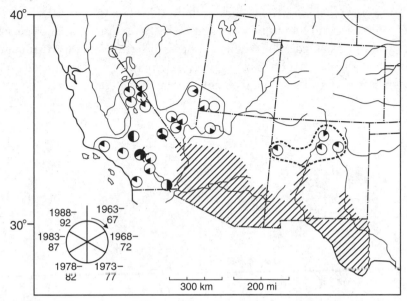

FIGURE 6.3. The pattern of diagonal lines shows the approximate breeding range of the western form of the summer tanager (*Piranga rubra cooperi*) as of the mid-1950s to early 1960s in the American Southwest. Localities of range expansion by pioneers and colonists through the 1990s are denoted by the symbols and dashed line. (From Johnson 1994).

the species in the region of interest, and establish priorities regarding development of vegetation communities and specific plant associations.

Synthesis

Reconstructing the likely animal species assemblages of a proposed restoration site is a valuable means of developing project goals. Unfortunately, the historic presence of a species does not indicate that it could reoccupy the site even if all necessary habitat conditions were restored. Many species have certain minimum area requirements that cannot be met on small restoration sites. Further, conditions on lands surrounding the restoration site may be unsuitable to allow recolonization. Thus, restoration goals for wildlife should be developed in light of both historical possibilities and current realities.

Many attempts have been made to develop guidelines for selecting species as a focus for restoration. In addition to ecological criteria, factors to consider often include economic, cultural, ethical, and esthetic dimensions. There is no single approach to developing a list of focal species for emphasis during project planning; there will be a multitude of reasons for selecting species. Where we run into trouble is when we try to use one or a few species as surrogates (indicators) of how other species will respond to threatening processes and management actions. The key element of any focal species approach is to clearly document the rationale used for all selection criteria.

Chapter 7

Design Concepts

Fundamental strategies for developing restoration plans and designing con
servation reserves have been widely debated for decades. Throughout this
book I have been developing the foundational issues of reserve design, including such core concepts as habitat quality and area, limiting factors, genetics, supplementation of populations (e.g., translocations, reintroductions),
and demography. In later chapters I discuss study designs (chapter 8) and
how to monitor results of restoration projects (chapter 9). In chapter 10, I
present four case studies that incorporate many of the concepts developed in
this (and other) chapters.

In this chapter I first continue to build the foundation for wildlife restoration by synthesizing the large topic collectively known as *habitat heterogeneity*, which includes the interrelated issues of patch size and type, connectivity, fragmentation, and corridors. These and other issues are fundamental to
the theory of reserve design (e.g., see Margules and Usher 1981; Margules et
al. 1988) and have been reviewed by others (e.g., Hilty et al. 2006; Morrison
et al. 2006; Lindenmayer and Hobbs 2007). Here I will focus on the application of principles of reserve design to the restoration of wildlife and wildlife
habitat.

Habitat Heterogeneity

The heterogeneity of resource patches in landscapes has been discussed by
different authors in various ways, and was summarized by Morrison et al.
(2006, 262–66). Among the aspects of resource patch heterogeneity discussed
by Morrison et al. are the richness and diversity of patches, dynamics such
as short- and long-term changes in patch richness, connectivity between

TABLE 7.1

Components of habitat heterogeneity in landscapes

Component	Description
Patch-type richness and diversity	Number and relative area of habitat types for a species within a landscape
Patch dynamics	Incursion and melding of patches over time as a function of disturbance events and successional growth of vegetation
Patch connectivity	Degree of adjacency of patches with similar conditions in a landscape
Patch isolation	Distance from one type of patch to the next (or *nth*) nearest patch of the same type
Fragmentation	Breaking up of contiguous environmental or habitat patches into smaller, more disjunct, and more isolated patches of different types
Corridors	More or less linear or constricted arrays of environments or habitats in a landscape serving to connect larger patches
Permeability	Degree to which an organism can move among patches within a landscape
Edge effect	Incursion of microclimate and vegetation into a patch, typically forested, from a disturbed edge or opening
Edge contrast	Degree of difference in vegetation structure between two adjacent patches

Source: Morrison et al. 2006 (table 8.2).

patches, and isolation and fragmentation of environments; a summary of their definitions are presented in table 7.1 and discussed in this chapter in the context of wildlife restoration.

Habitat heterogeneity is the degree of discontinuity in environmental conditions across a landscape for a particular species. Because habitat is a species-centric term, a particular environmental condition may constitute habitat for one species and a barrier for another. Environmental conditions can include vegetation composition and structure, soils, and hydrology, as well as human disturbances. Discontinuities in environmental conditions occur as *ecotones*, or relatively sharp breaks in environmental conditions, or as *ecoclines*, or broader gradations in conditions over areas of greater geographic extent. Discontinuities can occur naturally, as with changes in soil type or edges of water bodies, or anthropogenically, as with agricultural lands or roads. The cause of the discontinuity is less important to wildlife occurrence and health than is the nature and extent of the discontinuity (Morrison et al. 2006, 262–66). Clearly the type and extent of habitat heterogeneity, in relation to what target species are likely to tolerate, must be considered when developing plans for restoration of an area.

Fragmentation

Fragmentation refers to the degree of heterogeneity of habitats across a landscape, with a focus on the isolation and size of resource patches available. Note that fragmentation is necessarily a species-specific condition. The term *habitat fragmentation*, however, is engrained in the ecological literature and used widely to refer to virtually any sort of heterogeneous condition. Many authors (e.g., Bogaert et al. 2005) have referred to *landscape fragmentation*, which is strictly incorrect, as it is environments or resources (habitats for specific species) that become fragmented within landscapes, not entire landscapes per se. Thus, the following truism must be remembered in designing all restoration projects:

> The species-specific concept of habitat selection, which occurs across spatial scales (chapter 4), must be the focus of restoration with regard to fragmentation (and the broader issues of habitat heterogeneity). Likewise, the concept of species assembly rules (chapter 5) is directly relevant here because fragment size, shape, and the quality of the habitat therein will serve as a filter determining in part if a species can exist in a target area.

Various kinds or degrees of species-specific habitat heterogeneity can be described. In the extreme case, resource or vegetation patches can be isolated into islands surrounded by vastly different and, for specific species, unsuitable conditions. The response of species and assemblages of species to island situations has been the subject of much ecological study, which I will discuss further.

Partial isolation of habitats also pose challenges in research and management. Even though a substantial decline in overall abundance might not be evident, partial isolation can effect population viability by incrementally lowering the numbers of animals per unit area of unsuitable and suitable environments in a landscape, lowering dispersal, and lowering the effective size of the breeding populations.

Another kind of heterogeneity is temporal fragmentation, sometimes called ecological continuity. *Temporal fragmentation* refers to the degree to which a particular environment (e.g., successional stage of a vegetation type) occupies a specific area through time. If an old field ecosystem is perturbed, such as with widespread naturally or human-induced conversion, and then allowed to regrow, many of the original species associated with the original condition might be lost. Thus, regional and site histories are important to interpreting community composition and species occurrence. Much work remains to determine sensitivity of wildlife species to temporal fragmentation both within resource patches and across landscapes, although models of individual response to habitat patch configurations might provide a useful tool (Morrison et al. 2006, chapter 10). Restorationists are usually well versed in determining site histories when planning for recovery of plant communities; this same need to discover site histories applies to planning for animal species.

Heterogeneity and fragmentation also can refer to subtle discontinuities in environmental conditions within a fragment rather than to changes in just gross vegetation structure and successional stage; Morrison et al. (2006, 265–66) discussed these topics in detail. We would expect, for example, small-bodied animals (e.g., arthropods, amphibians, rodents) to react to changes in canopy cover within a fragment (patch of habitat) in a different manner than would larger-bodied animals (e.g., coyote, deer). Smaller animals might spend their entire life cycle within a fragment, whereas larger animals can more easily move between fragments. Once again we see the need to consider the issue of heterogeneity, and ultimately restoration activities, on a species (or at least group of like species) basis. Numerous mathematical indices have been developed to measure habitat heterogeneity (see review in Morrison et al. 2006, 266–70).

Thus, fragmentation is a useful concept if the response of animals is placed into an appropriate spatial hierarchy and temporal pathway. The changes that take place in the environment at several scales of resolution can be summarized as follows (Angelstam 1996):

- Fragmentation in a broad sense that takes place on the between-patch scale (what many would refer to as the landscape scale)
- Fragmentation of different plant associations that take place within a vegetation type
- Alteration of habitat quality within small areas

There are numerous, largely interacting, factors to consider when evaluating the effects of fragmentation on animals. In restoration, planning each of the following factors needs to be considered on a species-specific basis:

1. Absolute loss of habitat area (less patch interior)
2. Increased edge
3. Increased distances for movement of animal between patches

4. Increased penetration of predators, competitors, and nest parasites
5. Changes in microclimate with changes in patch area and edge

Note that this list includes the physical loss of area and changes in patch shape (items 1–2), as well as impacts on animal behavior (3–5) and environmental conditions (5). Changes in microclimate can also impact prey abundance (e.g., insect prey for birds), soil moisture and temperature, and understory plant vigor.

Bolger et al. (1997) studied the response of birds in a matrix of remaining chaparral vegetation in a residential landscape in southern California, finding that 6 species showed negative (area-sensitive) responses, 4 species positive responses, and 10 species showed no apparent response to area size. In an Illinois grassland, 5 species were area sensitive, 3 species responded in a positive manner to edge, and 6 species were restricted to specific vegetation configurations and were not area sensitive (Herkert 1994). Freemark et al. (1995) concluded that the species present in any particular landscape setting showed negative, positive, and no responses to area size. They found areas of <10 ha become unsuitable for many species, patches of 50–60 ha contain up to one-half of the regionally area-sensitive species, and that patch size often needed to reach 100–300 ha before area-sensitive species were present. The first message for restoration from Freemark et al. is, then, that patch size will largely drive what one can expect regarding species occupancy. Freemark et al. also showed, however, that sensitivity to fragmentation and area size was reduced when >30% of the surrounding area was forested. Thus, the conditions surrounding a patch will influence what can occupy the patch, which adds another element to restoration planning and goal setting.

Andren (1995) reviewed the literature and found that 82% of 22 studies conducted in agricultural landscapes showed that nest predation risk in forest fragments increased with proximity to edges and decreased with fragment size. According to Ibarzabal and Desrochers (2001), most studies showed that increased nest predation risk caused by edge effects is more serious for birds nesting in agricultural areas than those in forested landscapes. An updated review by Lahti (2001), however, found that of 54 study sites surveyed, 13 exhibited an edge effect, 31 did not exhibit one, and 10 showed mixed results. Lahti concluded that the type of edge, in terms of the nature of the adjoining plant communities, was not a good predictor of nest predation; but that an edge effect has been slightly more often indicated in landscapes with high fragmentation. Flaspohler et al. (2001) also reviewed the literature and concluded that creation of openings in forested landscapes reduced nest success for some ground-nesting species but not for canopy-nesting species.

Thus, the review shows that multiple interacting factors must be assessed to determine the dynamics of the impacts of predators on wildlife. I agree with Lahti's (2001) recommendation that researchers focus on species-specific predator behavior and its relation to habitat and landscape features rather than searching for a general relationship between edge and predation. Lahti concluded that predictions can be made and tested when the dominant predators in an area are known. Thus, when the dominant predators have been identified and their patterns of predation are known, land areas could be classified into (nest) predation types. Such information would be valuable when considering species most at risk of predation in an area, and the probable effect of changes in predator population sizes and behaviors (Lahti 2001). Thus, differences in the predator community in different landscapes support recommendations that edge effects be evaluated on a region-by-region basis (Flaspohler et al. 2001). These authors suggested that a general theory emerging is that avian nest success is influenced by landscape structure and nest predator community composition at relatively local scales. These conclusions match those made earlier by Martin (1993), who showed that vegetation type (i.e., shrub and grassland versus forested) and the degree of disturbance largely drove nest predation.

If the restoration goal is to conserve specific animal species (e.g., rare or legally endangered species), knowledge on how predators respond to habitat manipulations (restoration) aimed at the focal species is needed if positive impacts on the population(s) are to be achieved. For example, manipulating groundcover (density or species composition) might be used to manage snake populations in an area where such factors are of relatively minor importance to bird activity. Indirect benefits of such a holistic approach to management include an ability to predict when predators will be likely to exert substantial pressure on nesting birds, and a likely reduction in the need to attempt predator control. Direct removal of predators and competitors might also be needed to allow the target species to occupy and persist in an area. The message for all land managers and restorationists is that no general statements can be made concerning how environmental conditions will influence populations of predators that negatively impact target species for restoration. Careful evaluation of the predator assemblage and how it will respond to restoration should be included in project planning; such planning can reduce the probability of direct removal of predators following restoration.

As I discussed earlier, different species will be affected by different aspects of fragmentation. For mobile species with large home ranges, total habitat area is probably more important than interior area (patch size). In contrast, persistence of less mobile animals with small home ranges is likely related to patch isolation. Davidson (1998) developed recommendations for incorporating fragmentation issues into research and planning:

- The choice of spatial scale (both extent and grain) greatly affects development of restoration plans. There is no one correct spatial scale for analysis; efforts to minimize fragmentation at one scale could increase fragmentation at another scale.
- Analyses are affected by the number of patch types and patch definitions. For example, evaluating fragmentation of an oak woodland will likely result in different conclusions than an analysis that broke the same woodland into blue oak and coast live oak woodlands.
- Perimeter:area ratios should not be used as measures of fragmentation. These ratios change unpredictably and do not capture important aspects of fragmentation such as isolation.
- It is preferable to measure different aspects of fragmentation separately (see the numbered list earlier in this chapter), rather than trying to select one best measure or combining them into a single index.

Disturbance Ecology: Dynamics of Habitats in Landscapes

As reviewed by Morrison et al (2006, 270), the study of disturbance ecology focuses on dynamics of habitats in landscapes. By the definitions we use in this book, *landscape* can vary from square meters to square kilometers, depending on the ecology and movement ability of the species under consideration; habitats are species-specific and embedded within these landscapes of varying size. Disturbance ecology includes the study of the spatial and temporal dynamics of soil structure, fire ecology, vegetation succession, meteorology, climatology, and paleoclimatology. The effects of disturbance on wildlife can be evaluated according to their frequency, intensity, duration, location, and geographic extent. Thus, disturbances to a species habitat can be incorporated into spatially explicit models of population demography to project effects on species distribution and viability (Morrison et al. 2006, 270).

Four general categories of disturbances have been identified according to their intensity and geographic extent (figure 7.1):

- Type I disturbances: major environmental catastrophes that are relatively short term, intense, and that affect large areas. They include volcanoes, major fires, floods, and hurricanes (or typhoons in the Pacific Ocean).
- Type II disturbances: locally intense environmental changes from

		Geographic area affected	
		Widespread (>1000 ha)	Local (1-1000 ha)
Degree of disturbance	High	Type I *Major environmental catastrophe* (volcanoes, major fires, hurricanes)	Type II *Local environmental disturbance* (wind, ice storms, insects, disease)
	Low	Type Iii *Chronic or systematic change over wide areas* (predators, competition, forestry)	Type IV *Minor environmental change* (local fires, developments)

FIGURE 7.1. Four types of disturbances shown by degree, or intensity, and geographic area affected. (From Morrison et al. 2006, figure 8.5)

events such as wind storms, ice storms, and local outbreaks of defoliating insects; they often create gaps in forest or woodland canopy.

- Type III disturbances: chronic or systematic changes over wide areas, and include slow alteration of native landscapes for human habitations, ecological succession, and long-term climate change. Wildlife relations to Type III disturbances include changes in species abundance and ecosystems from changes in regional climate; shifts in climate can greatly alter the distribution of vegetation and wildlife over broad areas.
- Type IV disturbances: include minor and local environmental changes, such as low-intensity and local events such as spot fires, low-density rural developments along edges of natural landscapes, and gap dynamics of vegetation canopies. In particular, small vegetation gaps can be caused by natural plant death or by biotic mortality agents such as insect defoliators, plant pathogens, and plant diseases, and also by fire and weather conditions such as ice or windstorms.

The response of wildlife to each type of disturbance within a project area will vary by species, which complicates our ability to make broad recommendations for restoration. Of course, Type I disturbances will, in general, have much wider and catastrophic impacts on species relative to Type IV disturbances. A filter approach, as I described in chapter 5, can be applied to species within the context of the type of disturbance. As vegetation patches change in time so do responses by animals as they use patches for breeding, foraging, refuge from predators, resting, and dispersal. Some wildlife species likely evolved in concert with native disturbance regimes and take optimal advantage of resources distributed through space and time in shifting patches. Thus, the further that human activities alter native patch disturbance dynamics, the greater may be the discontinuity with the evolved habitat selection behaviors of some species. In some cases, even if a suitable environment is present but is greatly altered in patch distribution pattern and temporal occurrence, an associated wildlife species may be excluded.

Across watersheds and other broad spatial extents, disturbances affecting vegetation patch structure and composition tend to alter such ecosystem processes as surface water discharge, nutrient runoff, organic matter input to soils, net productivity, and microclimate. In turn, these changes can influence species composition associated with soil, ground surface, plant canopy, and other substrates (Morrison et al. 2006). Thus, restoration activities must include recovery of the foundational hydrological process so that the desired vegetation assemblages can be developed. Additionally, changes in vegetation patches can alter energy balances of individual organisms, such as by changing food content or values in foraging patches, thereby tipping the balance of foraging efficiency and affecting successful reproduction. Managing native vegetation conditions in landscapes that are subject to relatively frequent, intensive Type II and IV disturbances, such as stand-replacing fires, and in which native vegetation occurs only in small, isolated patches, can be a great challenge to the restorationist. In such cases, restoration of a native *animal community* is not possible; rather, decisions must be made to focus on specific animal species (and assemblages thereof).

Management Lessons

Noting that the dynamics of vegetation and environmental factors consist of a complex medley of changes occurring on multiple "schedules," Morrison et al. (2006, 275) asked "What management lessons can be drawn from this brief review of major categories of disturbances and dynamics of resource patches in landscapes?" They concluded the following:

- Landscapes must be assessed individually to determine which disturbance types occur, the local site histories, and the likely vegetation responses to any disturbance regimes caused or altered by management activities.
- Wildlife can play a major role in affecting disturbance regimes and how habitats respond to disturbances, such as through predation or transportation of disturbance agents (e.g., forest pathogens) and herbivory influence on vegetation otherwise susceptible to disturbances.

- Management may wish to more fully study how activities change native disturbance regimes and how wildlife may respond behaviorally (functionally) and demographically (numerically) to such changes.
- The specific future responses of most ecosystems that incur disturbances at multiple scales of space, intensity, and time are not very predictable. Rather, what can be better predicted, at least as probabilities or as frequencies, are the disturbance regimes themselves. In this sense, management can then craft a set of desired future dynamics, perhaps to reconstruct or mimic native disturbance regimes in which some species may have evolved optimal habitat selection behaviors.
- Studies of disturbance regimes can help guide land and resource management. For example, to what extent does management of vegetation emulate effects of various kinds of natural disturbances?

As I introduced in chapter 6, indicator or umbrella (protection of umbrella species will inadvertently protect a wider range of species) have been proposed as a means of overcoming limited data on sensitive species in project planning. Indicators of any kind must be carefully and narrowly defined to have any predictive value (Landres et al 1988; Morrison et al 1992). Flather et al. (1997) showed that one taxa usually fails to predict the response of other groups to environmental change. Although the use of indicator species is declining, planners still use (and misuse) indicator taxa because they believe that there are no other alternative methods (Roberts 1988). Mountain lions (*Felis concolor*) and golden eagles (*Aquila chrysaetos*) have been used as umbrella species in southern California because both are sensitive to habitat fragmentation. But, because both the lion and eagle can tolerate a degree of vegetation alteration within a fragmented landscape, they can continue to use areas that have become unacceptable to the majority of rare species.

Wildlife-habitat relationship (WHR) models have been used to estimate wildlife occurrence in the absence of information. Morrison et al. (2006) reviewed this topic in detail; two of their most consistent suggestions are the clear identification of objectives and the validation of models at each step of the model's development and use. Without detailed knowledge of occurrence, it is often difficult to estimate the error associated with species distribution, much less the actual distribution. Even slight errors in distribution, combined across the hundreds of species that may be considered in conservation planning, suggest that extreme caution must be undertaken when using surrogates like vegetation type as a predictor of species richness or diversity. Such crude habitat models can help generate initial lists of species potentially occupying a planning area. However, these habitat relationships break down when applied to the small (a few to several hundred hectare) size of most restoration projects.

As reviewed by Hilty et al. (2006, 38), the desired number of patches and their size are encompassed in the SLOSS (single large or several small) concept. That is, should we focus on one large reserve or several smaller reserves in an area? Proponents of the single large reserve emphasized survival of area-demanding species and survival following catastrophic events. Whereas proponents of multiple reserves argued that each reserve is likely to have a unique set of species and that more species can therefore be conserved by setting aside multiple reserves, and that concentrating individuals of a target species in a single patch risked extinction due to a catastrophic event. The SLOSS debate has abated because of recognition that the best plan is to set aside as many reserves as possible and that they should be as large as possible. That said, restoration will have an initial as well as a long-term, reoccurring role in maintenance of habitat conditions within small or large areas.

Corridors

Population viability is usually enhanced when individuals are able to move between populations (recall the discussion of metapopulations in chapter 3). At some spatial scale, all locations are heterogeneous (or patchy). What may appear to be homogeneous to a large mammal may be quite heterogeneous at the spatial scale of the amphibian. The connectivity among patches will often determine if a species can survive in a specific location and ultimately within a larger region. Hilty et al. (2006, 89) used the term *connectivity* to refer to the extent to which a species or population can move among landscape elements in a mosaic of habitat. This definition necessitates linkages among species at appropriate spatial and temporal scales. Corridors are one means of achieving connectivity.

As reviewed by Beier and Noss (1998), until recently most species lived in well-connected landscapes. However, human-induced impacts can disrupt the natural connectivity. A narrow path in the woods might prevent an amphibian from crossing, a condition that would have little impact on the movements of a larger and more mobile animal. Thus, the task of the restorationist is complicated by the species or group specific abilities of animals to move across potential barriers. A restoration plan should include a categorization of species by the potential impacts that various paths, roads, structures, changes in vegetation structure, and other features of the plan might have on species of interest. As such, many conservation biologists have been advocating the retention or development of corridors that help to link landscape patches.

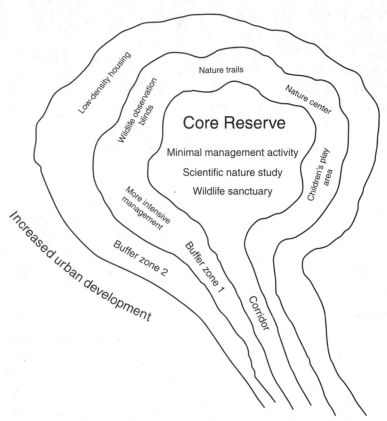

FIGURE 7.2. Typical design for an urban wildlife reserve, including buffers and corridors.

The intuitive appeal of corridors resulted in the widespread recommendation for their use in conservation planning and landscape design. A classic example of a reserve designed to promote population viability is shown in figure 7.2. Other elements of this design are discussed later, but note the corridor extending out of the central ("core") part of the reserve. The concept behind this basic corridor design is to allow movement of animal species of all sizes: first, as a *passage* route for species that can live in the core (or patch) but who are too large to have the majority of their home range contained within the more narrow corridor, such as medium and large mammals; second, as an extension of the core that allows some degree of residency in the corridor, if not breeding opportunities, such as small mammals and small to medium birds.

The use of corridors, however, is controversial. As reviewed by Hess (1994), in certain situations corridors can increase the chance of metapopulation extinction by promoting the transmission of disease. Thus, the probability of disease and other adverse factors (e.g., parasites) should be considered during the planning stages of all reserve networks. The presence of diseases that are known to infect species likely to use corridors should be evaluated prior to linking reserves. Strategies for containing the disease should be developed prior to establishing the linkages. Contingencies for treating epizootics include vaccination, removal of infected individuals, and temporary severing of the linkage.

Simberloff and Cox (1987) and Simberloff et al. (1992) questioned the value of corridors by noting that too little information is available to warrant wholesale adoption of corridors as a conservation action. In addition to disease, corridors can enhance the spread of fires, and can increase exposure of individuals to predation, domestic animals, and poachers. Thus, Simberloff and his colleagues recommended that each potential application of corridors be evaluated on its own merits because species-environmental interactions differ in time and space. Here again I emphasize that corridors per se might have great value, but they must be carefully evaluated and planned on a species-specific basis. Additionally, how a linkage for one species might negatively impact other species must be considered.

Simberloff et al. (1992) suggested that managing the landscape as a whole, rather than each reserve as an independent unit, could reduce the need for corridors. For example, maintaining patches of vegetation surrounding a core area in different seral stages could allow needed movements without establishing and maintaining a linear corridor. Another alternative is the development of *stepping stones* that are close enough together to allow animals to move between patches without the establishment of direct linkages (I discuss various types of corridors in more detail later). Mills (2007, 218) noted that we must go beyond asking "whether corridors work" and instead ask, "is the intended role of a corridor to induce emigration from a fragment, or simply to direct movements?" Unfortunately, there is not much evidence that corridors actually do the former, but there is strong evidence that they do the latter (Haddad et al. 2003). Mills (2007, 218) went on to note that if augmenting dispersal is the goal, then you need to determine how much dispersal is needed for the target species in the project region, and how best to achieve it. Linear corridor across the landscape, or even a simple passage under a highway, may be an effective management action.

Beier and Noss (1998, 1249) made an important statement by defining corridors with regard to "specific wildlife populations." They noted that evaluating the utility of corridors ". . . only makes sense in terms of a particular focal species and landscape." This conclusion follows naturally from our previous (chapter 4; Hall et al. 1997) development of the habitat concept as a species-specific phenomenon. As such, generalizations about the utility of corridors are difficult to make.

In the species-specific context of corridors, then, we can define *corridor* as a linear habitat, embedded in a matrix of dissimilar habitats, which connects two or more larger blocks of habitat (adapted from Beier and Noss 1998). *Passage* is defined as travel via a corridor by individual animals from one habitat patch to another. Beier and Noss (1998) explicitly excluded from their definition of corridor linear habitats that support many species but do not connect large habitat patches, such as narrow riparian strips embedded in agricultural fields. Although I understand their rationale, they have apparently not considered the spatial nature of patchiness in making this distinction. As introduced above, individuals with small home ranges but relatively long dispersal distances could be perceiving patchiness at the forest floor scale. Here, for example, changes in soil moisture or ground cover could substantially change the relatively fine-scale patchiness of such an animal's landscape. Thus, planning for corridors potentially requires designs at multiple spatial scales, depending on the overall goals of the study and species involved.

As discussed by Diefenbach et al. (1993), restorationists need to carefully study all aspects of the environment that animals might encounter. Readers can consult Morrison et al. (2006) for a more thorough review of corridors and empirical evidence (or lack thereof) of animals dispersing through and otherwise using corridors.

Corridor Management and Research Needs

Various suggestions have been made for the design of corridors (e.g., Hilty et al. 2006, chapter 5). In figure 7.3 I describe several of the typical designs for corridors. The classic depiction of a corridor (figure 7.3A) is a continuous passageway of some width that links two or more core habitat areas. Because a continuous linkage might not be possible or even required (e.g., in the case of some birds and other animals), the stepping-stone concept of corridors (B) has been proposed. The stepping-stone concept can also be expanded to include actual linkages between the steps (C), with the stepping-stone areas perhaps serving as refugia where animals can rest and perhaps feed but not actually reside. Finally (D), we can consider the situation where the corridor is a lower quality area relative to the core habitat, but one in which animals can move about but not actually breed or survive indefinitely. In this latter situation the area surrounding the core habitats could be viewed as sinks and the cores as sources. Certainly other combinations are possible, such as plac-

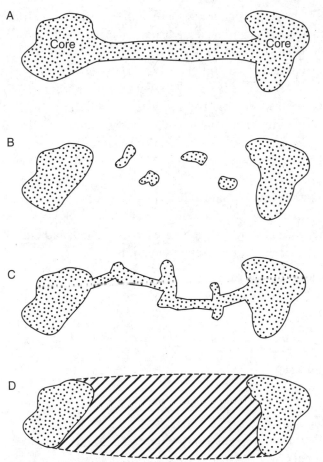

Figure 7.3. Schematic diagrams of general types of corridors. (A) The classic design with continuous linkage (or passageway) between core habitat areas. (B) The situation where linkage is broken into patches of habitat, or stepping stones, between core areas. (C) The stepping stones of B have now been linked with a narrow but continuous passageway. (D) The core areas are connected by a continuous matrix of relatively poor habitat were animals can move but not gather sufficient resources for anything more than temporary occupancy (e.g., resting).

ing the stepping stones of B into the matrix of lower quality habitat (the corridor) of D. If the entire intervening matrix can be made more permeable to movements of target species, then the connectivity payoffs will be much greater than any single corridor could be (Mills 2007, 218). As summarized by Smallwood (2001), the lands surrounding a corridor can be classified according to the functions they individually and collectively serve for specific animal species, such as the quality of the habitats on a scale of low to high.

Recall from our earlier discussion that there is no best corridor configuration. Rather, the design and establishment of corridors must be based on analysis of species-specific habitat and resource requirements, dispersal ability, and overall behavioral responses to a patchy environment. I am not suggesting that every species needs a different type of corridor. Rather, I am only stating that each species must be evaluated to determine the likelihood of it using a corridor, including use relative to movements outside of the corridor. If corridor use is unlikely, then alternative means of ensuring occupancy of a specific location must be considered (e.g., periodic augmentation of individuals; see chapter 3).

Hilty et al. (2006, 97) provided a summary list of hierarchical levels that should be evaluated when planning a corridor, arranged from levels of biodiversity, the scales at which corridors are implemented, and the potential goals that can result from corridor implementation:

- Levels of Biodiversity
  - Individual (of a species)
  - Deme (of a species)
  - Species
  - Community
  - Landscape
- Spatial Scale (of Linkage)
  - Local (e.g., underpass)
  - Regional (e.g., river corridor)
  - Continental or cross-continental (e.g., mountain range)
- Potential Goals
  - Daily movement (e.g., access to daily resources)
  - Seasonal movement (e.g., migration)
  - Dispersal (e.g., genetic exchange, mate finding)
  - Habitat (e.g., wide greenway corridor)
  - Long-term species persistence (e.g., adaptation to global warming)

Hilty et al. (2006,150) advised that restoration planners should be aware that all connections may not be equal, but rather that they play various roles in the larger network of which they are part. Corridor-planning efforts should consider entire planning areas, such as watersheds, and not develop a plan one corridor at a time.

Hilty et al. (2006, 177) described the ways in which corridor design can help avoid the pitfalls that I discussed earlier, thus increasing the chance that connectivity goals will be achieved. They reminded us that although general recommendations about design are difficult because of the numerous different situations where corridors could be implemented, recent research has offered some recommendations that could have general applicability for establishing and maintaining connectivity. In particular, spatial scale and focal species selection are two important components of design. Focal species need to be identified so that their habitat, dispersal, behavior, and other requirements are considered. The quality of habitat within an ecological corridor can determine whether the species of interest will use the corridor. Some species will not use a supposed ecological corridor because of low habitat quality. Understanding the periodic dispersal of a species may also inform the temporal scale that should be addressed so that a population does not become isolated (Hilty et al. 2006, 181–83).

Hilty et al. (2006, 184–97) presented a detailed description of the multitude of factors that must be considered when determining if corridors are a viable option within a restoration project; I summarize some of their key points here. Corridors will usually promote movement of some individuals of certain species, but the decision to use corridors must not be made in a haphazard fashion. Poorly designed corridors will act as filters that tend to allow passage of generalists species, such as those that use multiple habitat types, while specialist species will be unlikely to pass through. Thus, intraspecific and interspecific interactions are important factors to consider in connectivity planning, and social species might not use corridors unless they can move in groups. Some species have physical limitations that must be considered when designing corridors. All such potential barriers should be identified in the planning process so as to avoid corridors that look good on paper but in fact do not adequately serve species of interest. It may be necessary to replace human-created barriers such as fences or poorly designed culverts with wildlife-friendly designs.

Research supports the importance of continuous corridors as opposed to corridors that are bisected by roads or other barriers, especially for more sensitive species. Generally, if the habitat within a corridor is too fragmented, passage of species through the corridor may effectively be stopped. Gaps and barriers, such as roads, should be avoided where possible. What is perceived as a gap varies from species to species, and can range from a few meters to hundreds of meters.

The dimensions of corridors are, of course, important in determining what species occur within the corridor and the potential speed with which those species pass through the corridor. The length and width of corridors naturally are central to the ultimate utility of a corridor. In general, shorter corridors are more likely to provide increased connectivity than longer corridors. Corridors that are too long might not contain some species due to increased distance from core habitat. The preponderance of data indicates that wider corridors are generally more effective for maintaining connectivity. Again, corridor dimensions must be made on a species-specific basis.

Beier and Noss (1998) provide two central areas of research needed to clarify the value of corridors. Remember, however, that the value of corridors must be evaluated on a species-specific basis. They suggested that (1) experiments using demographic parameters as dependent variables, even if unreplicated, should be done to demonstrate the demographic effects of particular corridors in particular landscapes; and (2) observations of movements by naturally dispersing animals in already fragmented landscapes can demonstrate the conservation value of corridors if actual travel routes in both corridors and the surrounding matrix are analyzed.

As noted by Levey et al. (2005), even simple studies of behavior can provide great insight into how animals will react to the presence of a corridor. Morrison et al. (2006, chapter 7) detailed the use of behavior in studies of wildlife ecology, which can be implemented in restoration projects that will likely need to rely on some type of artificial connectivity to meet wildlife goals.

Buffers

As developed by Meffe and Carroll (1997), reserves may use land-use zoning to influence activities surrounding the reserve and make them more compatible with project goals. The classic diagram of a buffer applied to a reserve is depicted in figure 7.4. Here, activities within the buffer zone would be de-

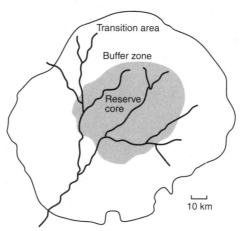

FIGURE 7.4. Schematic diagram of a zoned reserve system showing the possible spatial relationship among a core area, a buffer zone, and a transition area. (From Meffe and Carroll 1997)

signed to minimize impacts on the core area. For example, if logging was allowed in the transition zone, then perhaps only camping would be permitted
in the buffer. Or, hunting might be allowed in the transition zone but prevented in the buffer and core. It can be argued, of course, that the core area
should simply extend to the edge of the transition zone, thus providing at
least the opportunity for a greater extent of the core. This is a valid argument,
but one that is often difficult to implement because of limitations in existing
core areas, and because of political considerations. The buffer can, in
essence, be extended linearly from the reserve to connect with an adjacent reserve, thus forming a passage corridor (but see earlier discussion on corridors). Later (Populations and Restoration: Management Implications) I discuss the use of buffers in planning for wildlife restoration, including the case
of mitigating disturbance due to human activities.

Effects of Isolation

With small populations in isolated situations, isolation can increase the likelihood of adverse genetic effects, such as fixation of deleterious alleles, increasing homozygosity, and overall decline in allelic diversity of the gene
pool, as caused by genetic drift. Inbreeding depression—including depressed
fertility and fecundity, increased natal mortality, and decreasing age of reproductive senescence—is one manifestation of small effective population size.
Small colonizer populations are subject to founder effects that set the stage
for loss of genetic and phenotypic diversity with subsequent isolation from
outbreeding (recall our discussions in chapter 3).

At the assemblage (community) level, recent isolation can result in a decline in species richness over time. This is termed *faunal relaxation*, when
wildlife species are lost from an isolated environment, and it has been documented for oceanic islands isolated from mainlands by submergence of
land bridges, for patches of forest isolated by slash-and-burn or clear-cut timber harvesting, and for patches isolated by human development. For example, the proportion of native plant cover in a chaparral fragment decreased
over time in southern California. The edges of fragments are encroached
upon by exotic plants, refuse, and other impacts, thus reducing the quantity
and quality of vegetation and the habitat of many species within the fragment. Eventually, the fauna reaches an equilibrium species richness where
local extinction and emigration of species equals immigration and colonization. Species can actually be added to such degraded fragments, although
they are often exotics (e.g., house mouse, European starling, and numerous
species of plants).

Isolation of reserves and parks has been a concern for some biologists who suspect that relaxation effects are causing declines in native wildlife in protected areas (Newmark 1986). In some cases, legal and ecological boundaries of protected areas, such as national parks, do not necessarily align, although it is imperative in such analyses to determine the actual status of the environment and populations in lands adjacent to such protected areas. Newmark (1995) analyzed extinction of mammal populations in national parks in western North America and concluded that extinction rates have exceeded colonization rates, and that extinction rates are greater in smaller park units. He also found that the major factors affecting greater rates of extinction of mammals were small initial population size, and lower age of maturity (which is positively related to shorter population generation time).

Are Isolation and Fragmentation Always Bad?

My discussion of potential adverse effects of isolation and fragmentation of environments should not be taken to mean that remnant patches of natural environments have little or no conservation value. Isolation of environments and populations is not inherently bad per se, and there are circumstances where isolation is an advantage for conservation. As discussed earlier, one advantage is avoiding spread of disease (Hess 1994), parasites, and pathogens. Another advantage is the establishing of several founder populations in sites with different disturbance dynamics and thus different likelihoods of success. In this case, overall persistence of the metapopulation is enhanced if there is little correlation of potentially disastrous environmental disturbances among the population centers, and if populations are large enough to avoid genetic problems or if there is occassional gene exchange (outbreeding) among populations. Yet another advantage of isolation is the maintaining of relictual faunas naturally isolated by changes in climate, vegetation, or landform.

In many cases, naturally developed, isolated faunas are best left isolated from exotic species, especially introduced predators and competitors. Species at the edges of their range might occupy some isolated and very different environments through long-distance colonization events or because of relictual distributions in refugia. Peripheral environments may prove important for long-term dispersal of the species and for evolution of unique morphs, subspecies, new species lineages, or entire species complexes. Often, it is important to carefully describe the spatial and temporal scale of isolation, as well as the causes, to determine appropriate management actions to maintain the condition or to ameliorate undesirable isolation problems caused by human activities. Likewise, fragmentation of habitats in landscapes is not necessarily always undesirable. Remember that at some scale, nearly all environments and species-specific habitats are fragmented. Environments or resources are often naturally fragmented through time, as with seasonality of seeds and fruits, foliage, arthropods, and other items needed by animals. In some circumstances, natural fragmentation in space and time of environments or resources can lead to the evolution of new morphs or life forms. As with isolation, it is important to determine the causes and effects of fragmentation of environments to help direct appropriate management action.

In fact, remnant patches may be all that is left in a landscape, watershed, or geographic region from which to rebuild a more natural biota or ecosystem. Thus, remnants can have disproportionately high conservation value per unit area than do large natural areas, depending on the landscape context and management needs. Such conservation measures may be highly efficient, in that they would entail only a small land area but would have great benefits. Remnant patches of native environments offer vital service by conserving sources of associated plants and animals and providing stepping-stone connectivity of a habitat for a species throughout a landscape. Remnants can also provide valuable learning experiences for restoration management.

What should be done about isolation or fragmentation of habitats for specific species that is clearly caused by human activities? The simple answer is to provide habitat linkages, including corridors or dispersed environments or resources, and to block habitats in the future. But these are blanket solutions that do not necessarily always meet multiple conservation objectives, or fit the capability of the land or ownership patterns.

The Landscape Matrix as a Planning Area

A valuable planning approach involves integrating management goals across broad geographic areas (i.e., landscapes) that include multiple ownerships and thus likely many land allocation categories. As developed in Morrison et al. (2006, 309–10), such landscape planning entails recognizing reasonable expectations for the conservation contribution by individual owners and thus landscapes. For example, most agricultural lands will not be managed to maintain a full array of species, habitats, and ecological functions on their own. Agricultural lands can, however, provide some of the landscape elements needed to maintain regional biodiversity, or provide safe environments for some dispersing species.

Lindenmayer and Franklin (2002) outlined how matrix forests can serve key roles in biodiversity conservation by (1) supporting populations of some species, (2) regulating and even affording the movement and dispersal of some organisms, (3) buffering sensitive areas and reserves, and (4) maintaining the integrity of aquatic systems. Thus, in conjunction with corridors and reserve areas, the forest matrix can be managed to contribute to a regional approach of conservation.

Marcot et al. (2001) showed how a broadscale approach to tropical forest conservation could contribute to conservation across multiple spatial scales, called a *protected area network* (PAN): (1) large, existing or potential national parks and other major reserves, which may include protected land allocations such as botanical waysides, wilderness areas, wildlife sanctuaries, and research natural areas; (2) corridors and lands serving to connect the first two areas; (3) small patches of key wildlife resources within lands otherwise allocated to intensive resource extraction such as timber concessions; and (4) individual components, substrates, or elements of high wildlife value within intensive resource-use areas. Thus, PAN elements 2, 3, and 4 can include the matrix lands discussed above by Lindenmayer and Franklin (2002). Tscharntke et al. (2002) extended the approaches of Lindenmayer and Franklin and Marcot et al. (2001) to small fragments of grasslands in cropland landscapes for conserving insect assemblages.

Major principles of reserve design can be summarized as follows (from Noss et al. 1997):

- Species well distributed are less susceptible to extinction than are species confined to small locations.
- Larger blocks containing larger populations are better than small blocks.
- Blocks of habitat close together are better than blocks far apart.
- Habitat in continuous blocks are better than fragmented habitat.
- Interconnected blocks of habitat are better than isolated blocks.
- Populations that fluctuate are more vulnerable than stable populations.
- Disjunct or peripheral populations are likely to be more genetically impoverished and vulnerable to extinction, but are also more genetically distinct than central (core) populations.

The challenge in restoration planning is to calculate the area needed in various cover types, given the expected frequency, intensity, location, and distribution of disturbances. The approach should entail calculating (1) the area of specific habitat conditions required for meeting conservation goals, (2) the frequency of disturbance and area affected by disturbance events, and (3) the postdisturbance recovery time for producing desired conditions again. Geographic Information Systems (GIS) simulation and analysis may be useful (e.g., Herzog et al. 2001; Schelhaas et al. 2002).

Populations and Restoration: Management Implications

Given the dynamics of populations and metapopulations in heterogeneous environments, guidelines for habitat management were developed by Morrison et al (2006, 112–15) and included the following concepts, which are directly applicable to restoration planning.

Guidelines for Species Richness and Overall Biodiversity

Guidelines can be applied at three scales, and can be derived from understanding how populations vary as a function of habitat complexity at each scale:

- within habitat patches, or alpha diversity, such as with the observed correlations between foliar height diversity and bird species diversity presented in chapter 4;
- between habitat patches, or beta diversity; and
- among broader geographic areas, or gamma diversity.

This hierarchy links directly with the filtering concept (or mechanism) developed in chapter 5, whereby the species present in a location are a result of the limitations imposed by abiotic and biotic factors and other constraints. Beta diversity is often maintained by moderate disturbance regimes which provide for variations across space—typically within subbasins—in vegetation elements, substrates, and abiotic characteristics (including soils) of the environment. Beta diversity appears at different spatial scales for species with different body size, home range area, and vagility. Gamma diversity is often controlled by climate, landform, geographic location, and broadscale vegetation formation features. Most restoration occurs at the within-patch scale, which I discuss next.

Guidelines to Maintain Within-Patch Conditions

It may be desired to ensure that specific habitat patches remain viable environmental units. In this case, the size, and the topographic location, adjacency of other patches, and susceptibility to disturbances such as floods or fires, can influence within-patch conditions. Within-patch conditions usually are mediated primarily by patch size and within-patch dynamics, but also by the context of the patch (that is, the kinds of vegetation conditions in adjacent patches); the type and intensity of natural disturbances within the patch; and particularly the kind, frequency, and intensity of management activities within the patch. As noted throughout this book, within-patch conditions must be developed to target specific animal species if the outcome of restoration is to meet project goals.

Guidelines to Maintain a Desired Occupancy Rate of Habitat Patches

A collection of habitat patches may be occupied by organisms over time depending on the size and quality of the patches, the type and quality of the intervening matrix environment, and the spatial juxtaposition of the patches. In dynamic environments such as wet meadows and fire-adapted forests, the location of patches will change over time under disturbance regimes and presence of species that provide physical surfaces and substrates.

Habitat patches can often be mapped and key links identified as part of a set of habitat features important to metapopulation dynamics. As discussed earlier, stepping stones or other habitat corridors do not necessarily have to consist of primary habitat (that is, to provide for all life history needs to maximize realized fitness) in order to provide value to connect populations across a landscape; refer to figure 7.3.

Seemingly unoccupied patches may also serve as occasional sink habitat (chapter 4) to house nonreproductive floaters not yet assimilated into the breeding segment of the population. Floaters are often important for maintaining populations and often occupy marginal habitats. Additionally, we do not think that all suitable habitat is occupied because of grouping (conspecific attraction; chapters 3 and 4) behavior that is witnessed even in species we do not usually equate with colonial behavior. Mills (2007, 218) also noted that species whose dispersal depends on the presence of conspecifics may require more intensive management than just managing the matrix; he mentioned the use of nest boxes, song playback, or even white paint to simulate droppings.

Guidelines for Habitat Configuration

Configuration of habitat patches may be directly manipulated by management activities, an indirect result of other human land-altering activities, or a result of stochastic and unpredictable natural disturbance events. Habitat patches can be managed to remain reachable by dispersing and migrating individuals, and organisms moving within their home ranges. Feeding and resting habitats can be provided within daily dispersal distances or home range areas. Secondary or marginal habitats can be provided as peripheral but close to primary habitats to help serve as sinks for surplus or floater individuals in good reproductive years; the stepping stones depicted in figure 7.3 can serve in this role.

As developed in chapter 3, expanding the area of habitat allows the numbers of individuals of a species to increase along with an increase in levels of demographic organization (see figure 3.4). Spatially unconstrained populations are characterized by multiple, interconnected metapopulations that occupy a region. As noted by Smallwood (2001), the space needed by animals must be considered if we are to link population viability with restoration ecology. Smallwood (2001; figure 3.4) went on to merge his analyses of the areas needed to support functioning demographic units to project areas, including the area likely to be lost to land conversions and areas to be gained by restoration. Figure 7.5 shows how you can predict the response of populations to changes in the size of available area if you have information on the area needed to support different demographic units. Smallwood gathered the data

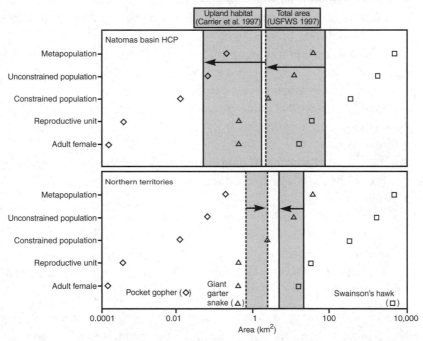

FIGURE 7.5. The areas needed to support functionally significant demographic units of animal species can be compared to project areas, including areas to be lost to land conversions and areas to be gained by restoration. The vertical lines correspond to habitat area (km²) that are upland (dashed lines) and both upland and wetland (solid lines) at the start and end of projects, as indicated by the respective origins and directions of the arrows. Arrows pointing left indicate loss of habitat space, and arrows pointing right indicate gains. (From Smallwood 2001, figure 4)

for the species depicted in figure 7.5 from that available in the literature, underscoring the importance of conducting thorough literature reviews when developing restoration plans for wildlife (see also chapter 6). In figure 7.5 we see that the areas needed to support functionally significant demographic units of animal species can be compared to project areas, including areas to be lost to land conversions and areas to be gained by restoration. The vertical lines correspond to habitat areas (km²) that are upland (dashed lines) and both upland and wetland (solid lines) at the start and end of projects, as indicated by the respective origins and directions of the arrows. Arrows pointing left indicate loss of habitat space, and arrows pointing right indicate gains.

As I have developed throughout this book, and synthesized in chapter 5, we must consider the full range of factors that potentially limit the occurrence of species in a proposed restoration area. Such limiting factors and related issues, and potential means of mitigating them, include the following:

- Disturbance caused by human activities.
  —Control human access or timing of access (e.g., no entry into a sensitive area during breeding season).
  —Establishing buffers around key, sensitive areas (e.g., roosting areas).
- Disease
  —Do not allow animals to concentrate in small areas.
  —Consider treatment of the environment or treatment of selected animals.
- Size of area
  —Functions as a factor limiting occupancy of certain species; consider possibility of linkages (corridors).
- Seasonality
  —Consider availability of water, roosts, and other resources; such resources might need to be artificially established and maintained.
- Biotic factors, including predation and competition
  —Consider direct control of exotic, or in some cases, native (e.g., cowbirds) animals.

These and other limiting factors thus play a direct role in helping establish goals and subsequently establishing specific management practices for a restoration project.

Synthesis

Conservation biologists have expended considerable time and effort in developing concepts that might promote species persistence. Although the concepts make intuitive sense, it is important to remember that they are concepts, and as such, usually have minimal empirical support. In large part this is because of the difficulty in replicating even small reserves, and the multitude of interacting environmental factors present. Thus, restoration projects that implement ecological and conservation concepts should be viewed as tests of these concepts. If designed properly (see chapter 8), such projects can provide invaluable guidance for future projects. This makes publication of projects results—regardless of the success in attaining project goals—an essential part of the growth of the field of wildlife-habitat restoration.

Habitat heterogeneity is the degree of discontinuity in environmental conditions across a landscape for a particular species. Because habitat is a species- or organism-centric term, a particular environmental condition or gradient may constitute habitat for one species and a barrier for another. The species-specific concept of habitat selection, which occurs across spatial scales (chapter 4), must be the focus of restoration with regard to fragmentation (and the broader issues of habitat heterogeneity). Likewise, the concept of species assembly rules (chapter 5) is directly relevant here, because fragment size, shape, and the quality of the habitat therein will serve as a filter determining in part if a species can exist in a target area. Thus, fragmentation is a crude concept unless the response of animals is placed into an appropriate spatial hierarchy. The changes that take place in the environment at several scales of resolution include (1) fragmentation in a broad sense that takes place on the regional scale, (2) fragmentation of different plant associations that take place within a vegetation type, and (3) alteration of habitat suitability within small areas. To best understand effects on wildlife, disturbances should be depicted according to their frequency, intensity, duration, location, and geographic extent. In this way, disturbances to habitats can be incorporated into spatially explicit models of population demography to project effects on species distribution and viability. As vegetation patches change in time so do responses by animals as they use patches for breeding, foraging, refuge from predators, resting, and dispersal. Some wildlife species likely evolved in concert with native disturbance regimes and take optimal advantage of resources distributed through space and time in shifting patches.

The intuitive appeal of corridors resulted in the widespread recommendation for their use in conservation planning and landscape design. However, corridors must be discussed and then described on a species-specific basis. The utility of corridors only makes sense in terms of a particular focal species within a particular landscape. This conclusion follows naturally from our previous (chapter 4) development of the habitat concept as a species-specific

phenomenon. As such, generalizations about the utility of corridors are difficult to make.

Perhaps the most prudent approach to planning for wildlife habitat is to
consider the geographic region as a whole and then decide which elements
within it best contribute to specific conservation goals. The first challenge, of
course, is to articulate those goals unambiguously and precisely, so it is clear
which species, assemblages, habitats, and vegetation and environmental conditions are of conservation value and interest, which usually include those
that are scarce, isolated, disjunct, or declining. A true landscape planning approach then entails integrating management goals across ownerships and
land allocation categories.

Empirical studies and models of populations in patchy environments can
help provide realistic expectations for occupancy rates within and among
habitat patches. This has important implications for designing inventory,
monitoring studies, and interpreting results of such studies. Configuration of
habitat patches may be directly manipulated by management activities, an indirect result of other human land-altering activities such as growth of towns
and roads, or a result of stochastic and unpredictable natural disturbance
events. We can attempt to guide direct management activities such that habitats are configured to afford rescue effects of habitat colonization by organisms over time.

Chapter 8

A *Primer on Study Design*

A key role of conservation science is predicting the response of flora and fauna to environmental change within a context of increasingly complex interactions due to the human element. Unfortunately, we are not yet very good at predicting the response of animals to *natural* environmental changes, let alone making these predictions with the added complication of planned or unplanned human intervention.

Thus, it is easy to argue that adherence to scientific methods that provide a known and repeatable level of knowledge is essential if we are to advance our understanding of the environment, and thus be in a better position to predict the consequences of our actions. It is important that we avoid basing management decisions on preconceived notions and built-in biases that derive from either a lack of study or poorly conceived studies. All science requires adherence to rigorous scientific methods; only good science should ever be applied. In large part, ecology as a discipline has failed to advance rapidly (in its ability to provide reliable knowledge), not because it is inherently flawed, but rather because its practitioners have failed to treat it as rigorous science (sensu Romesburg 1981; Peters 1991).

My goal for this chapter is to summarize the principles required for gaining reliable information, in the context of the study of restoration of wildlife. This chapter, adapted from Morrison (1997) and Morrison et al. (1998; 2006, chapter 4; 2008), could, however, easily be adapted to the study of any plant or animal. I briefly mention some statistical tests that accompany various study designs, but I do not delve into analytical details.

Scientific Methods

Knowledge is discovered through the application of scientific methods. There is not, however, any single best scientific method. Rather, several methods are suitable for differing purposes (Romesburg 1981). Romesburg identified three main scientific methods. First, *induction* applies to the finding of associations between classes of facts: for example, making the observation that certain animals are usually observed within the interior of an oak (*Quercus* spp.) woodland in California and then correlating the abundance of these animals with distance from the woodland edge. The more times (trials) this association is made, the more reliable it becomes; you are *inducing* a relationship. Thus, induction is reliable if done properly, but it does not provide insight into the process causing the relationship. Management decisions made following induction often fail because users attempt to extend the relationship well beyond its original intent, for example, by applying the relationship developed in the California oak woodland to a woodland in the eastern USA.

Second, *retroduction* refers to the development of research hypotheses about processes that are explanations or reasons for observed relationships. In our previous example, we might hypothesize that the association between animals and woodland edge was due to the inability of predators to penetrate deeply into the woodland, thus providing protection to animals in the interior portions of the woodland. Unfortunately, retroductive explanations often provide unreliable information because there are usually many possible alternative explanations. In this example, although our initial explanation seems reasonable, the relationship might actually be due to favorable microclimatic conditions in the woodland interior. A management decision to begin predator control to enhance the species of interest would fail (whereas the correct decision would probably be to increase woodland size).

This problem with retroduction (and actually induction as well) brings us to Romesburg's third category, the *hypothetico-deductive*, (H-D) method. H-D compliments retroduction in that it starts with an hypothesis—usually developed through retroduction—and then makes testable predictions about other classes of facts that should be true if the research hypothesis is true. H-D is a method of determining the reliability of our ideas. In our example, we would design an experiment to test the influence of interior distance from the woodland edge, microclimatic conditions, predator abundance, and likely several measures of woodland structure on some measure of the animal population (e.g., abundance, breeding success).

Most studies in ecology are based on inductive or retroductive methods.

This is one reason why many researchers have been calling for studies that seek to determine *why* relationships are occurring (Gavin 1989). If the underlying processes are understood, then we have a better chance of applying results obtained in one area or time to another area and time. Again, this does not diminish the usefulness of inductive-retroductive studies. Rather, it only means that we should not expect them to do more than they are designed to do; and, we should avoid the temptation to make our conclusions and design our management policies on potentially biased, unreliable information.

I want to emphasize that the previous discussion of the scientific method does not lessen the value of observational studies or natural history observations. Indeed, the only way one can reliably begin an ecological study is to understand the ecological system within which the study will take place; and the best way to begin to gain an understanding of ecology is to go out and observe nature. Thus, I view the scientific method as a series of steps that build upon each other in a hierarchical fashion, from observation to induction to retroduction to development of testable hypotheses.

Terminology

I will first explain some of the key terms used in study design. Clearly, one must be able to communicate with experts in study design and statistics if you are to understand how to develop rigorous studies.

An *element* is an item on which some measurement is made, such as an animal, roost site, snag, or other item of interest. The *sampling unit* is a collection of elements, usually taken over some defined space and time. The *sampling frame* is the complete list of sampling units. This is a fundamental term used in study design, and it represents how you organize your collection of elements and sampling units. It ultimately defines your ability to draw inferences from your study. The relationship between elements, sampling unit, and sampling frame is depicted in figure 8.1.

The *sampled population* is simply the elements contained in the sampling frame. As noted, the sampled population determines the *population of inference* for a study. Finally, the *target population* is the specific area, location, group of animals, or other defined item that is the focus of your study. The researcher justifies the target population based on the sampled population (unless they are essentially the same). Target population is the entire grid area; but the sampled area is in the upper right corner—a clear mismatch of target and sample (as shown in figure 8.2). The goals of a study should be linked to the target population, and they need to be explicitly stated.

Sampling hierachy

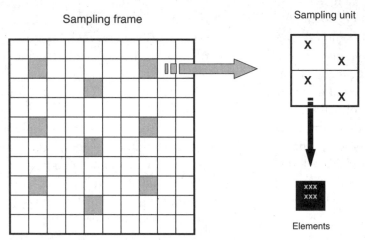

Sampling frame

Sampling unit

Elements

FIGURE 8.1. Relationship between elements, sampling unit, and sampling frame.

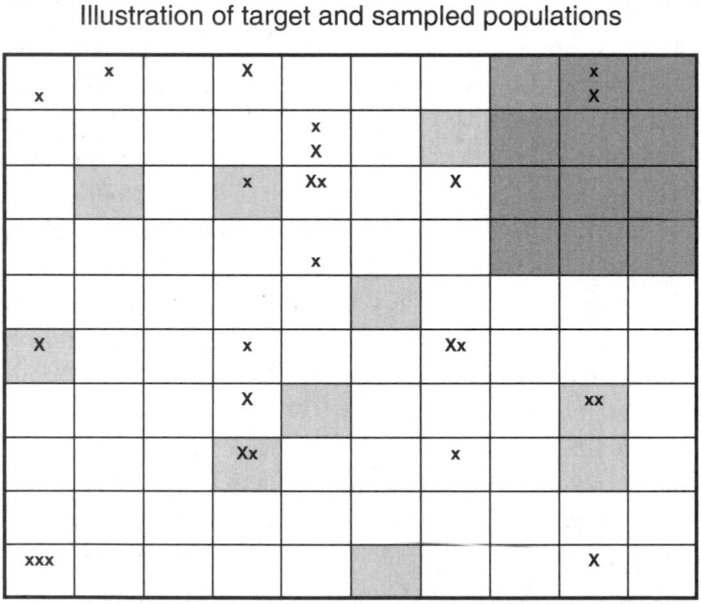

Illustration of target and sampled populations

FIGURE 8.2. Target population is the entire grid area; but the sampled area is in the upper right corner—a clear mismatch of target and sampled area. A design using the lighter squares would be more appropriate.

Sources of Variation

Temporal and spatial variation must be considered in the context of establishing your sampling frame. It also then relates to your necessary sample size (discussed later). It is important to your design so that the likely temporal and spatial variation inherent in your target population is an integral part of your design. In figure 8.3, note the difference between the two sites at time 2 and time 4. Spatio-temporal variation is an ecological process that introduces variation, which is not the same as variation introduced through sampling (assuming no observer bias).

There will be variation among your sampling units (e.g., plots). The variation within the units combines to increase the variation across the units, again impacting sample size needs. As the spatial extent of a study increases the overall variability in the samples also tends to increase; the issue (discussed later) of preliminary data collection helps you anticipate these issues. Note here you are only sampling from part of the areas (grids, plots), which introduces variation among the sampling units.

Within each sampling unit, there will be variability in the elements (e.g., animals, plants). This variability accumulates across elements and is reflected in variability for each sampling unit, and thus among-unit variation. Enumeration variation is usually due to incomplete counts.

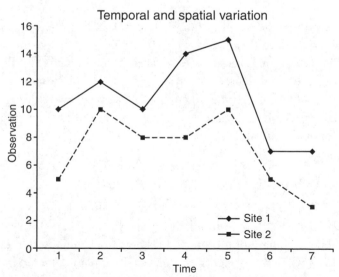

FIGURE 8.3. Illustration of the importance of designing a study that accounts for the likely temporal and spatial variation inherent in the target population. Note the difference between the two sites at Time 2 and at Time 4. (From Thompson et al. 1998, figure 1.12)

Ultimately, then, your ability to identify a biological or statistical effect is based on the variability among your elements. The central point is that these variations determine what you can conclude biologically given your sample size; best to go into a study with all of this resolved (this is power and sample size analysis, which we will cover later).

Monitoring as Research

Resource managers often state monitoring is something different than research, and as such, requires less rigor in its design and implementation, which leads to poorly designed studies and ultimately poor management decisions (Morrison and Marcot 1995). As defined by Green (1979, 68) a *monitoring study* has the purpose of detecting a change from the present conditions. The resulting data can provide a baseline against which future changes (impacts) can be measured. However, the precision at which such an impact can be measured will be dependent on how well developed the monitoring study was.

In developing a monitoring study, researchers must decide the level of change to be considered an impact prior to initiation of the work, which in turn determines the sampling intensity (e.g., number of study plots) and frequency (e.g., time between sampling). For example, suppose a restorationist wants to determine the change in abundance of a focal species that is thought to be increasing its geographic range. The amount of change considered to be biologically significant must be determined. Then, specific steps can be taken to design the sampling design and protocol needed to detect this change with a desired degree of certainty. I discuss the design and implementation of monitoring studies in chapter 9. Elzinga et al. (2001) and Feinsinger (2001) presented very useful summaries of sampling and monitoring methods for plants and animals.

Principles of Study Design

Green (1979) developed basic principles of sampling design and statistical analysis of relevance to environmental studies that should be followed in any study. The development of the study goals flow from initial statement of the problem (e.g., will native rodents eat the seeds of a plant targeted for restoration?), to the stating of the specific null hypothesis (e.g., the density of seeds does not significantly reduce the mass of a focal native rodent species). It is absolutely critical to the success of any study that this step be well developed, or all that follows might be for naught. In establishing the goals for a study, it is critically important to establish the spatial and temporal applicability of the results; recall our discussion of sampling frame and target population earlier in this chapter. For example, if the goal is to determine the best means of reducing an exotic species that has a wide distribution, then establishing all study plots in a restricted space and time will be unlikely to have broad applicability (i.e., the inference will be narrow because of variations in environmental conditions). Such decisions will determine the spatial distribution of sampling areas (the sampling frame) and the temporal nature of sampling.

Additionally, the confidence one desires in the results must be determined during planning. Asking if a reduction in canopy cover reduces breeding success by $50 \pm 10\%$ requires much more rigorous sampling than asking if it reduces it by $50 \pm 25\%$ because the latter goal has a wider allowable variability ($\pm 25\%$) than the former ($\pm 10\%$). Similarly, asking if canopy cover changes breeding success is different than asking if it reduces success; the former is looking for an increase or a decrease, while the latter is just looking for a decrease. Thus there is a pivotal interplay between the question you ask, the sample size needed to reliably answer it, and thus the generality you can expect from your results.

The term *significant* or *significance* has caused much confusion in ecology. We typically use the term significance in terms of biological, statistical, and social significance. Unfortunately, most authors do not separate statistical from biological significance, which leads to results of minimal absolute difference often being given ecological importance based on a statistical test.

When ecologists say that something is *biologically significant*, we mean that there is enough difference, or a strong enough relationship exists, to make us believe that it matters biologically. Because we would seldom expect exactly no difference between two biological entities, we need to specify how much difference we think would matter to the entities of interest (Cherry 1998; Johnson 1999). For example, given a large enough sample size, a difference in foraging rate of 10% between males and females of a species might be statistically significant (e.g., $P < 0.05$); but does it matter to survival or reproductive success? Although the H-D method is a valuable approach within the scientific method, setting a null hypothesis of "no difference in foraging rate" is thus trivial and likely to be rejected. In contrast, setting a null hypothesis of "males will forage 20% quicker than females" incorporates a biological expectation into the statistical method. Of course, one must be able to justify the magnitude of difference—here 20%—which would be based on the literature or previous study of the situation (Morrison et al 2008, chapter 1).

Conducting a preliminary analysis or pilot study is an important but often overlooked aspect of any study. *Pilot studies* are needed when one has little concrete notion of the best sampling design, sampling intensity, or even sampling methods. As stated by Green (1979, 31), "Those who skip this step because they do not have enough time usually end up losing time." Preliminary work allows you to verify that your sampling procedures are actually sampling the segment of the population or environment that you intended, allows evaluation of necessary sample sizes, and allows modifications in sampling techniques before you become locked into a particular methodology. I usually advise that one should first "study the study" before charging into establishing a magnitude of effect and your sampling frame. Because few studies are so unique that the methods being used have not been implemented before, most data collected during the preliminary or pilot portion of a study will be available for analysis with the overall data set.

There are several prerequisites for an *optimal study design*: (1) The impact (a general term used for change: e.g., construction, treatment) must not have occurred, so that before-impact baseline data can provide a temporal control for comparing with postimpact data; (2) the type of impact and the time and place of occurrence must be known; and (3) nonimpacted controls must be available (figure 8.4, sequence 1). As noted by Green, an optimal design is thus an areas-by-times factorial design in which evidence for an impact is a significant areas-by-time interaction. Given that the prerequisites for an optimal design are met, the choice of a specific sampling design and statistical analyses should be based on your ability to (1) test the null hypothesis that any change in the impacted area does not differ from the control, and (2) re-

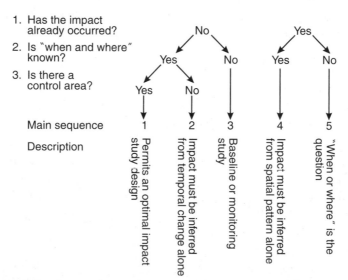

FIGURE 8.4. The decision key to the *main sequence* categories of environmental studies. (From Green 1979, figure 3.4)

late to the impact any demonstrated change unique to the impacted area and to separate effects caused by naturally occurring variation unrelated to the impact (Green 1979, 71).

Optimal and Suboptimal Study Designs

It is often not possible to meet the criteria for development of an optimal design. Impacts often occur unexpectedly, time or funding is not available to implement the desired design, or control areas are unfeasible or logistically impossible. Many if not most restoration projects fall into the *suboptimal study design* category because of the following reasons:

- Treated (restored) sites are usually selected nonrandomly.
- Replication of restoration sites is often not possible.
- Pretreatment data are usually scanty or nonexistent.
- Control areas are difficult to establish.

If no control areas are possible (figure 8.4, sequence 2), then the significance of the impact must be inferred from temporal changes alone (discussed later). If the location and timing of the impact is not known (i.e., it is expected but cannot be planned; e.g., fire, flood, disease), the study becomes a baseline or monitoring study (figure 8.4, sequence 3). If properly planned spatially, then it is likely that nonimpacted areas will be available to serve as controls if and when the impact occurs. This again indicates why *monitoring* studies are certainly research, and will allow the development of a rigorous experimental analysis if properly planned. Thus, the field of impact assessment offers guidance on how to quantify treatment effects in as rigorous a manner as possible, given the suboptimal situation; these designs are developed in detail later in Impact Assessment.

Unfortunately, the impact has often occurred without any preplanning by the land manager. This common situation (figure 8.4, sequence 4) means that impact effects must be inferred from among areas differing in the degree of impact; study design for this situation is discussed later in this chapter. Finally, situations do occur (figure 8.4, sequence 5) where an impact is known to have occurred, but the time and location are uncertain (e.g., the discovery of a toxin in a plant or soil).

Experimental Design

Experiments can be placed into two major classes: *mensurative* and *manipulative*. Neither type is necessarily more robust than the other; it all depends upon the goal of the study. Manipulative experiments have the advantage of

being able to test treatments and evaluate the response of something to varying degrees of perturbation; and, do so in a manner usually within your control. Mensurative studies—also termed observational studies—measure differences in treatments without implementing a designed treatment (or
experiment). Critical in the development of any experiment is proper consideration of replication, the use of controls, randomization, and independence.
Hurlbert (1984) presented an excellent review of the steps necessary for development of a strong experimental design; much of the material in this section is summarized from that article (see also Morrison et al. 2008).

Everyone is familiar with the general need for experimental controls,
whether the study be mensurative or manipulative. *Controls* are required because biological systems exhibit spatial and temporal variation. As noted by
Hurlbert (1984), if we could be certain that a given system would be constant
in its properties over time in the absence of an experimental impact, then
separate controls would be unnecessary (i.e., pretreatment data from the to-
be-treated site would be an adequate comparison). Additionally, in many experiments, controls allow separation of the effects of different aspects of the
experimental procedure. *Similar* is most often used within the context of
study design to denote identification of sampling units, where sites are selected based on similarity in key environmental characteristics. In experi-
mental/manipulative designs, sampling units are selected (in an appropriate
fashion; discussed later) and then treatments randomly applied; nontreated
units are termed controls. In impact assessment studies where such controls
are not possible, units selected for comparison (with impact unit[s]) are usually termed *reference* sites because they are referencing the conditions on the
impact sites. I do not use the term with regard to study design herein because
of the confusion this would cause with the term *reference conditions*, as is
commonly used in restoration. Selecting many control sites, even when there
is only one treated site, is recommended. Multiple control sites produce a
much better understanding of what is occurring naturally (for comparison
with the treated site) than would be available if only one control was used. I
discuss the selection of (multiple) control sites in chapter 9.

Another definition of controls includes all of the design features associated with an experiment. Thus, *randomization* controls for (reduces) experimenter biases in the assignment of experimental units to treatments. Randomization reduces potential bias by the experimenter, thus increasing the
accuracy of the estimate. *Replication* controls for among-replicates variability
inherent in any study. Replication reduces the effects of random variation (often referred to as *noise*) or error, thus increasing the precision of an estimate.
Interspersion (of study plots) controls for regular spatial variation in the exper-

imental units. The term control also refers to the homogeneity of the experimental units, to the precision of treatment procedures, or to the regulation of the physical environment in which the experiment is conducted. As emphasized by Hurlbert (1984), the adequacy of an experiment is based both on your ability to control the physical conditions during the experiment and the use of an adequate number of treatment controls (e.g., replicated control plots).

Independence refers to the probability of one event occurring that is not affected by whether or not another event has or has not occurred. For statistical analyses, it should be assumed that the error terms are independently distributed. Departures from independence occur from correlations in time and/or space of the experimental samples (Sokal and Rohlf 1981). A somewhat common misconception is that the application of nonparametric statistics relieves this assumption; it does not.

Blocks and factors (discussed in more detail later) are methods that are used to try to control for extraneous or disturbing variables. Although blocking or factoring can be done after data collection, this usually substantially lowers sample size and thus your ability to recognize biological or statistical significance. It is critical to understand the loss of degrees of freedom as blocks or factors are added. Consider the following simple but realistic example:

		Sample size
Habitat use of deer:		
all ages and sexes combined		75
sexes separated	Male	35
	Female	40
ages separated		
Subadult		
	Male	15
	Female	23
Adult		
	Male	20
	Female	17

Thus, the likely reasonable sample size of 75 is quickly reduced to <25 for each element in the block or factor (depending on the analysis used). And as we will study, how the samples are gathered could negate a reliable inference if the samples—regardless of the number—are not collected in a manner appropriate to the study goals. Note the reduction in sample size with blocking, making a priori planning essential. Also, unequal sampling across blocks can bias overall results. Thus, although you can block and examine factors retrospectively, it is far better to plan ahead (be prospective).

In the sections that follow I provide a brief review of common study design and associated statistical analyses; some of the material is summarized from Morrison et al. (2008), who provide an in-depth development of study design and analysis as applied to studies in wildlife research. All restorationists should possess a fundamental knowledge of study design and statistics to be able to, at a minimum, understand and evaluate study designs, data descriptions and presentations, and conclusions of research studies.

Single-Factor Designs

Single-factor designs are those in which one type of treatment or classification factor is applied to all experimental units, such as all elements (e.g., animals, plants) in a unit. The treatment may, however, be of different levels (e.g., three severities of burning in a single vegetation type). Different treatments are expected to act independently. We will cover this class of simple designs as a foundation for all other designs that we cover. There are two goals here: First, to understand the basic designs and concepts; and second, to be able to refer back to these basic designs as we move into more complicated designs.

Unpaired and Paired

Unpaired designs depict an inability to match the environmental conditions of the treatment plots with similar reference plots (figure 8.5). This design usually results in a relatively high variance that might be overcome by a large sample size given that variance does not continue to increase with increasing sampling (e.g., such as including samples from an increasingly large spatial area). Note that not pairing is discussed in the context of a failure because you end up with high variance relative to what would be possible if pairing was accomplished (there are, of course, situations in which no pairing is appropriate; such as in simple descriptions). Typical statistical tests used for unpaired designs are the *t*-test and Mann-Whitney.

In contrast, *paired designs* depict an ability to match the environmental conditions of treatment and control plots. The analytical advantage is that the usual statistical tests address the difference between pairs and not the absolute values of each pair, which usually results in a low variance relative to the unpaired situation. Pairing is actually the most basic form of blocking (blocking is discussed later). The difficulty with pairing is being able to justify

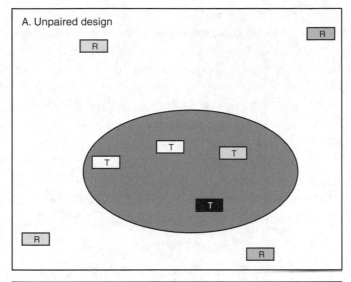

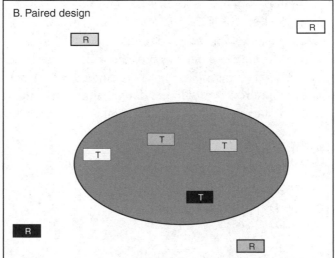

FIGURE 8.5. (A) Unpaired design where sampling plots (small rectangles) are randomly placed within the treated (T; inner oval) and reference or control (R) areas. (B) Paired design where each T is matched with a similar R.

that the pairing was based on randomized techniques and not observer bias. As such, pairing is usually done prospectively; retrospective pairing requires a clear rationale. Of course, you may always *unpair* a paired design and conduct statistical analyses on the absolute values. Paired *t*-tests of nonparametric equivalents are used for analysis.

COMPLETELY RANDOMIZED

One of the most basic designs, the *completely randomized design* allocates sampling units in a completely randomized manner across a study area. This design actually has limited application in wildlife research because it usually is applied in situations where you have no knowledge of the underlying distribution of elements (figure 8.6). This design is probably most appropriate during a preliminary study that is intended to gather data on the distribution and abundance of unknown or little known elements. Note that in figure 8.6 treatments are applied randomly or samples are taken randomly, depending on the application. Typical statistical analyses include the *t*-test, Mann-Whitney, or one-way anova (or nonparametric equivalent, Kruskal-Wallis).

RANDOMIZED COMPLETE BLOCK DESIGN

In the *randomized complete block design*, blocking is incorporated into the completely randomized design; note that by definition (complete) all treatments are in all blocks (figure 8.7). *Blocking* is an extension of pairing to >2 groups (i.e., paired data can be analyzed under anova). The goal of blocking is to make individual plots as similar as possible within a block, and for the blocks to be as dissimilar as possible. Thus, blocking is used when you have a specific purpose in mind for controlling variance. Although numerous blocks can be assigned, problems involving stratification arise as does the practical issue of sample size allocation among blocks. The anova family of statistics is the primary analytical tool.

Completely randomized design
$T = 3$ treatments

Treatment

A	B	C
A	C	B
B	A	C
C	B	A

FIGURE 8.6. Basic format for the completely randomized design.

Randomized complete block design
$T = 3$ treatments, $B = 4$ blocks

Block   Treatment

Block			
1	A	B	C
2	A	C	B
3	B	A	C
4	C	B	A

FIGURE 8.7. Basic format for the randomized complete block design.

Block design; field application

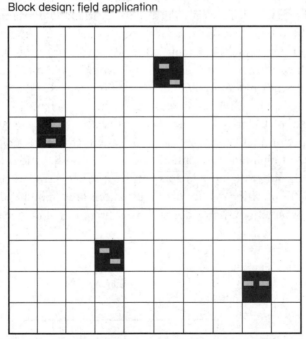

FIGURE 8.8. Application of the randomized complete block design. Black squares are blocks and shaded rectangles are plots.

Figure 8.8 depicts a simple blocking arrangement as applied to a field situation where the study goal is to determine the abundance of an element (e.g., rodents, ant mounds, birds, plants) across a region. Blocking could be used because not all sections (squares in the figure) are available for (random) sampling and you wish to study variance within and between blocks rather than just generate an overall (all eight plots) variance. Blocking could also be implemented when a treatment is applied, or when there is a marked difference in conditions, such as vegetation density, plant association, and elevation. Analysis of variance in blocking allows separation of between-block and within-block variance. This figure can also be used as an example of pairing.

There are other applicable designs beyond the scope of this book, including incomplete block design and Latin squares design (see Morrison et al. 2008).

Multiple-Factor Designs

Factorial designs are common in both experimental and especially in observational research in wildlife. Unlike blocking, here the objective is to examine several possible interactions among factors. Thus, you are looking at the influence of potential factors on your response variable. For example, the influence of different age classes and sex through time (years) on foraging rate. Interpretation becomes increasingly difficult as you use more than two to three factors; in the preceding example we would have three main effects (age, sex, and time) and four potential interactions (age × sex, age × time, sex × time, age × sex × time). All analyses involve the anova family.

Advantages of factorial designs include the ability to have one analysis rather than performing a one-way anova for each factor. Also, factorial analyses allow for testing interactions among factors (figures 8.9 and 8.10).

The effects of two or more factors can be assessed simultaneously using factorial designs in which treatment combinations are applied to the observational units. A benefit of such a design is that there might be decreased cost.

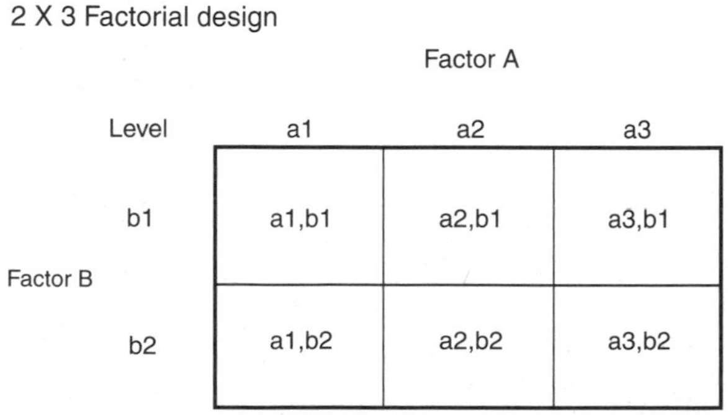

2 X 3 Factorial design

Factor A

Level	a1	a2	a3
b1	a1,b1	a2,b1	a3,b1
b2	a1,b2	a2,b2	a3,b2

Factor B

FIGURE 8.9. A 2 × 3 factorial design where factor a has two levels and factor b has two levels (e.g., factor a = habitat type and factor b = sex of study species).

2 X 2 X 3 Factorial design

Factor B	Factor C		Factor A	
b1	c1	a1,b1,c1	a2,b1,c1	a3,b1,c1
	c2	a1,b1,c2	a2,b1,c2	a3,b1,c2
b2	c1	a1,b2,c1	a2,b2,c1	a3,b2,c1
	c2	a1,b2,c2	a2,b2,c2	a3,b2,c2

FIGURE 8.10. A $2 \times 2 \times 3$ factorial design where factor a = 3 levels, factor b = 2 levels, and factor c = 2 levels (e.g., factor a = habitat type, factor b = sex of study species, and factor c = age [juv., ad.]).

In a two-factor design, your interest is in two classes of treatments (with ≥1 level of treatment per treatment class possible) (figure 8.9). A simple example is 2 elevations × 3 habitat types. It is preferable to have balanced (orthogonal) designs, although unbalanced designs—and even designs without replication—are possible. In a multiple factor design, you combine ≥1 class of treatment with ≥1 class of experimental units (figure 8.10). Extending the previous example gives 2 elevations × 3 habitat types × 2 age classes.

As introduced earlier, interactions indicate how the response variable varies by combinations of the factors (e.g., animal abundance by each combination of the elevation-habitat type categories). Note that interactions are written as *factor1* × *factor2*, where the "×" denotes "by" (so, factor1 by factor2, or in our example, elevation by habitat type). A graphical representation of interactions between factors is shown in figure 8.11. In the top panel we see that there is an effect of age—because learning time changes across ages— and an effect of sex—because the two lines are separated. But, there is no interaction between age and sex (i.e., sex × age interaction not statistically significant) because the same pattern is seen in the two lines. In the bottom panel, however, we see that while the age and sex effects remain (as in the top panel), they do not hold for all combinations of age and sex: note where the two lines merge on the left of the figure (this is where there is no effect of age or sex on learning time).

HIERARCHICAL DESIGNS

Hierarchical designs have some levels of one factor occurring in combination with the levels of one or more other factors, and other distinctly different

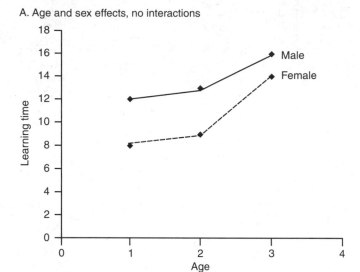

A. Age and sex effects, no interactions

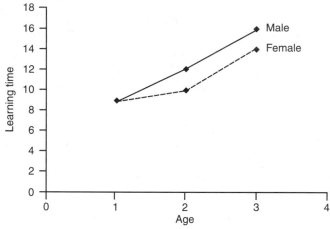

B. Age and sex effects with interaction

FIGURE 8.11. A graphical representation of interactions between factors. A. No interaction between age and sex for learning time. B. An interaction between age and sex for learning time.

levels occurring in combination with other factors. This design has application to restoration but is compromised by lack of replication and level of independence among treatments and replicates. It does represent a way to apply a treatment to a large area with plots and subplots receiving additional or different treatment. For example, applying fertilizer over an entire project area and then using different tree planting densities within plots in the project area and different tree species within subplots. These designs are also used to address some questions where a hierarchy occurs naturally, such as eggs within a nest and sexes within the individuals of the brood.

Split-plot designs are a form of nested factorial design where the study area is divided into blocks, the blocks are further subdivided in large plots, and these large plots are then subdivided into smaller plots called split plots. This design results in an incomplete block structure, which is relatively difficult to analyze.

Other Designs

The following designs are not unique study designs per se, but rather techniques that increase your ability to reach a conclusion in situations where replication in space is minimal or nonexistent. Thus, the following designs have broad applicability to wildlife and restoration studies.

The *sequential design* can be applied where a sequence of treatments are applied to the same element or plot, with the caveat that the treatment does not cause a permanent change or that the change that occurs is planned (e.g., an increasing intensity of the same treatment). This design has the advantage that among-plot variance is minimized (because you are not changing plots); but it is usually used in situations where sample size is minimal. This also applies to observational studies where you can justify that your repeated presence does not influence the activity of the elements under study (this even applies to studies of plants where trampling might be an issue; or other studies where leaving human scent could be a factor).

The *cross-over design* can be applied where the treatment is not permanent, but can be crossed-over between control and treatment plots. For example, in testing the influence of cowbird removal on host-nesting success, treatments and controls can be switched between years (assuming no substantial depression on cowbird populations occurs on the treated plots in year 1; e.g., see Morrison and Averill-Murray (2002).

Impact Assessment

Impact is a general term used to describe any change that perturbs the environment, whether it is planned or unplanned, human induced, or an act of nature. In this section we concentrate on unplanned—and thus unforeseen—perturbations, and planned studies that are constrained by few or no replicates; as discussed earlier, the latter applies to most restoration studies. Often treatments are applied to small plots to evaluate one resource, such as plants, and you have been funded to study animal responses. In such situations, the initial plots might be too small to adequately sample many animal species. Or, there might be no treatment involved, and the project focus is to quantify the ecology of some element within a small temporal and spatial scale. The suite of study designs that fall under the general rubric of impact assessment are applicable to studies that are not viewed as having caused an environmental impact per se. Designs that I cover later, such as after-only gradient designs, are but one example.

As introduced earlier in this chapter, Green (1979) outlined the basic distinction between an optimal and suboptimal study design. Here are highlighted fundamental aspects of and additional examples within optimal and suboptimal designs of impact assessment. Morrison et al. (2008) present a detailed description of impact assessment studies as applied to wildlife science.

Disturbances

Planned or unplanned disturbances can impact the environment in three primary ways: pulse, press, and those affecting temporal variance (Bender et al. 1984; Underwood 1994).

Pulse disturbances are those that are not sustained after the initial disturbance; the effects of the disturbance, however, may be long lasting. *Press disturbances* are those that are sustained beyond the initial disturbance. Both pulse and press disturbances can result from the same general impact. The magnitude of the pulse disturbance will determine our ability to even know that an impact has occurred. For example, figure 8.12 depicts mild (B) and relatively severe (C) pulse disturbances; the former would be difficult to detect if sampling was less frequent (i.e., if sampling had not occurred between times 6 and 8) and/or the variance of each disturbance event was high. Figure 8.12 depicts mild (C) and relatively severe (D) press disturbances. The former would be difficult to distinguish from the variation inherent in the control sites.

Disturbances affecting temporal variance are those that do not alter the mean abundance, but change the magnitude of the occillations between sampling periods. These changes can increase (see figure 8.12A) or even decrease (see figure 8.12B) the variance relative to predisturbance and/or control sites. A pulse disturbance can thus resemble a disturbance affecting the temporal variance.

Identifying a disturbance is thus problematic because of background variance caused by natural and/or undetected disturbances (i.e., a disturbance other than the one you are interested in). For example, note the similarities between the conclusion of no effect in figure 8.13A, temporal variance (mi-

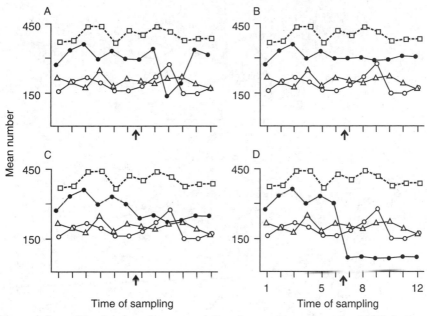

FIGURE 8.12. Simulated environmental disturbances in one location (solid dots), with three controls, all sampled six times before and after the disturbance (at the time indicated by the arrow). (A–B) The impact is an alteration of temporal variance after the disturbance; temporal standard deviation times 5 in A and times 0.5 in B. (C–D) A press reduction of abundance to 0.8 (C) and 0.2 (D) of the original mean. (From Underwood 1994, figure 3)

nor pulse) in figure 8.13B, and a pulse disturbance in figure 8.13C. Here we see that the temporal variance in many populations creates noise that obscures more subtle changes (such as environmental impacts), and the lack of concordance in the temporal trajectories of populations in different sites (Underwood 1994). These disturbances clearly show the critical role that control sites play in being able to separate treatment effects from background stochasticity.

The duration of the impact study will be determined by the temporal pattern (length) of the impact. Although by definition a pulse disturbance will recover, a press disturbance usually recovers relatively slowly as either the source of the impact lessens (e.g., a chemical degrades) or the elements impacted slowly recover (e.g., plant growth, animal recolonization). Also note that an impact can change the temporal pattern of an element, such as the pattern in fluctuation of numbers (see figure 8.3). A change in temporal patterning could be due to direct effects on the element or through indirect effects that influence the element. Direct effects might result from a change

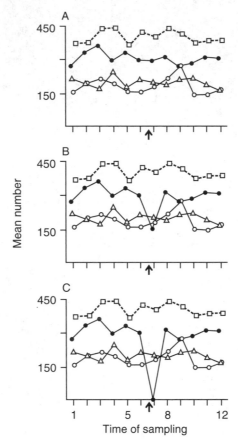

FIGURE 8.13. Identifying a disturbance is problematic because of background variance caused by natural and/or undetected disturbances. Note the similarities between the conclusion of *no effect* in A, *temporal variance* (minor pulse) in B, and a *pulse disturbance* in C. (From Underwood 1994)

in activity patterns, because the impact agent modified sex or age ratios (i.e., differentially impacted the sex-age structure); indirect effects could result because the impact agent severely impacted a predator, competitor, or food source of the element being monitored.

Recovery and Assumptions

As summarized by Parker and Wiens (2005), impact assessment requires making assumptions about the nature of temporal and spatial variability of the system under study. They (see also Wiens and Parker 1995) categorized assumptions about the temporal and spatial variability of a natural (nonimpacted) system as in steady state, spatial, or dynamic equilibrium (figure 8.14).

A *steady-state system* is typified by levels of resources, and the natural factors controlling them that show a constant mean through time (A). Hence, the resource at a given location has a single long-term equilibrium to which it will return following perturbation (if it can, indeed, return). Such situations usually only occur in very localized areas. In A, the arrow denotes when the state of the system (solid line) is perturbed to a lower level (the dashed line).

Spatial equilibrium occurs when two or more sampling areas, such as impact and reference, have similar natural factors and, therefore, similar levels of a resource (B). Thus, in the absence of a perturbation, differences in means are due to sampling error and stochastic variations. Look closely at the dashed line in A versus the dashed line in B; the primary difference between figures is that multiple areas are considered in B.

Dynamic equilibrium incorporates both temporal and spatial variation, where natural factors and levels of resources usually differ between two or more areas being compared, but the differences between mean levels of the resource remain similar over time (C). In such systems *recovery* occurs when the dynamics of the impacted areas once again parallel those of the reference (control) area. Note in C that the reference (solid line) line fluctuates around the mean (also solid line), while the impacted area (dashed line) drops well below the natural (although lower than the reference) condition (lower solid line).

Parker and Wiens (2005) also presented an example of when ignorance of the underlying system dynamics can lead to erroneous conclusions on recovery. In figure 8.14D, we see considerable natural variation around the long-term, steady-state mean. In this figure the horizontal solid line represents the long-term mean of the fluctuating solid line. If this long-term variation is not known or not considered, the perturbed system might erroneously be deemed recovered when it is not; for example, at point a in the figure. Conversely, the system might be deemed to be impacted when in fact it has recovered (point b; note that the dashed line is now tracking the solid line that represents the natural state).

The assumptions surrounding all three of these scenarios about system equilibrium also require that the perturbation did not cause the resource to pass some threshold beyond which it cannot recover. In such situations a new equilibrium will likely be established, for example, when an event such as fire, overgrazing, or flooding permanently changes the soil. Under such situations the system would recover to a different state.

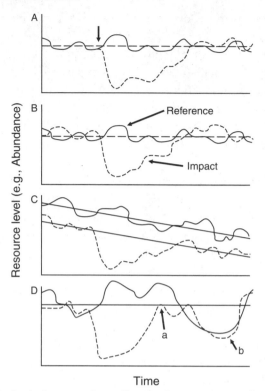

FIGURE 8.14. Ecological assumptions affecting the assessment of recovery from an environmental accident. (A) Steady-state equilibrium. The vertical arrow indicates the accident, and the dotted line the state of the affected system; the solid curve indicates the dynamics of the system in the absence of a perturbation, and the solid horizontal line the steady-state mean level of the resource. Recovery occurs when the impacted system returns to the mean steady-state equilibrium. (B) Spatial equilibrium. In this case, two different areas have similar long-term dynamics. The unaffected system serves as a reference for the impacted system, and recovery occurs when the impacted system returns to the point where its dynamics are again similar to those of the reference site. (C) Dynamic equilibrium. The reference and impact areas have different levels of the resource (spatial variation), but their temporal dynamics are similar (in this example, a long-term decline). Recovery occurs when the dynamics of the impacted system once again parallel those of the reference system, even though levels of the resource differ between the areas. (D) In this example, there is considerable natural variation about the long-term, steady-state mean. If this long-term variation is not considered or is unknown, the impacted system may erroneously be deemed recovered when it is not (point a) or may be considered still to be impacted when its dynamics in fact match those of an unimpacted system (point b). (From Parker and Wiens 2005, figure 1)

Applications to Restoration

Green (1979) developed the *before-after/control-impact*, (BACI), framework for application to environmental impact situations where little or no replication was possible. The BACI design is basically a simplified application of a manipulative experiment that has only one impact and one control site. In BACI, an impact is inferred from a difference in the pattern between the single treated (impacted) site and the single control site. Although such designs are, statistically, an areas by times factorial design (with evidence for an impact being a significant areas by times interaction), the lack of true replication usually negates any meaningful statistical analysis.

Because there is usually no option for additional treatment—especially then the treatment is an undesired environmental or catastrophic impact—the best option is to increase the number of control sites, which will generally strengthen your ability to infer a treatment effect. Your ability to infer is increased because you have gained a more thorough understanding of the natural or nontreated response of the target element (recall our earlier discussion on understanding the dynamics of the system being studied). Although restorationists usually have limited ability to increase the number of treated sites, additional control sites can usually be found.

The manner in which system dynamics interact with the design of an impact study are summarized in table 8.1. *Baseline* is defined as a study that compares pre- and post-data from the impact area only. This is analogous to Green's (1979) main sequence 2, where the impact is inferred from temporal variation only. Because natural factors usually vary through time, however, results from baseline studies are seldom sufficient to determine if recovery has occurred. *Single year studies* compare impact and control areas but within a single year. These designs approximate spatial equilibrium through the use of multiple sampling areas, which requires a close matching of natural conditions across sites (e.g., the paired design discussed earlier). Recovery occurs when impact and control sites are similar. *Multiyear studies* reduce the effects of temporal and spatial variation by removing naturally varying temporal effects; here multiple control sites are very helpful. If the impact and control area(s) are in a dynamic equilibrium, recovery occurs when differences in annual means become constant (trend lines become parallel, as explained earlier for figure 8.14C).

Morrison et al. (2008, chapter 6) reviewed several modifications of the BACI designs that have been developed to address the limitations of the basic BACI. These designs, such as pairing, are equivalent to the basic experimental designs we developed earlier.

TABLE 8.1

Three design strategies for assessing recovery from environmental impacts on biological resources in temporally and spatially varying environments

Attributes	Baseline	Single year	Multiyear	
			No reason to reject/suspect assumptions	Reason to reject/suspect assumptions
When to use	Temporally invariant taxa	Spatial equilibrium achievable, short recovery period	Temporally variant taxa, long recovery periods, information on recovery process desired	Temporally variant taxa, long recovery period, taxa on multiple recovery periods, information on recovery process
Data needs	Pre- and postimpact only	Impact and reference sites, covariates	Time series for impact and reference areas or for gradient	
Comparison	Pre- vs. postimpact	Impact vs. reference, matched pairs, gradient||	Impact vs. reference and gradient over time||	
Equilibrium assumption	Steady-state	Spatial	Dynamic	Reject or suspect assumptions
Breakdown in assumptions	Temporal variation confounds with recovery	Spatial variation confounds with recovery	Temporal variation differs for impact and reference categories	NA
Statistical methods	t test: student's, paired; BACI§	ANCOVA, paired t test, gradient||	Level-by-time, trend-by-time, repeated measures	Gradient|| (with or without covariates), impact/ref, others
Conditions needed for recovery	Equal pre- and postmeans	Impact and reference means equal, no impact on gradient||	Difference in means constant, gradients|| constant	Failure to reject multiple assessments of impact effect
Advantages	Rreference sites not required (though useful)	Single year of data, extrapolation reasonable	Nonrandom site selection	

Disadvantages	Equilibrium assumption not reasonable, pre-impact data required	Recovery snap-shot, co-variables needed, matched sites for matched pairs	Multiyear data required, difficult to extrapolate from non-random samples	
Comments	Use with multi- or single-year studies, provides partial information on recovery process	Corroborate with contamination and toxicity (triad approach)	Use preimpact data, validate assumption	Verify with habitat changes, use α level > 0.05

Note: NA, not applicable.
Reasons may include zero means. Entries that span the last two columns pertain to both situations.
Methods addressed in Wiens and Parker (1995).
§BACI uses prespill data at impact and reference sites and relies on the assumption of dynamic equilibrium.
Gradients are dose-response regressions of biological resources vs. gradients (i.e., continuous measures) of exposure.

Designs classified under suboptimal apply to the impact (restoration treatment) situation where you had no ability to gather preimpact (pretreatment) data or plan where the impact was going to occur. Such situations are frequent and involve events such as chemical spills and natural catastrophic events, and simply when funding only becomes available shortly before initiation of the project; hence no opportunity was afforded for gathering pretreatment data from treated or control sites.

Restorationists are regularly called upon to restore sites that have been heavily impacted by fire, flood, years of agriculture, invasion by exotic plants, and a host of other natural and human-induced factors, but for which no direct pretreatment data are available. Development of desired conditions, as introduced in chapter 6 and integrated into each of the case studies in chapter 10, must be used to design the restoration project (including developing baseline conditions). Such *after-only impact designs* also apply, however, to planned events that resulted from management actions, such as timber harvest, road building, and restoration activities, but were done without any study plan. Here I highlight some features of after-only designs that have applicability to restoration projects.

Single-Time

Because you are seldom restricted to a single sampling period following an impact, this group of designs is seldom applicable. The primary situation in which single-time designs would apply is when you have a sharp pulse impact followed by a rapid recovery. In such cases it is unlikely that anyone would be too concerned with the actual impact because recovery is rapid. In restoration it is more likely that a study would become a single-time design because you identified recovery after your first sampling period, thus negating the need for additional work. Therefore, this design would seldom apply in restoration projects because longer-term monitoring is usually needed to claim project success (e.g., for plant or animal viability on a site).

Morrison et al. (2008) explained the basic designs applicable to after-only situations, including (a) impact-reference, (b) matched pair, and (c) gradient. The gradient approach is especially applicable to localized impacts because it allows you to quantify the response of elements that are at varying distances from the impact. Further, you might be able to identify the mechanism by which the impacted site recovers, such as through a gradual lessening of the effect of the impact along the gradient from distal to proximal ends. The ends of the gradient serve, in essence, as nonimpacted reference sites. Figure 8.15 depicts a simple but effective gradient design.

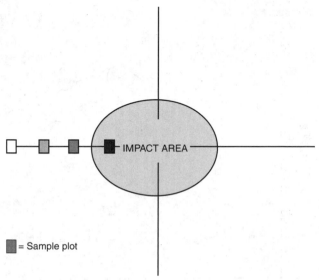

FIGURE 8.15. Representation of a gradient design. Note how the measurement of the impact lessens with distance from the source.

Interactions

This is perhaps the most applicable design to many after-only situations in resource management, including both planned management activities and unplanned catastrophes. Here you establish multiple control sites matched with the impacted site, and gather samples over a period of time, the length of which depends on the impact and the elements under study. Here the measure of interest (e,g., animal presence, plant vigor) will vary in magnitude through time; impact is inferred when the pattern of change in the element differs significantly from that of the control sites. Statistically, such a comparison is a factorial anova where a significant time interaction infers an impact (as graphically shown in figure 8.16A); no impact is inferred from figure 8.16B. As described by Morrison et al. (2008), such analyses can be performed with either categorical (e.g., low, medium, high) or continuous (chemical concentration in ppm) measurements of the impact.

Power and Sample Size Analyses

A fundamental goal of research is to generate the appropriate amount of information to answer the project goal or address the specific hypothesis. The goal is not to generate the maximum amount of information, or the amount of information possible given the time and funding available. It is actually

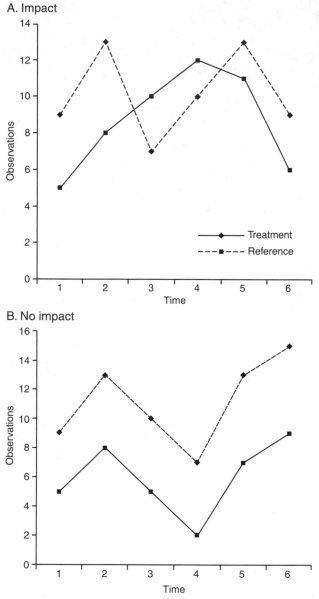

FIGURE 8.16. Illustration of where an impact occurs (A) and does not occur (B).

possible to gather too much information in the sense that doing so has wasted resources that could have been used elsewhere. There are specific tools available to help guide the gathering of the appropriate amount of data needed to address a specific study question, including especially statistical power analysis, which I discuss in this section.

Data can be obtained from hand-written notes of qualitative observations to those obtained using sophisticated measuring devices. Whether data are in a qualitative form such as "I observed that species A used oaks more than pine," or of a more quantitative form such as "I found a significantly ($P <$ 0.01; t-test) greater use of oak ($x = 25.5\%$, SD = 5.20) than pine ($x = 11.0\%$, SD = 3.15)," we have been provided information on an observed phenomenon. Although I suspect that most people would put more faith in my second example, in reality, given the information provided, there really is no reason to trust one observation over the other. If I had added "On 75 occasions I observed. . . ." to the first example, and $n = 6$ to the second, I would certainly put my trust in the former even though the latter does present a statistically significant difference. I say this because, without justifying one's sample size, no data set should be assumed to be valid regardless of the associated P-value. Finding a significant P in no way justifies a conclusion; recall our earlier discussion of statistical versus biological significance. Erroneous and often contradictory conclusions may be reached with variations in sample size, because alpha levels vary as sample size increases (see also Morrison 1984a,b).

To understand power and sample size determination we must first review a fundamental aspect of statistical theory, namely error type. The probability of committing a *Type I error*, termed alpha, is rejection of the null hypothesis when it is actually true. In contrast, *Type II error*, termed beta, is failing to reject the null hypothesis when it is in fact false. Lower probabilities of committing a Type I error are associated with higher probabilities of committing a Type II error, and the only way to minimize both errors is to increase sample size. Improving the power of a test has special importance to land managers. Looking at *power* another way, it tells us the likelihood that we will falsely reach a conclusion of no effect do to a treatment (or impact). As developed later, power analysis is most useful in the design phase of a study to help guide the sample size needed to render a rigorous (and thus believable) conclusion. Although we can all provide reasons (make excuses) for failing to gather an adequate data set, there is no reason for failing to provide quantified justification for the samples that were collected, and thus the impact of the sample on our conclusions.

Determining Power

Statistical power is determined by four interrelated factors: power increases as sample size, alpha-level, and effect size increase; power decreases as variance increases. Understanding statistical power requires an understanding of Type I and Type II errors, and the relationship of these errors to null and alternative hypotheses. Any biometry book will provide a good refresher of these topics, which should be reviewed before continuing with this material. Morrison et al. (2008) also provide a more detailed discussion as directly applicable to wildlife studies; here I summarize some of their material.

Effect size is defined as the difference between the null hypothesis and a specific alternative hypothesis. If a null hypothesis is one of "no effect," the effect size is the same as the alternative hypothesis. As developed earlier in this chapter, however, we are usually interested in some effect size that has biological meaning. As I have repeatedly emphasized, because the detectable effect size decreases with increasing sample size, in most studies a finding of a statistically significant difference has no biological meaning (for example, a difference in canopy cover of 5% over a sampling range of 30%–80%). As such, setting a biologically meaningful effect size is the most difficult and challenging aspect of study design. Setting an effect size with biological meaning is often referred to as *the magnitude of biological effect*. Effect size is not a population parameter but is a hypothetical value set by the researcher (although it is certainly based on biological knowledge). This point is important in power analysis because it means that we must set power prospectively rather than retrospectively (retrospective power analysis can be conducted but requires extreme care; this will be discussed later).

Figure 8.17 depicts the relationship between power and effect size to detect a population trend; note that as effect size increases statistical power increases. That is, for a given sample size it is easier to statistically detect effect when the change is large than it is when the change is small.

By setting effect size or just your expectation regarding results (e.g., in an observational study) a priori, the biology drives the process rather than the statistics. That is, using statistics to first help guide study design, and then again to compliment our interpretations, is the proper procedure. The common practice of collecting data, applying a statistical analysis, and then interpreting the outcome misses the needed biological guidance. What you are doing is deciding to accept whatever guidance the statistical analyses provide and then trying to come up with a biological explanation. Even in situations where you are doing survey work to develop a list of species occupying a particular location, stating a priori what you would expect to find and the relative order by abundance provides a biological basis for your interpretation of results.

Power analysis is used in three basic scenarios:

- To determine the number of samples necessary to achieve a specified power given the effect size, alpha, and variance that you have established (scenario 1)

An accident assessment design

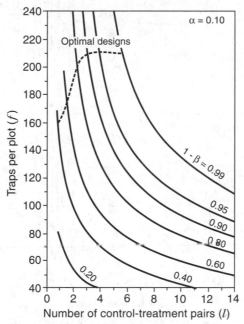

FIGURE 8.17. An example of power contours for the test of an impact hypothesis. (From Skalski and Robson 1992, figure 6.8)

- To determine the power of a statistical test likely to result when the maximum number of samples that you think can be obtained are gathered (scenario 2)
- To determine the minimum effect size that can be detected given the power, alpha, variance, and sample size you have established (scenario 3)

Morrison et al. (2008) detailed how each of the above scenarios can be applied in ecological research. In summary, the advantage of power analysis is in the insight you gain regarding the design of your study.

Conducting power analysis after data collection—termed *retrospective power analysis*—is generally not recommended (Steidl and Thomas 2001). Despite the controversy, retrospective power analysis is a useful tool in management and conservation. Retrospective power analysis should never, however, be used when power is calculated using the observed effect size. In such cases the resulting value for power is simply a reexpression of the *P*-value, where low *P*-values lead to high power and vice versa (Morrison et al. 2008).

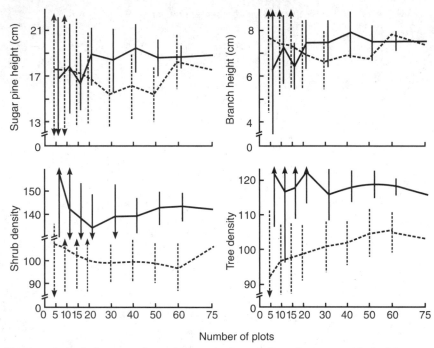

FIGURE 8.18. Influence of sample size on the stability of estimates (dashed horizontal lines) and measurements (solid horizontal lines) of bird-habitat characteristics. Dashed or solid vertical lines represent 1 SD from point estimates for visual estimates and direct measurements of the variables, respectively. (From Block et al. 1987, figure 2)

Sequential Sample Size Analysis

A primarily nonstatistical, graphical method of evaluating sample size during data collection, as well as justifying the collected sample size, is called *sequential sample size analysis* (Morrison et al. 2008). Under this method you can plot the values of any variable of interest as the sample size increases. For example, calculating means and variance as every 10 vegetation plots are gathered. You can justify ceasing sampling when the means and variance reach an asymptote; see figure 8.18. Likewise, you can take increasingly large random subsamples from a completed data set, calculate the mean and variance, and determine if the values reached an asymptote.

Synthesis

Knowledge is discovered through the application of scientific methods. There is not, however, any single best scientific method. I view the scientific method as a series of steps that build upon each other in a hierarchical fashion, from observation to induction to retroduction to development of testable hypotheses.

A common misconception among resource managers is that monitoring is something different than research, and as such, requires less rigor in its design and implementation. This misconception is widespread and has lead to many poorly designed studies and subsequent misguided management decisions.

Experiments can be placed into two major classes: mensurative and manipulative. Neither type is necessarily more robust than the other; it all depends upon the goal of the study. Manipulative experiments have the advantage of being able to test treatments and evaluate the response of something to varying degrees of perturbation, and do so in a manner usually within your control. Critical in the development of any experiment is proper consideration of replication, the use of controls, randomization, and independence. Project directors should avoid the temptation of rushing into study initiation without first developing all phases of the study, including likely statistical analyses. Even crude estimates of necessary sample sizes are essential if field efforts are to be maximized—there is no reason to under- or oversample.

The development of the study goals flow from the initial statement of the problem to the stating of the specific null hypothesis. It is critical that this step be well developed, or all that follows might be for naught. In establishing the goals for a study, it is critically important to establish the spatial and temporal applicability of the results. Such decisions will determine the spatial distribution of sampling areas and the temporal nature of sampling. The statistical literature, and an increasing amount of the wildlife literature, provides in-depth guidance on designing, conducting, analyzing, and interpreting field studies. I highly recommend that restorationists and wildlife scientists increase direct communication on wildlife studies, especially during the design phase.

The terms *significant* or *significance* have caused much confusion in ecology. We typically use the term significance in terms of biological, statistical, and social significance. Unfortunately, most authors do not separate statistical from biological significance, which leads to results of minimal absolute difference often being given ecological importance based on a statistical test. When ecologists say that something is *biologically significant*, we mean that there is enough difference, or that a strong enough relationship exists, to make us believe that it matters biologically. Because we would seldom expect exactly no difference between two biological entities, we need to specify how much difference we think would matter to the entities of interest.

Many if not most restoration projects fall into the design category of *suboptimal* because (1) the treated (restored) area is not randomly selected and (2) pretreatment data are often scanty or altogether unavailable. Thus, the field of impact assessment offers guidance on how to quantify treatment effects in as rigorous a manner as possible, given the suboptimal situation.

Chapter 9

Monitoring: Field Methods and Applications

As I introduced in chapter 8, a cornerstone of science is the gaining of reliable knowledge so we can improve our understanding of nature. Restorationists understand that their efforts are improved over time through practice and experience. Thus, the better we document our activities, the better we will be at understanding the project outcomes, and the easier it will be to convey this understanding to the public, land managers, and scientists. Recall that I opened chapter 2 with a discussion of the need for restorationists to understand the theoretical underpinnings of ecology. One of the primary ways we accomplish these dual goals of understanding and conveying an outcome falls under the general rubric called *monitoring*. Monitoring is, however, a largely misunderstood process. I must be clear: monitoring is a scientific endeavor, and as such, it should be pursued with care and rigor. Monitoring is research, and as such, must be conducted with the rigor we expect from other scientific investigations.

In this chapter, I will first describe the process of monitoring as applied to restoration of wildlife and wildlife habitat. The principles discussed apply, however, to most ecological applications. I will then develop the proper application of commonly used sampling methods. Each description of a method summarizes the strengths and weaknesses, sampling intensity necessary to achieve various levels of confidence regarding the results, and suggested improvements on standard techniques and potential ways of combining methods to improve a survey.

There are numerous methods available to sample animals, and most were designed to sample a particular aspect of a population. The specific sampling method or methods chosen for a study must be applicable to the goal of that study. For example, point counts are not superior per se to spot mapping to assess birds, and area-constrained surveys are not superior per se to pitfall traps for assessing herpetofauna. Each method has strengths and weaknesses that must be matched with the project goals. Although the literature provides guidance on the applicability of a particular method to specific project goals, there is little guidance on the frequency and especially the intensity of application necessary to achieve reliable results (recall that techniques to guide sampling frequency and intensity were outlined in chapter 8).

Definitions

An *inventory* is used to assess the state or status of one or more resources. It can provide information on environmental characteristics, such as the distribution, abundance, and composition of wildlife and wildlife habitats. An inventory is typically confined within a specified area or set of areas to provide a basic understanding of the wildlife present. A common goal of an inventory is to generate a list of species occupying the area(s) and perhaps provide a basic understanding of abundance and specific location. As introduced in chapter 8, however, inventories are not synonymous with "quick-and-dirty" surveys. It can often take many months of sampling to generate an adequate species list (depending on the goal of the specific project).

Inventories associated with restoration projects are typically done to establish a baseline data set on existing conditions (see chapter 6). Verifying presence is relatively straightforward: if you detect the species by sight, sound, or sign, then you have established presence. However, failure to detect a species does not necessarily mean it was absent when you sampled, or that it might use the area during a different time period. As noted, the sampling intensity and methods used to determine presence are critical if confidence is to be placed in a determination of no occurrence on the area; I develop the issue of determining detection probabilities later.

Monitoring is usually defined as a repeated assessment of the status of some quantity, attribute, or task within a defined area over a specified time period (Thompson et al. 1998, 3). There is confusion between monitoring and the term inventory, and the related terms *baseline* monitoring or *assessment* monitoring; these latter terms are actually inventories. Repeated samples can be taken to produce an estimate (inventory) for a particular time period (e.g., month, season, year); thus, the winter can be the time period. For example, an inventory may be conducted to determine a point estimate for animal abundance during the winter, or breeding success during the spring. This estimate tells us nothing about the change in these parameters over several time periods. Repeating the inventory at another time period generally results in a monitoring study. Thus, monitoring measures or indexes dynamics, whereas inventories index the state of the resource. Thompson et al. (1998) reviewed other uses of the term monitoring.

Inventory and Monitoring of Wildlife

The variables measured during monitoring might be identical to those used in an inventory. Other variables, however, are unique to monitoring in that they measure rates such as survival that require repeated sampling. The length of monitoring will be determined by the goals of a study; for example, the initial three years following completion of a restoration project. If monitoring of a population continues long enough, the population will likely be subjected to the usual range of environmental conditions (e.g., drought) present in the study region. The use of control sites thus becomes increasingly important as the length of a monitoring study increases so we are able to separate treatment effects from natural variation (see chapter 8).

Monitoring has been placed into four overlapping categories. Here I briefly review this terminology to assist the reader in interpretation of other literature:

Implementation monitoring is used to assess whether or not a directed activity has been carried out as designed. For example, a restoration project was designed to establish 40% willow cover at 3 m height within four years of implementation. Implementation monitoring would be done to evaluate if the goal had been accomplished.

Effectiveness monitoring is used to evaluate if the stated action met its objective: that is, if the reason for the 40% willow cover was to provide nesting locations for target bird species. Effectiveness monitoring would determine if the birds were actually nesting.

Validation monitoring is used to determine whether established management direction provides guidance to meet its stated objectives. For example, did a mitigation plan actually result in recovery of a species in the area?

Compliance monitoring is done when mandated by law or statute to ensure that the project met legal requirements. For example, compliance monitoring is done to determine if allowable "take" of an endangered species has been exceeded.

The methods and guidance provided in this chapter can easily be adapted to address all four of these monitoring categories.

Sampling Considerations

Inventory and monitoring projects require an adequate sampling design to ensure accurate and precise measurements. Thus, adequate knowledge is required of the behavior, general distribution, biology and ecology, and abundance patterns of the resource of interest (Thompson et al. 1998). Once these basic properties of the resources are understood, a sampling methodology appropriate for the study goals must be developed. Basic sampling issues were developed in chapter 8.

Resource Measurements

Resources can usually be measured directly or indirectly. *Direct measures* are variables that link clearly and directly to the question of interest. If direct measurements exist and are feasible to obtain, then they are preferable over indirect measures because direct measures establish a causal link between factors. For example, measuring the abundance of prey being used by a foraging animal directly links the animal with a critical measure of its environment.

Indirect measures are, however, widely used in inventory and monitoring studies because they tend to be easier to gather. Indirect measures attempt to establish a surrogate for the direct, causal link between variables of interest. In the example of prey abundance used above, indirect measures of prey abundance, such as tracks, burrows, and feces, might be sought because of the difficulty and expense involved with obtaining actual prey abundance.

A broad class of indirect measures are known as *ecological indicators*, which are surrogates of the ecological state of a resource of interest (i.e., they indicate condition). This concept was originally proposed by Clements (1920) to explain plant distribution based primarily on soil and moisture conditions. Because quantifying the ecological requirements of animals on a species-specific basis is extremely difficult, many investigators have tried to develop indicators (indirect measures) of the phenomenon of interest. Relationships between animals and the environment are not as strong for animals as they are for plants because of the mobility of most animals (Block and Brennan 1993; Morrison et al. 2006). The use of indicators is controversial and should be attempted only after careful evaluation; many sources provide direction on the proper use of indicators (e.g., Morrison 1986; Landres et al. 1988; Morrison et al. 2006).

Habitat is often monitored as a surrogate for monitoring the animal directly (i.e., to index population trends for a species). Costs of monitoring a population to have an acceptable statistical power can be high (Verner 1984). Additionally, little information exists for most species such that a strong link between habitat and population trend can be established. As developed in chapter 4, habitat is a complex concept that entails much more than the structure and floristics of vegetation. Modeling of suitable habitat based on remote sensing data can be successful but usually is only capable of predicting presence or absence of the species of interest. When information beyond presence-absence is needed (e.g., abundance, reproductive success), data must usually be gathered on increasingly finer spatial scales.

Selection of Sampling Areas

A fundamental step in any study is to clearly define the target population and the sampling frame (chapter 8). This first step establishes the extent to which inferences can be extrapolated to areas outside your immediate study area (which would have value in any restoration study). In most restoration projects, however, the physical size of the area is predetermined by factors outside your control. Relatively small (i.e., <100 ha) areas are unlikely to support viable populations of many vertebrates unless the surrounding area is also suitable for inhabitation or at least provides passage routes. Therefore, if the study goal is to maintain population viability over the long term, then the sampling area would likely need to include locations outside the immediate restoration site.

The mobility of many wildlife species can confound the inferences drawn from the project area. In most ecological studies, a somewhat arbitrary decision is made to define the population potentially affected by a project (see chapter 3 for detailed discussion of populations). Rather, you are usually just sampling (dealing with) a portion of a population. The primary assumption is often that animals within this investigator-defined area are the ones most likely to be affected by the project. Not measured, however, are the interacting effects that can impact animals beyond the defined population area. Thus, the results of the monitoring may be questionable because factors outside the project area can be affecting the animals within the area.

In some cases, the project area is small enough that a *census*—defined as a complete count—of the area is possible. More often, however, the entire project area cannot be surveyed, thus requiring the use of sampling plots (i.e., subsample the project area).

The primary considerations involved with establishing plots are their shape and size, the number of plots needed, and placement of plots within the project area.

Determining shape and size is complicated by factors such as the methods used to collect data, biological edge effects, distribution of species under study (e.g., clumped or random), biology of the species (e.g., nocturnal or diurnal), and logistics involved with data collection. Thompson et al. (1998, 44–48) summarized the primary considerations and tradeoffs in choosing a plot design. For example, long and narrow plots may allow for more precise estimates, but square plots will have less edge effect. They concluded that no single design was optimal for all situations, and they suggested trying several designs in a pilot study (see chapter 8).

The number of sample plots and the distribution of those plots within the study area depend on several sampling considerations including sample variance, and the distribution and abundance of the species. As developed in chapter 8, sample size should be defined by the number of plots necessary to provide precise estimates of the parameter(s) of interest. It is critical to remember that monitoring plans, like any research study, require that specific effect sizes be established prior to initiation of the work. Thompson et al. (1998) and Morrison et al. (2008) reviewed many of these sampling issues.

Study Duration

The duration of a monitoring study will be based on several interacting factors, including project objectives, field methods, ecosystem dynamics, ecology of target species, funding, and logistical feasibility. A primary consideration for inventory and monitoring studies is the temporal qualities of the process being measured, such as plant succession, length of breeding cycles, and generation time (Franklin 1989), which are influenced by biotic and abiotic factors (Parker and Wiens 2005; see chapter 8). Thus, it will be necessary to collect data over a period of time sufficient to allow the population(s) to experience a range of environmental conditions; these dynamics can have a long periodicity. For example, in pinyon pine woodlands of the Great Basin of the United States, major pine crops occur only every 6–10 years, and rodent populations in this region fluctuate many fold in response to this phenomenon (e.g., Morrison and Hall 1998).

In summary, four primary phenomena necessitate long-term studies: slow processes such as plant succession and many vertebrate population cycles; rare events such as fire, flood, and disease; subtle processes where short-term variability exceeds the long-term trend; and complex phenomena such as intricate ecological relationships. The reality of budget constraints, however, usually inhibits such long-term sampling. Therefore, innovative approaches are often required to attempt to achieve unbiased results from suboptimal monitoring duration. There are several main alternatives to long-term studies (see Strayer et al. 1986):

- *Retrospective studies* have been used to address many of the same questions as long-term studies. A key use of retrospective studies is to provide baseline data for comparison with modern observations. They can also be used to characterize slow processes and disturbance regimes, and how they may have influenced selected ecosystem attributes. Perhaps the greatest value of retrospective studies is in characterizing changes to vegetation and wildlife habitats over time. For example, dendrochronological studies can provide information on frequencies and severities of historical disturbance events, which can be used to reconstruct ranges of variation in vegetation structure and composition at various spatial scales. Other potential data sets for use in retrospective studies include databases from long-term ecological research sites, forest inventory data bases, pollen studies, and sediment cores.
- *Substitution of space for time* is achieved by finding samples that represent the range of variation for the parameter(s) of interest to infer long-term trends. For example, forests in several seral stages can be studied instead of waiting for a forest to grow. Morrison and Meslow (1984) studied the impact of herbicides on birds on clear-cuttings sprayed two years and five years earlier to infer potential impacts (see discussions on impact assessment in chapter 8).
- *Models* are conceptualizations of how an ecological process might behave under various conditions. Models range from simple word or pictorial descriptions of a process of interest to sophisticated mathematical constructs. For example, estimates of demographic parameters such as survival, reproductive output, and recruitment, can be used to project the size of the population into the future. Various types of habitat models are used to link a measure of animal presence or abundance with one (univariate) or multiple (multivariate) measurements of the animal's environment (see chapters 3 and 4). Especially popular techniques are simple and multiple regression, discriminant analysis, and logistic regression. For example, Morrison et al. (1994) used multiple regression to develop a relationship between animal abundance and vegetation measurements as a guide to restoration prescriptions (e.g., plant species composition and structural classes). Verner et al. (1986), Patton (1992), and Morrison et al. (2006) provide detailed descriptions of these and related models.

Adaptive Management

Based on results of preproject inventories, modifications of initial project plans can be made to enhance the outcome of the restoration or other management activity. Once the project has been initiated, monitoring will then indicate if the desired conditions (chapter 6) are being met. It is critical, of course, that rigorous baseline data be available. This system of planning, sampling, and modification has been formalized within the concept of adaptive management. *Adaptive management* (or *adaptive resource management* [ARM]) is centered primarily on monitoring the effects of land-use activities on key resources, and then using the monitoring results as a basis for modifying those activities to achieve project goals (Walters 1986; Lancia et al. 1996). Adaptive management is an iterative process whereby management practices are carefully planned, implemented, and monitored at predetermined intervals. If outcomes are consistent with predictions, then the project continues as planned. If outcomes deviate from predictions, however, management can (1) continue, (2) terminate, or (3) change, depending on project goals.

Morrison et al. (2006, 406–12) summarized the application of adaptive management for habitat modification, explaining that the adaptive management approach entails (1) identifying areas of scientific uncertainty, (2) devising field management activities as real-world experiments to test that uncertainty, (3) learning from the outcome of such experiments, and then (4) recrafting management guidelines based on the knowledge gained. Management guidelines can include testable hypotheses where monitoring and adaptive management studies equate to conducting the experiment, and revision of the management guidelines equates to reevaluation of the study results in terms of testing the validity of the initial hypothesis (Morrison et al. 2006, 407).

Adaptive management is not synonymous with *trial-and-error* approaches to restoration. Trial-and-error approaches attempt to fix a problem after implementation and lack an action plan prior to initiation of a project. In contrast, adaptive management provides a structure whereby clear goals and objectives are established and monitored, and specific actions are planned at the outset of the project for responding to deviations from projected interim and final goals. This general process has been summarized in a seven-step process that includes feedback loops that depend largely on monitoring results (figure 9.1). The primary feedback loops in figure 9.1 are between steps 5–6–7, 2–7, and 7–1. The 5–6–7 loop is the shortest and implies that management prescriptions are essentially working and need only slight adjustments. The primary obstacle in this loop is the time lag between project im-

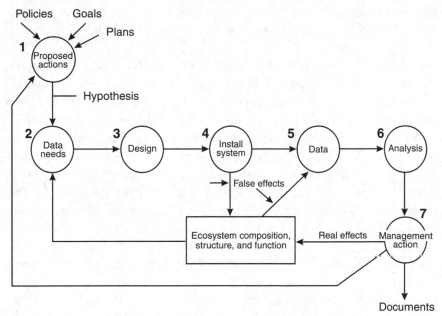

FIGURE 9.1. A seven-step adaptive management system. False effects are data that reflect high installation impacts, observer bias, location effects, and observer-caused disturbances during measurement. (From W. M. Block, personal communication.)

plementation and the point where the project has advanced sufficiently to monitor for effectiveness. Consequently, the loop is often severed, and feedback is never provided. The second loop, 2–7, indicates that monitoring was poorly designed, the wrong variables were measured, or monitoring was poorly implemented. The 7–1 feedback loop is when a decision must be made regarding the course of future management and monitoring activities. Here, if monitoring was done correctly, then informed decisions can be made for future management directions. If monitoring was done poorly, then another opportunity was lost to provide a scientific basis for resource management. Adaptive management principles are incorporated into the case studies presented in chapter 10.

For adaptive management to be effective, specific *thresholds* of key resources of interest must be established prior to initiation of the project, which can include stocking densities of plants, foliage cover, and the presence or a specific abundance of a target animal species. Violating such a threshold—either high or low—will then *trigger* a specific management action. Thus, the concept of thresholds and triggers are a central component of adaptive management and represent a predetermined level that when violated will lead to specific responses. For example, a restoration plan might stipulate a target for cowbird density of less than one female/ha within five years of vegetation restoration so that breeding songbirds can successfully reproduce. If this threshold was not achieved, then specific actions are prescribed at initiation of the study (here, trapping and removal of cowbirds). Without establishing a priori thresholds, there would be no way to design an appropriate monitoring program for cowbirds. Additionally, there may have been no reason to even attempt certain restoration activities under this hypothetical scenario if political considerations would have negated cowbird control as an option.

Sampling Principles

In this section I briefly outline general sampling principles and considerations that apply to all inventory and monitoring studies, as well as other sampling programs. As developed in chapter 8, pilot studies should be included as an integral part of project design. In fact, the first few months of most projects become pilot by default because of the errors made and lessons learned when trying to implement the design. Following the guidance here will assist with developing a rigorous study design.

Observer Bias and Training

Biases influence not only the way we design a study, but they are inadvertently inserted into field sampling when multiple observers are used to gather data. Observer-based variability may lead to incorrect conclusions because results include artifacts, spurious relations, or irreproducible trends (Gotfryd and Hansell 1985, 224). For example, *intraobserver reliability* is a measure of the ability of a specific observer to obtain the same data when measuring the same behavior on different occasions, and thus measures the ability of an observer to be precise in his or her measurements (Martin and Bateson 1993, 32–34). Assessing intraobserver reliability in field studies is difficult because animals seldom repeat their behavior in exactly the same fashion. One test is to videotape animals and then repeatedly to present (in some random fashion) individual sequences to the observer. The results of such trials can then be used to estimate the degree of observer reliability.

Interobserver reliability measures the ability of two or more observers to obtain the same results on the same occasion and often inserts substantial variability into the dataset (Martin and Bateson 1993, 117). Not surprisingly, differences in experience among observers accounts for much of the interobserver variability noted in field studies. Initial training, followed by regular retraining, can help minimize intra- and interobserver variability. Additionally, standardized data forms, accompanied by sampling protocols that define each variable and the appropriate recording methods, are highly recommended.

Researchers conducting behavioral studies should be aware that their presence and activities likely influence an animal's activities, which introduces often substantial bias into a study. Because wild animals are constantly vigilant for predators and competitors, the presence of an observer likely heightens its awareness. After all, it is likely that the animal knew you were there long before you ever saw it! Such high awareness or responsiveness is termed *sensitization*. The waning of responsiveness is termed *habituation*, and is considered to be a form of learning (Immelmann and Beer 1989). Animals that appear to become habituated to the presence of observers have thus adopted a modified pattern of behavior that allows them to keep the observer under surveillance. The biases associated with estimations of animal abundances should also be considered carefully in habitat studies because many of our analytical procedures correlate animal numbers with features of the environment (see Morrison et al. 2006). Clearly, a study that has low bias among habitat characteristics can be ruined by biased count data.

Types of Information

The most basic information of interest for inventories is a list of the species present in an area over a specific period of time, usually a season (i.e., breeding, winter). The number of different species present is termed *species richness*. Studies of species richness provide basic information and help with the design of a more detailed study. A survey often begins by generating a species list by reviewing the literature and distribution maps, and by observers conducting walking surveys throughout the study area. Because of the cryptic nature of many species, rigorous techniques are almost always needed, however, to gather a complete species list.

Population monitoring is a large and varied field of study used to quantify and monitor trends in abundance or various demographic parameters over time. Because population size changes over time due to weather, food availability, disease, catastrophes, and many other factors, it is difficult to separate these interacting factors from various human-induced influences. Substantial interyear variation in animal numbers can mask a longer-term trend, thus delaying the implementation of remedial actions (i.e., complicating implementation of adaptive management). A further cornerstone of population monitoring is the use of consistent methods over time. Repeatable counts need not be accurate in the sense that they represent the absolute or true number present. Rather, they need to use the same technique and intensity of application (given appropriate sampling intensities and associated detection probabilities).

Habitat assessments are used in monitoring to quantify changes in the amount and condition of habitat, and to predict the impacts of land-use practices, over time and space. As developed in chapter 4, habitat is species specific and incorporates many components that can limit the size and distribution of animal populations. Habitat can serve as a preliminary focus for monitoring the abundance of habitat over larger spatial extents (landscapes), although verification of occupancy by animals should also be conducted.

Sampling Errors

The true value of the phenomenon we are trying to determine—density, species richness, or habitat—will usually be unknown to us. The difference between the true value and our estimate is termed *error.* Error is, in turn, composed of natural variation and (usually observer or sampling) bias. A study with low variation is considered to have high precision, and one with low bias is considered to have high accuracy. *Precision* measures how close our estimates are to one another, regardless of how closely they approximate the true value (e.g., a tight cluster of gunshots that widely miss the target center have high precision but low accuracy). The results of a study can be biased for numerous reasons, including especially those developed earlier concerning intra- and interobserver biases, but also because of inadequate effort, differences in vegetation and other environmental features (figure 9.2), animal behavior (e.g., noisy versus quite species), and numerous other factors.

We can, however, measure the precision of our estimates. Precision can also be increased by increasing sample size if measurements are carefully

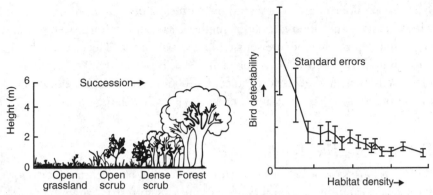

FIGURE 9.2. Birds are more conspicuous in open areas than in dense woodland. The hypothetical species is equally abundant across the succession, but might appear more abundant in the grassland and young trees where it is more easily detected. This effect is particularly serious if the bias arises from the same source as the object of study (such as the effect of forest succession on bird communities). (From Bibby et al. 2000, figure 2.7)

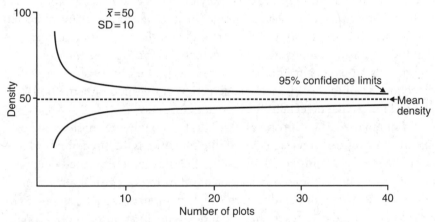

FIGURE 9.3. Relationship between the number of plots sampled and the confidence that can be put on the result for a particular hypothetical population. (From Bibby et al. 2000, figure 2.2)

planned (figure 9.3). As a general rule, however, precision does not increase directly with an increase in sample size but rather increases in proportion to the square-root of sample size. Thus, it would take 30 samples to double the precision obtained from 10 samples; doubling it again would take another 120 samples. Here again we see why the success of a monitoring study requires determining the necessary precision prior to initiating field sampling (i.e., use of pilot studies). The common sources of bias in counting animals, which include the observer, count method, sampling effort, weather, and vegetation types, have been discussed for herpetofauna (Heyer et al. 1994), birds (Ralph and Scott 1981; Ralph et al. 1995; Bibby et al. 2000), and mammals (Wilson et al. 1996); Morrison et al. (2006; 2008) provide detailed summaries of sampling methods and procedures for minimizing errors.

Detection: Are They Really Absent?

Biologists have expended considerable effort on estimating abundance. However, when monitoring population dynamics of terrestrial mammals, fish, birds, or plants through time or space, it is not appropriate to assume that you detected all individuals of interest or that each species is equally detectable. Thus, if you do not observe a species in an area, is it because the species was not present or because you were simply unable to detect its presence (Morison et al. 2008)? Not detecting a species, or all individuals of a species, can be due to extraneous noise (e.g., running stream, highway traffic) that masks songs and calls, or failure to see an individual because of its location (e.g., high in a tree, in a tree cavity). Population indexes based on uncorrected counts (i.e., just what you saw or heard) rely on the unrealistic assumption that the organisms under study are all detected equally across multiple vegetation types, observers, or time periods. Thus, in the case of such imperfect detection, we must directly estimate detection or use modeling methods that account for varying detectability of the target species.

As summarized by Morison et al. (2008), probably the most well-known historical survey techniques in the wildlife literature is the Breeding Bird Survey (BBS) for land birds and road-spotlight counts for white-tailed deer. Historical survey approaches have relied on absolute or raw counts—the number of birds or deer seen or heard at each survey point or along a transect—to be an index of abundance, such that the index is assumed to be proportionally related to the true population size. The index value is then used to monitor changes in trends through time. However, uncorrected indices must assume that the probability of detection remains constant between all survey sites, observers, weather conditions, species, and time periods; this assumption is seldom if ever valid. Additionally, it is likely that the magnitude of the influence of factors such as weather on the index will vary in intensity through time. Changes in vegetation structure as restoration proceeds will most certainly influence our ability to detect animals on a project area (e.g., as shrubs grow and become taller and denser) and substantially bias conclusions on trends based on uncorrected counts.

We have historically dealt with unequal detection by assuming that detection probability varies randomly across time, space, and treatments, and therefore, on average, detection will be the same. As noted earlier, an assumption of constant detectability is unreasonable and should not be used as a basis for monitoring trend (Morrison et al. 2008). A means of accounting for varying detectabilty has been to identify the variables that cause detection probability to vary (e.g., weather, vegetation growth) and model them as predictors of the counts; this is a viable option but relies on a statistical relationship and requires rigorous knowledge of the disturbing variables. The preferred method, therefore, is to estimate the detection probabilities of target species along with those factors influencing variation in detection directly because it relies on weaker assumptions of the attribute of interest (population size); this approach does require substantially more effort than the other approaches.

Detection probabilities are usually obtained by initially oversampling; for example, conducting eight point counts or running traps for six nights for a target species when you know from the literature that a fewer number are probably sufficient. Rather simple techniques are then available, including the use of available software, to calculate appropriate sampling regimes based on project objectives. A thorough development of calculating detection probabilities is beyond the scope of this book, but can be found in MacKenzie et al. (2006).

Wildlife Sampling

Questions about the animals present in an area, including their distribution, abundance, and vigor form a key component of any restoration project. Even if the project is not focused on restoration of particular animal species, it is likely that animals will either inhibit (e.g., gophers and other rodents) or promote (e.g., butterflies) plant restoration. Thus, all restorationists should have at least a basic knowledge of the techniques available to sample animal populations. The specific technique(s) selected and the intensity of their implementation is determined by project goals, and the need for species-specific surveys will be driven by the behavior of the target species. In the following sections, I outline many of the techniques available for the inventory and monitoring of vertebrates and supplement this basic material with sources of more detailed guidance.

Amphibians and Reptiles

Heyer et al. (1994) provided a detailed description of most of the sampling techniques available for amphibians and reptiles (collectively referred to as

TABLE 9.1

Standard techniques used to sample amphibians and reptiles, type of information gained, and relative time and cost

Technique	Information gained[a]	Time[b]	Cost[c]	Personnel[d]
1. Complete species inventories	Species richness	High	Low	Low
2. Visual encounter surveys	Relative abundance	Low	Low	Low
3. Audio strip transects	Relative abundance	Med	Med	Low
4. Quadrat sampling	Density	High	Low	Med
5. Transect sampling	Density	High	Low	Med
6. Patch sampling	Density	High	Low	Med
7. Straight-line drift fences and pitfall traps	Relative abundance	High	High	High
8. Breeding-site surveys	Relative abundance	Med	Low	Med
9. Breeding-site drift fences	Relative abundance	High	High	High
10. Quantitative sampling of amphibian larvae	Density/relative abundance[e]	Med	Med	Med

Source: W. R. Heyer et al. 1994 (table 4).
[a]Designations are hierarchical: techniques that provide density estimates also provide relative abundance and species richness. But if a technique provides only relative abundance, an additional technique must be used to provide density.
[b]Relative time investment.
[c]Relative financial cost: high = relatively expensive; medium = moderately expensive; low = relatively inexpensive.
[d]Personnel requirements: high = more than one person required; medium = one or more persons recommended; low = can be done by one person.
[e]Some methods included in technique 10 give relative abundance only, and some provide density estimates.

herpetofauna). Although their book focuses on amphibians, most of the methods are applicable to reptiles. They summarized the 10 most commonly used techniques, reproduced here as table 9.1, which vary in the type of information gained, and the effort needed in terms of time and costs. Obtaining a density estimate (numbers per unit area) and a complete survey of the species present is always expensive (recall the issue of developing detection probabilities, earlier). Obtaining a complete inventory of species in an area usually takes many months, and techniques listed with low or medium time are not a substitute for the complete inventory. Also note that the 10 standard techniques are not mutually exclusive. For example, focused surveys at breeding sites (e.g., ponds, creeks) can be used to supplement more general techniques, such as visual encounter surveys and pitfall trapping. Halliday (2006) and Blomberg and Shine (2006) presented brief but informative summaries of sampling techniques for amphibians and reptiles, respectively.

Scott (1994) detailed the general techniques that are available for generating species lists for an area, which involve searching the surface of the substrate, and turning over rocks, logs, and other cover. Such techniques have been used for both short- and long-term monitoring. The field techniques available can be used for sampling many of the species likely to be present in an area. However, secretive, canopy-dwelling, fossorial, and deep-water species often require more specialized techniques (as outlined in Heyer et al. 1994).

Short-term (a few days or weeks) sampling will not give much insight into the total number of species present at a particular site. Thus, longer-term and intensive sampling is necessary to establish baseline (prerestoration) conditions and then to conduct postimplementation monitoring.

Visual encounter surveys (VES) are those in which observers walk through an area for a prescribed period of time (i.e., time constrained) systematically searching for animals; time is usually expressed as the number of person hours. The VES is considered appropriate for both inventory and monitoring studies, although the caveats given above regarding secretive species apply. Although the VES is used to determine the species richness of an area and to estimate relative abundance of species, the method usually misses highly cryptic species. VES can also be done in a plot (technique 4), along a transect (technique 5), and in a user-defined patch of vegetation (technique 6). Crump and Scott (1994) provided details on implementation of various designs and derivations of VES, and Jaeger and Inger (1994), and Jaeger (1994a, b) described the related techniques of quadrat, patch, and transect sampling, respectively.

Audio strip transects (technique 3; see Zimmerman 1994 for details) are used to count calling animals (usually frogs). The width of the transect is varied depending on the detection distance of the various species' calls, and the counts are then used to estimate parameters such as relative abundance of calling individuals, species composition, and breeding or microsite use. This technique is basically an adaptation of singing bird surveys that have well-developed methodologies and analytical techniques.

Pitfall traps are used commonly to sample herpetofauna (and also rodents). Pitfalls provide estimates of species richness and relative abundance, and are particularly effective in capturing secretive, fossorial species because the traps can be left open for extended periods of time. A *pitfall trap* is a container placed in the ground so that its open end is flush with the surface; animals are captured when they fall into the trap. Pitfall traps are usually constructed from small cans, plastic buckets, or PVC pipe, and are 40–50 cm deep and 20–40 cm wide (Jones et al. 1996). A short (e.g., 50 cm tall) fence, called a *drift fence*, is often located to radiate out from the pitfall and thus help direct animals into the trap. As depicted in figure 9.4, a drift fence is composed of fencing material such as a thin aluminum sheet, hardware cloth (woven wire), or plastic sheet. Depending on the target species and condition of the ground, the fence can be buried at various depths to prevent movement under the fence. Corn (1994)

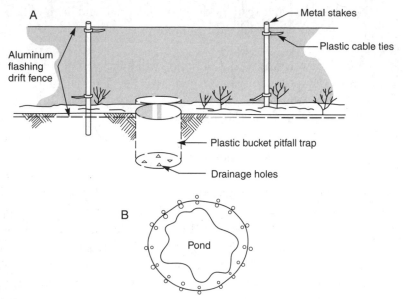

FIGURE 9.4. Pitfall trap with a drift fence established to help guide animals into the trap. (A) Illustrates the location of the fence relative to the trap. (B) An example of a continuous fence around a pond. (From Heyer et al. 1994, figure 17)

and Halliday (2006) provided a detailed description of trap design and placement, and basic survey methods.

The use of pitfalls became controversial because historically each trap was filled with some liquid so that animals would perish and be preserved, thus negating the need for frequent (at least daily) checking of trap contents. Pitfalls can, however, be used in a more humane manner that includes the provision of protective material in the trap bottom and frequent checking of trap contents.

Snakes are cryptic and often difficult to capture; they can usually escape from pitfalls. To assist with capturing snakes, a technique called a *funnel trap* can be used with good success. A funnel trap consists of rounded tubes or rectangles (window screen or hardware cloth) with an inwardly directed, funnel-shaped opening at one or both ends. When used with a drift fence, snakes are directed into the funnel but find it difficult to negotiate their way back out of the trap.

An artificial wooden cover, called a *cover board*, has been used to sample herpetofauna (see Fellers and Drost 1994 for details). This technique has limited applicability, however, and should be used only if the target species is known to use surface cover, or used in combination with other sampling techniques. Several key disadvantages are that the technique provides only an index of population abundance, not all species use artificial cover, the use of artificial cover can decrease as temperatures increase and the environment becomes dry, and cover boards are difficult to place in many vegetation types (e.g., shrubs, tall grass). It is highly unlikely that cover boards alone would provide a thorough assessment of the herpetofauna present in an area.

Night driving is a form of nonrandom line transect in which the transect is a paved road, which is usually used to sample herpetofauna that are attracted to the heat radiating from the road as the evening air cools. The technique by itself, however, cannot provide reliable estimates of abundance for most species for an area. Driving can also occur shortly after dawn to locate animals that were killed by passing cars the previous evening. For example, Morrison and Hall (1999) located new species for the Inyo and White mountains of eastern California through road driving that were not recorded during three seasons using pitfalls and VES sampling. Implementation of this method is, of course, constrained by the availability of suitable roads; results are also biased by the obvious nonrandom placement of the sampling locations (roads).

Birds

In this section I outline the basic techniques available for counting birds. Several detailed works critically review sampling methods, including Ralph and Scott (1981), Ralph et al. (1993), Bibby et al. (2000), and Gibbons and Gregory (2006). Much of this section is summarized from the excellent coverage of counting techniques given by Bibby et al. (2000).

As developed in chapter 8, a properly designed study begins with a clear elucidation of study goals. Specifically, will results apply to a large geographic area or a small plot? Are simple presence-absence data sufficient, or is a density estimate necessary? What magnitude of error is acceptable? Bibby et al. (2000) provides a detailed explanation of the standard sampling techniques, including sample data forms, recording methods, and methods of data interpretation. Here I will highlight the key components of these standard techniques.

Territory or Spot Mapping

The territorial behavior of many bird species provides the basis for what is termed the territory or *spot mapping technique*. The conspicuous behavior of many species, especially vocal passerines, is the basis of this technique. As is commonly known, singing males display from various locations to attract and breed with a female, which establishes an area usually defended from other males of the species (i.e., conspecifics); this area forms the territory. Thus, the number of territories in an area provides an estimate of the density and location of birds. These data can then be used to monitor the number and location of birds through time, and also to correlate changes in the number and size of territories with environmental conditions (see chapter 4).

Spot mapping is usually the most time consuming of the standard bird-counting techniques and is usually applied only to small areas (<20 ha) in studies where detailed information is needed on the location of birds, territory size, and habitat use. Studies of rare and threatened species often incorporate this method as part of an assessment of nesting behavior, demographics, and behavior. To use this method, an observer repeatedly crosses the study area, recording the specific location and behavior of each bird (of the target species) seen on a scale map of the area (see Bibby et al. 2000, chapter 3, for a detailed description). Six to ten visits to the study area have been established as a standard if accurate details on territory boundaries are needed. This method can also be used, however, to locate approximate territory boundaries to assist with later location of nests and fledglings.

Line Transects

Line transects are a commonly used technique for assessing the distribution and abundance of birds over large areas (i.e., >20 ha) of relatively uniform terrain. Transects need to be widely spaced (usually >200 m) to avoid double counting of individual birds. Detecting birds while walking, however, requires excellent birding skills. Thus, this technique is sensitive to bias from observer quality. Transects can be used year round, but bird detectability changes substantially between seasons due to bird behavior (especially changes in vocalizations), weather, and foliage cover.

Bibby et al. (2000, 70) depicted various specific field designs for transects (figure 9.5), with the specific type used depending on the goals of the study and the structure of the vegetation in the study area. For example, if abundance indices are needed in open (e.g., grassland, low shrub, marsh) vegetation, then either no distance measuring, or recording within a fixed belt, should suffice; see A or B1 or B2 in figure 9.5. However, if density estimates need to be calculated, then more specific measurements of distance must be determined; see C or D.

Transects are usually not used to develop relatively fine-scale assessments of bird habitat use because an adequate number of bird detections need to be

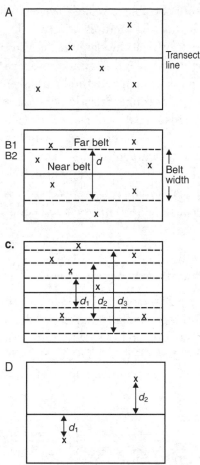

Figure 9.5. (A) No distance measuring; all birds are counted. This method is simple but different species are counted on different scales because of differing detectabilities. Five birds (x) have been recorded. (B1) Fixed belt. All birds are counted within a predetermined fixed belt (near belt). This lowers the total count but removes distant records of the more conspicuous species. In this case, four birds have been recorded and three birds have not been recorded. (B2) Two belts. All birds are counted but attributed to one of two belts. This is an effective method that is very simple to use in the field. Relative densities can be estimated. Four birds have been recorded in the near belt and three in the far belt. (C) Several belts. Birds are attributed to one of several belts of fixed width (d_1–d_3). This is harder to do in the field because distances have to be estimated to greater precision. It is often more satisfactory to use the methods given above or below. Counts in the first four belts were 1, 3, 2, 1. (D) Distances are measured to all birds. Distances are perpendicular to the route even if the bird was ahead when detected. This is the hardest method to use in the field but generates the best data for estimation of densities. Birds were recorded at distances d_1 and d_2. (From Bibby et al. 2000, figure 4.3)

made along the transect to be able to relate to vegetation and other environ-mental conditions—this usually requires at least 100 m of transect. Thus, transects are best suited for relatively large-scale assessments of bird abun-dance and habitat relationships. Although a transect can be subdivided into smaller (e.g., 50 m) segments, such a method partially defeats the purpose of the transect technique; point counts provide a more useful method of devel-oping finer-scale, bird-habitat relationships.

Determination of distances are often necessary because of the substan-tially different detectability between species (figure 9.6). Although this does not present a problem comparing a species within a vegetation type, as dis-cussed earlier, it complicates comparing a species across vegetation types and through time. Bibby et al. (2000, 90) summarized key assumptions associated with transect counts.

Point Counts

Point counts can be viewed as transects of zero length conduced while mostly stationary. Point counts were initially developed for use in rough terrain where simultaneously walking and observing birds was difficult and poten-tially dangerous (Reynolds et al. 1980). Point counts have, however, gained wide use in all vegetation types, including open terrain. By remaining at each

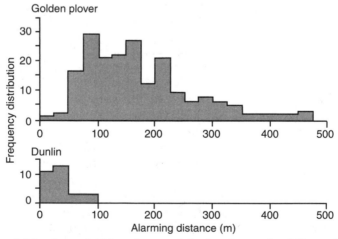

FIGURE 9.6. Numbers of golden plover and dunlin, detected at different distances from the observer. Golden plovers are noisy and conspicuous, especially as they re-act to an observer and give alarm at greater distances than do dunlin. Dunlin are cryptic and sit tight, so are not detected beyond 100 m. This difference between spe-cies needs to be taken into consideration in the design of appropriate breeding wa-ter-sampling methods. (From Bibby et al. 2000, figure 4.4)

point, more time is available per unit area surveyed to identify birds. Additionally, because observers are stationary, there is no noise associated with walking. Because observers must remain near the count center, most birds are recorded based on vocalizations rather than sight.

Distance estimations are usually used in point counting in a manner analogous to transects. Thus, the previous discussion of using no distance estimation, a fixed distance (e.g., 50 m radius plot), or individual distance estimations apply here. However, because detectability of birds begins to decline rapidly at 50–75 m from the observer, a distance of 50 m is often chosen to establish a fixed plot. But because of differences in detectability, using fixed plot radii is usually unacceptable when comparing results between different vegetation types.

Bibby et al. (2000, 91–92) developed how point counts were often efficient relative to transects because many points could be counted per unit time. However, this does not address the issue of independence between points. Avian ecologists typically separate points just far enough to avoid double counting birds at adjacent points; a typical distance is 200–300 m between points. However, points only ~300 m apart are usually sampling the same environmental conditions within a small area, and thus should probably not be counted as independent samples. For example, in figure 9.7, panel A establishes one transects while panel B establishes six points. However, independence of the points would depend on the scale of the map.

Despite the potential problem of independence of points, this technique is preferred over transects in developing relatively fine-scale assessments of bird-habitat relationships. With points, habitat parameters are sampled at and around the point (often within the fixed circular plot) and are related to the birds counted there. Ralph et al. (1995) provided a detailed discussion of point-counting applications, and chapter 4 of this volume summarized many of the methods used to develop wildlife-habitat relationships, including habitat assessment techniques (see also Morrison et al. 2006).

Bibby et al. (2000) provided a useful comparison of mapping, transect, and point counts applied to the same hypothetical situation. They also provided specific recommendations for individual species and groups of species that require specialized techniques or adaptations of standard techniques (e.g., colonial nesting birds, raptors, nocturnal birds).

Mammals

As developed above for the other herpetofauna and birds, multiple techniques are necessary to thoroughly access mammalian diversity in a study area. Mammalian diversity can be high even in small areas, which presents a

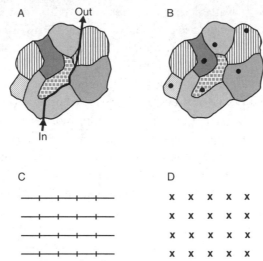

FIGURE 9.7. (A) In a fine-grained environment, such as wood, a transect following an access route might not be very representative. It would not be easy to divide the bird records into vegetation types. Indeed, in this example, two of the vegetation types have not been sampled at all. (B) In the same place, point counts could be set out at random or systematically so as to represent the full range of conditions present in the wood. Each point could also have the conditions recorded around it. (C) In open country, transects could be set out in a way to cover more of the ground and divided into sections for recording birds and habitats. (D) The equivalent design for point counts would theoretically record fewer birds but would take about the same time to execute. However, if birds were flushed ahead of the observer, as is generally the case in open ground, this would be a poor design because the observer walking up to a point would scare all the birds away. (From Bibby et al. 2000, figure 5.1)

challenge to the observer that must be mastered if an adequate assessment of the mammalian assemblage is to be obtained.

Wilson et al. (1996) provided a detailed description of most mammalian sampling techniques; much of the material I will review is summarized from articles within that publication. Krebs (2006) gave a brief summary of count techniques for mammals.

Voucher Specimens

Small and medium-sized species are often difficult to identify, especially when captured and released alive in the field. Identification can be difficult because subtle morphological characteristics, including those of skull and other bones, are required to separate species. Additionally, young and sub-adult animals are often difficult to identify in hand. Thus, specimens must be taken so that an archival record of the study can be developed. Such voucher specimens physically and permanently document data by (1) providing for confirmation of the identities of mammals accumulated and used in a study, and (2) assuring that the study can be repeated, reviewed, and reassessed accurately (Reynolds et al. 1996). Voucher specimens are also frequently needed in studies of herpetofauna.

Observational Techniques of Nonvolant Mammals

Many observational techniques can be used to determine presence and abundance of mammals in a study area, including those involving detection of animals by sight, sound, and physical evidence (especially tracks and dens). Here I will review several standard sampling techniques. Rudran et al. (1996) provided descriptions on several more specialized techniques such as aerial and night surveys.

Transects

Line and strip transects are conducted in a similar manner as done for herpetofauna and birds. Most of the assumptions of the techniques and the analytical methods for estimating mammalian abundance are the same as for other vertebrates.

In the line transect technique, an observer travels along a line recording either the animal itself, its call, or sign. Depending on the animal (and sign being used for detection), traveling can include walking, driving, or riding a horse. *Strip transects* follow line transects in assuming that all animals (or animal sign) within the strip are seen. For sign, the strip has traditionally been narrow, such as the width of the observer's outstretched arms; tracks, burrows, and scat are often counted in this manner. For observations of individual animals, the width of the strip is determined by the visibility of the animal and the denseness of the vegetation.

Road counts are often used to study larger mammals, especially ungulates (including nocturnal spot-lighting surveys). Many people use this technique because travel through a study area is relatively easy, and large areas can be covered rapidly. As outlined above for herpetofauna, counting from roads provides biased data and this technique should be avoided.

Quadrat or Plot Sampling

This method uses square or rectangular plots to sample animals or their sign. Quadrat sampling usually involves small, randomly (or stratified randomly) placed plots (1 m^2 to 100 m^2) that are surveyed intensively for sign (burrows, tracks, scat) and can be useful for obtaining an index of animal activity over time.

Observational Techniques for Bats

There is increasing interest in learning more about bat ecology, including a better understanding of their interactions with humans. However, because of their nocturnal habits, little information is available on the distribution and abundance of bats. Most field guides provide only broad descriptions of bat distribution and habitat use. Like birds, many bats migrate between breeding and wintering areas and require roost and feeding sites while traveling. Bats that winter in an area require locations to hibernate—hibernacula—that have specific microclimates and are free from human disturbance. Thus, bats should be carefully considered for emphasis in restoration projects.

Basic sampling methods for bats include (1) direct roost counts, (2) disturbance counts, and (3) nightly emergence counts. Kunz et al. (1996) and Jones et al. (1996) detailed general sampling techniques and specific capture techniques, respectively; here I highlight some of these methods.

Roost Counts

Direct roost counts are done by one or more observers who systematically count all visible bats in a roost. For bats that roost by hanging from horizontal surfaces (e.g., ceiling of a building, mine adit, or cave), such counts can obtain a nearly total count. However, many bats roost in crevices that obscure an observer's view. Thus, direct counts can be used to monitor bat abundance within a specific roost that is small and has an uncomplicated physical structure. Bats will also move between roosts (e.g., between tree cavities) within a project area, which means a thorough assessment of potential roost sites is needed.

Disturbance counts are used to at least partially overcome the problem of bats roosting out of sight or moving between count occasions. In this technique, bats are forced to take flight during the day and are counted as they become airborne. The success of this method depends on the sensitivity of the bats to disturbance and the skills of the observers. The count of bats will be unreliable if some fail to fly, do not leave the roost, or continually leave and reenter the roost. This method should only be used during favorable weather, and should never be used during breeding because of disturbance to pregnant and nursing females. Only experienced bat ecologists should conduct disturbance counts.

Nightly emergence counts are conducted as bats leave diurnal roosts; Kunz et al. (1996) described a related technique termed nightly dispersal counts. Emergence counts are an effective way to count bats that roost in inaccessible locations, in roosts with a complicated internal structure, and for minimizing disturbance to the roost. Because of the difficulty in counting bats as they emerge, multiple observers are usually required. For use in monitoring, observer effort, observer skill, positioning of observers, weather conditions, and time of observation must be standardized. Some observers have used still photography and videography to record emerging bats; night-vision devices can enhance observations.

Hibernaculum counts are done in winter to quantify hibernating bats. Extreme caution must be used to minimize disturbance to hibernating bats because disturbance can cause bats to awaken, which can lead to the death of individual bats. Because of the importance of adequate hibernaculum to maintenance of bat populations, it is, however, essential that these roosts are located and protected (Szewczak et al. 1998; Kuenzi et al. 1999).

ULTRASONIC DETECTORS

Ultrasonic sound detectors are a useful technique for detecting the presence of bats based on echolocation calls, and with the appropriate software, recordings can be used to make species identification. These devices are frequently used to access species richness at nonroost sights during the night such as water holes and riparian foraging areas. Distinguishing ecolocation calls of different species is based on species-specific features such as frequency composition and changes in frequency with time (e.g., O'Farrell et al. 1999; Szewczak 2004; Weller et al. 2007); computer software is available to assist with call identification (e.g., SonoBat, AnaBat). Most calls are highly structured with frequency that range from 10 kHz to 200 kHz; most North American species range from 10 kHz to 60 kHz. A substantial amount of research is underway that aims to allow more reliable separation of species.

Echolocation calls cannot, however, be directly used as count estimates because we cannot know the number of unique individuals actually passing the detector relative to the number of calls produced (e.g., 1 bat passing 10 times or 10 bats passing 1 time). Ecolocations can provide an index of bat activity, and an index can be used as a measure of the relative difference in activity between recording locations (Kuenzi and Morrison 2003).

Capture Techniques

Many means exist for capturing mammals that have gained general accep-
tance by biologists. These commonly used methods offer an appropriate starting point from which modifications can be made as preliminary data are evaluated in relation to project goals. Generally accepted techniques for
capturing mammals were outlined by Jones et al. (1996); here I summarize
many of the most frequently used techniques.

There are several commonly used devices for capturing small (<150 g)
mammals, each with a specific application depending on the species involved and the purpose of the study. *Snap traps* are used for killing rodents
during rapid assessment of species richness, especially in remote areas that
are logistically difficult to reach. Snap traps are frequently used when the goal
is to estimate abundance through the removal method (see Lancia and Bishir
1996). Of course, snap traps should not be used when protected species are
suspected to be present. Although small mouse traps are available in most
hardware stores, they are not powerful enough to instantly kill anything but
the smallest rodents and should not be used.

Box traps are the most commonly used device for live trapping small
mammals. Box traps are used primarily for studies in which animals are not
killed, and the goal is to determine species richness and derive an index of
abundance. These traps are available in multiple sizes so they can be effective with animals of various size and weight. The most popular traps are manufactured by H. B. Sherman Traps. Most people refer to box traps as *Sherman's* in the same sense that tissue paper and photocopies are described by
popular brand names. There are, however, multiple manufacturers of small
mammal traps (see Wilson et al. 1996, appendix 9, for a comprehensive
listing).

Because box traps are usually left open overnight, animal survival is maximized only when specific steps are taken to protect the animals, including
providing adequate bait and insulative material (shredded paper or fiber bat-
ting (polyester, cotton, wool), and covering the trap and cover with grass and
a small amount of foliage. Traps should never be placed unprotected on the
soil surface. Protecting traps also reduces trap disturbance by predators. Field
workers have an ethical obligation to ensure that animal discomfort is minimized. Box traps are also used for medium (up to 5 kg) and large (>5 kg)
mammals. However, such traps are heavy and difficult to position in adequate numbers in the field.

Pitfall traps are an effective means for sampling very small (<10 g) mammals, especially shrews (*Sorex*). The design and placement of pitfalls for

mammals is the same as that described above for herpetofauna, including the use of drift fences. Jones et al. (1996) also summarized more specialized capture techniques that can be used for medium and especially large mammals, including nets, dart guns, and drugged bait.

Trapping Arrays

The placement of traps—the trapping array—varies, depending on the goal of the project. For inventories that seek to obtain species richness, traps can be placed along transects that bisect all of the vegetation types and other environmental features of interest (e.g., rock outcrops, brush piles). Transect length is also determined by project goals and the nature of the environment (e.g., patchy versus relatively homogeneous). As a starting point, Jones et al. (1996) recommend a transect of at least 150 m long with traps spaced every 10–15 m for small mammals. For larger species, transect length should be based on home-range size. In medium-sized species, trap spacing of ≥100 m is recommended. However, because of the wide variation in movements of larger species, standard distances cannot be provided. Parallel transects can be used to sample larger areas and become, in essence, large-scale trapping grids.

Typical rodent trapping arrays are placed in grids with 7 × 7 (49 traps) to 10 × 10 (100 traps) with a single trap at each grid point and 10–15 m intertrap spacing. Jones et al. (1996) recommended 10 × 10 arrays with 2 traps at each grid point. The number of traps at a grid point depends, however, on the size of the trap needed (e.g., small for mice and large for chipmunks) and animal abundance.

Sampling Length

The goals of most monitoring studies include determining the species richness and abundance. Because, in most restoration studies, an increase in richness and likely abundance is expected over time, obtaining reliable estimates of these parameters is essential for evaluating the progress and success of the project (recall the earlier discussion of adaptive management). Thus, standardizing the time that traps are open (i.e., number of consecutive days) over time is not appropriate in most situations. This is because it usually takes a longer trapping period to quantify species richness and abundance in relatively species rich areas. The majority of rodent studies use a three-day trapping session, with many fewer going for four to five days. In fact, Jones et al. (1996) recommended that traps be operated on a schedule that coincides

with up to five to six activity periods for the species of interest and also recommended a period of at least seven days for medium-sized mammals. I
agree, and recommend that preliminary trapping be used to determine what
the trapping interval should be (refer to my previous discussion of detectability). Jones et al. (1996) provided a useful description of how to remove small
mammals from traps and handle them during processing. Field personnel
should also become acquainted with diseases that are sometimes carried by
small mammals.

Techniques for Small Volant Mammals

Bats are captured for determination of species identification, sex-age ratios,
reproductive status, and to apply marks. Jones et al. (1996) summarized trapping methods, and Kunz (1988) gave detailed methodology. Here I briefly
describe several popular techniques, emphasizing biases and cautions in
their use.

Several factors should be considered when designing an inventory project
that involves capturing bats, including weather conditions, moon phase, season (e.g., migration, dispersal, breeding), vegetative structure (e.g., short vs.
tall canopy), and colony size. These factors will help determine the type,
number, and arrangement of capture devices. These factors should be standardized or otherwise accounted for when comparing one or more study sites
through time. Caution must be used when trapping near roosts because of
the problem of capturing a large number of bats.

It is difficult to assess the use of most areas by bats because bats concentrate foraging activities in areas of high insect abundance, such as near water
sources and over riparian trees. Thus, most trapping has been done near such
sites, biasing our understanding of bat activity. However, bats do forage away
from wet areas. For small project areas without any apparent attractants for
bats, the observational (including acoustic) techniques described earlier
should be used initially to determine if trapping is necessary.

Mist nets, often the same type as those used for birds, are the most commonly used method to capture bats. Mist nets are made of light nylon or polyester thread in 2–3 m × 6–12 m panels that are strung between two poles. The
nets are difficult to see and do not usually return echolocation calls. Traps are
usually placed in arrays around sites frequented by bats, such as roosts (caves,
buildings) and water holes, but can be placed in any location thought to be
frequented by bats (figure 9.8). Results are biased, of course, by the height
that the nets are placed. Nets can be raised on poles or a pulley system to capture high-flying individuals, such as those foraging along the top of vegetation

Figure 9.8. Example of a mist-net placement near water.

(Jones et al. 1996, figure 19). Although radio transmitters can be used with bats, the small size of most bat species precludes the use of transmitters that remain active for more than a few weeks and have minimal (<1 km) detection distance.

Harp traps are another popular capture device (figure 9.9) made of a rectangular frame crossed by a series of vertical monofilament lines. When a bat hits the wires, it falls into a bag beneath the trap from which it can be removed. These traps are usually used to capture bats in narrow locations (e.g., stream corridor, cave entrance), as opposed to the relatively larger sampling area usually covered by mist nets. Harp traps and mist nets are not necessarily equally effective in capturing bats.

Synthesis

The goals of this chapter were to give you a general sense of the topics to consider and the effort needed in designing inventory and monitoring projects, and to provide an understanding of the diversity of methods available for sampling animals in the wild. Inventories associated with restoration projects are typically done to establish a baseline data set on existing conditions; that is, before initiating habitat-altering activities. Monitoring is usually defined as a repeated assessment of the status of some quantity, attribute, or task within a

FIGURE 9.9. Example of a harp-net placement.

defined area over a specified time period. Inventory and monitoring projects require an adequate sampling design to ensure accurate and precise measurements. Thus, adequate knowledge is required of the behavior, general distribution, biology and ecology, and abundance patterns of the resource of interest. Relationships between animals and the environment are not as strong for animals as they are for plants because of the mobility of most animals. Thus, the use of indicators in monitoring is controversial and should be attempted only after careful evaluation.

Restoration project must begin with establishment of the sampling universe from which samples can be drawn and the extent to which inferences can be extrapolated to areas outside your immediate study area. In many restoration projects, however, the physical size of the area is predetermined by management or regulatory issues. Relatively small areas are unlikely to support viable populations of many vertebrates unless the surrounding area is also suitable for inhabitation or at least provides passage routes. If the study goal is to maintain population viability over the long term, then the sampling area may include locations outside the immediate restoration site. I emphasize that monitoring studies are simply a special type of research project and as such must employ a rigorous study design.

The duration of a monitoring study depends on study objectives, field methods, ecosystem processes, biology of the species under study, and logistical feasibility. A primary consideration for inventory and monitoring studies is the temporal qualities of the ecological state or process being measured. The duration of a monitoring study depends on study objectives, field methods, ecosystem processes, biology of the species under study, and logistical feasibility. There are several main alternatives to long-term studies: retrospective studies, substituting space for time, and modeling how an ecological process might behave under various scenarios.

The concept of adaptive management is centered primarily on monitoring the effects of land-use activities on key resources, and then using the monitoring results as a basis for modifying those activities to achieve project goals. It is an iterative process whereby management practices are carefully planned, implemented, and monitored at predetermined intervals. Adaptive management is not synonymous with trial-and-error approaches to restoration. Attempting to fix a problem after implementation is quite different than developing an action plan prior to initiation of a project. Thus, thresholds and triggers represent a predetermined level that when violated will lead to specific responses.

Biases influence not only the way we design a study, but additional biases are inserted into field sampling when multiple observers are used to gather data. Ignoring observer-based variability may lead to incorrect conclusions because results include artifacts, spurious relations, or irreproducible trends. The biases associated with estimations of animal abundances should also be considered carefully in habitat studies. This is because many of our analytical procedures correlate animal numbers with features of the environment. Obviously, a study that has low bias among habitat characteristics can be ruined by biased count data, and vice versa.

Questions concerning animal diversity include those related to vegetation types or specific areas, and those related to specific species or groups of species. The primary goal of vegetation- or area-based questions is to determine the species that occur in specific vegetation types or specific areas. Species-based studies may focus on one or more populations across space or time to determine, for example, geographic or ecological distribution. Species-based studies may be conducted as an inventory, but a thorough description of the species present usually requires that multiple techniques be used. The specific technique(s) selected and the intensity of their implementation is, of course, driven by project goals. Additionally, species-specific surveys will be driven by the behavior of the species of interest. Certain techniques, no matter how popular in the literature, may be unsuitable for determining even the presence of various species. I briefly outlined many of the techniques available for the inventory and monitoring of vertebrates, and provided supplemental recommendations for sources of more detailed directions.

Animal ecologists have expended considerable effort in developing rigorous sampling methods for vertebrates. I am often asked what the *best* sampling method for a particular species is. This question can only be answered when placed in the context of the particular goals of the study. That is, the sampling methods and sampling intensity must be matched with study goals. As developed in chapter 8, there is no justification for over- or under-sampling. As synthesized in this chapter, restorationists will find a well-developed literature on vertebrate sampling methods to assist with the design of their projects.

Chapter 10

Case Studies

I think all would agree that restoration requires the synthesis of many environmental and more specifically, ecological concepts. In organizing this book I first laid out basic definitions and discussed the philosophical underpinnings of restoration ecology (chapter 2). We then stepped through the concepts and applications to wildlife ecology and management of populations (chapter 3) and habitat (chapter 4). In chapter 5 we discussed the ways in which biotic and abiotic factors and other constraints are thought to determine the species present in a specific location. Chapter 6 discussed ways in which we can use old and new data to identify the conditions we desired to achieve through restoration. Chapter 7 reviewed issues of spatio-temporal scale, habitat hererogeneity, corridors, and general concepts of reserve design as applied to restoration. In chapters 8 (study design) and 9 (monitoring), we reviewed the principles of study design including applications of impact assessment to restoration, as well as a review of basic sampling methods for wildlife.

Thus, to this point in the book we have developed the fundamentals of wildlife ecology as applied to restoration. Although each chapter interrelated many concepts, none described how individual chapters could be linked into an overall synthesis. In this chapter, then, I present case studies that are examples of how wildlife restoration plans can initially be constructed. Rather than develop these case studies myself, I thought it would be much more insightful and informative if I requested that other people write the sections for this chapter. To accomplish this task I presented four ongoing wildlife restoration projects that my students and other coworkers and I am conducting to a combined undergraduate and graduate course in wildlife restoration I teach at Texas A&M University. The course was composed of advanced (MS and PhD) students in wildlife and range science and management. Based on class size, the students divided into four restoration teams and selected from the available projects, which focused on the following:

- An endangered songbird in montane meadows in the Sierra Nevada, California: The case study on restoring the willow flycatcher in the Sierra Nevada, California, develops a program that integrates physical changes in meadow hydrology, planting of vegetation used for foraging and nesting, potential manipulation of predators, and active attraction of birds to the restoration sites. The authors use the literature and data being collected from an ongoing study to develop specific, quantified goals for changes in meadow wetness and vegetation cover.
- A multispecies restoration effort in the Lake Tahoe Basin, California and Nevada: This study incorporates both vertebrates and invertebrates into planning for restoration of several watersheds in the Lake Tahoe Basin. Because of historic interest in the basin, some information was available in the literature to assist with development of desired future conditions for wildlife.
- Endangered songbirds on private lands in central Texas: Unlike the previous two case studies, this project involves primarily private lands and incorporates methods for providing monetary incentives to landowners to encourage participation in the program. Using the state and federally endangered golden-cheeked warbler and black-capped vireo, this study incorporates both restoration of the physical environment as well as management of several predatory species.
- An endangered kangaroo rat in central California: This case study concerns a very small and completely isolated species of rodent, the Fresno kangaroo rat. Building upon existing data, this project incorporates a multitude of restoration techniques to both enhance existing habitat and provide new habitat areas; issues of corridor design and implementation are also included in the design.

Although we used my earlier book (Morrison 2002) for the course, I provided the students with draft copy of some chapters of this book for reference. Although they developed their plans as separate team efforts, we held weekly meetings to discuss progress and ideas. The students were free, then, to develop the plans as they deemed most appropriate and not necessarily follow any of the plans that were being developed by my coworkers and me for each project. These plans are not engineering and construction plans; that would be clearly beyond the scope of this book and our collective expertise. Rather, the case studies present the steps and rationale used to develop the overall framework—including goals, objectives, and constraints—needed to initiate a wildlife restoration plan. Each plan follows a common format for consistency, and includes a monitoring component set in a general adaptive management framework.

The restoration plans developed by each team are presented below, with each team coauthoring their individual plan. Many of the ideas developed by the students will help guide and modify the actual restoration planning for each project. I think you will find these case studies fascinating reading and will note where specific topics developed elsewhere in this book come into play. These case studies also reflect the difficulties and tradeoffs required to make a restoration plan a reality. In practice, these plans would undergo additional revision as we worked with hydrologists, plant ecologists, engineers, and other individuals to turn the framework into on-the-ground action. But, these case studies are thought provoking and serve as a valuable template for your own initial planning efforts.

Four Case Studies

Each study should be considered a complete and separate presentation for study, review, evaluation, and critique. For classroom use, it would be especially valuable for instructors to have students review and critique the approaches used and suggest ways in which the studies could be improved. For purposes of practical application, however, these case studies are a useful model on how to incorporate into a practical plan many of the issues involving wildlife restoration that we have been discussing in this book.

RESTORING A RARE SONGBIRD IN THE SIERRA NEVADA

Theresa L. Pope, *Department of Wildlife and Fisheries Sciences,
Texas A&M University.*
M. Constanza Cocimano, *Department of Wildlife and Fisheries Sciences,
Texas A&M University.*
Annaliese K. Scoggin, *Department of Wildlife and Fisheries Sciences,
Texas A&M University.*
Erin Albright, *Department of Wildlife and Fisheries Sciences, Texas A&M University.*

The Sierra Nevada is scattered with high-elevation, wet meadows. Past and current management activities, including livestock grazing, water diversions, and road development, have led to meadows that are currently drier than in the past. However, these meadows provide habitat for breeding willow flycatchers (*Empidonax trailli*), a neotropical migrant whose populations have been in decline in the Sierra Nevada. Drier meadow conditions provide predators access to willow flycatcher nests. High predation rates have been a major factor in the decline of the species, by decreasing the nest success and thereby reducing the number of individuals in the population. The decline in habitat and increase in predation rates have led us to take action and propose a project that will restore willow flycatcher habitat, increase habitat quality, and deter nest predators.

In this restoration plan, we will introduce the willow flycatcher and provide information about threats to this species' persistence in the Sierra Nevada. We will describe the approach we are taking for this restoration project and the desired conditions we hope to achieve. We will also detail the design of the project and how we will monitor our progress and evaluate the project's success. The overall goal of the restoration project is to develop a plan that can be implemented across the Sierra Nevada to allow for the recovery of willow flycatcher populations and eventually lead to delisting the Sierra Nevada subspecies.

Willow Flycatcher Natural History

The willow flycatcher is a small, dark, greenish or brownish flycatcher found throughout most of the contiguous United States, except the Southeast and southern edge of Canada (Green et al. 2003). Until the 1970s, willow flycatchers and alder flycatchers (*Empidonax alnorum*) were considered a single species that was eventually split by the American Ornithologists Union (AOU) based on song characteristics and geographical distribution (Green et al. 2003; AOU Check-List 1983). There are four recognized subspecies of willow flycatcher (*E. t. adastus, E. t. brewsteri, E. t. extimus,* and *E. t. traillii*) and three of these subspecies are present in California. *E. t. extimus* is found in southern California, whereas *E. t. brewsteri* and *E. t. adastus* are considered the Sierra Nevada subspecies, found on the Pacific slope and the eastern slope, respectively (Unitt 1987; Browning 1993). All California subspecies were listed as state endangered by the California Department of Fish and Game in 1990 (Steinhart 1990). The southwestern subspecies (*E. t. extimus*) reached such a critical level that it was listed as federally endangered in 1995 (USFWS 1995).

Considered a neotropical migrant, willow flycatchers nest almost exclusively in shrubby vegetation near a permanent or seasonal water features in the United States and winters in tropical and subtropical areas from southern Mexico to northern South America. In the Sierra Nevada, willow flycatchers nest in montane wetland shrubs. The adults begin arriving by late May and territories are established and nest building begins by mid-June. After the nests are built, eggs are laid over a period of 1 to 4 days and are incubated for about 12 to 14 days. Nestlings take around 15 days to fledge and remain in the vicinity of the nests with their siblings for several more days. During the fledgling stage, adults continue to bring food to the offspring. The period of the departure of birds is spread from late July to mid-late September (Green et al. 2003). If a nest is unsuccessful, Sierra Nevada willow flycatchers often renest in the same territory, though the clutch size is usually smaller for renests. Willow flycatchers tend to nest and forage in clumps of willow (or other shrubs) in and around their territory. They forage by hawking: waiting at a perch with good visibility and then pursuing arthropod prey that comes near (McCabe 1991).

Management History and Threats

The greatest threat to Sierra Nevada willow flycatchers is loss, fragmentation, and modification of riparian habitat, which they use for breeding and foraging. Habitat degradation and alteration primarily affects the hydrology and characteristic vegetation of the meadows (Green et al. 2003). Changes in hydrology have made many meadows drier than they were historically. This decrease in meadow wetness favors the establishment of lodgepole pine (*Pinus contorta*), which constitutes habitat for mammalian predators such as chipmunks (*Tamias* spp.) and Douglas squirrel (*Tamiasciurus douglasii*). Mammalian predators are a major component in the species decline (Green et al. 2003). In addition, a decrease in water levels leads to a reduction in willow that is the substrate for willow flycatcher nests (Green et al. 2003).

Habitat degradation began with livestock grazing, because livestock trampled vegetation and streambeds. Livestock were also an attractant for brownheaded cowbirds (*Molothrus ater*), which are brood parasites that lay their eggs in nests of host species (Friedmann 1929), including willow flycatchers (Mathewson et al. 2006a, b), often leading to reproductive failure of the host. Other human-related activities followed, such as road construction, timber harvesting, water diversions, and recreational activities. Road construction has been particularly important because roads were placed along and through meadows. Structures to maintain the roads and water flow around roads (e.g., gullies) were constructed to intercept water flow, making the meadows drier (Green et al. 2003). Therefore, the main action that will help in the recovery and viability of the declining Sierra Nevada population of willow flycatchers is restoring the vegetation and hydrology of meadows these birds inhabit (Green et al. 2003).

Location

Willow flycatchers in the Sierra Nevada generally breed in wet, montane meadows and riparian thickets from 1200 to 2500 m in elevation (Serena 1982; Harris et al. 1987; Flett and Sanders 1987; Valentine et al. 1988; Bombay 1999; Bombay et al. 2003; Green et al. 2003). Territory size ranges from 0.06 to 0.89 ha, with an average of 0.34 ha. Foraging occurs as far as 100 m from the bird's defended territory. Nesting occurs in riparian deciduous shrubs, preferably willow or alder, at least 2 m in height. A foliar density of greater than 25% for the lower portion of the shrub (below 2 m) and 75% for the above portion is optimal. Precipitation in the area is generally from snow, with as little as 36 cm per year on the eastern slope, to as much as 205 cm per year on the western slope (Bombay et al. 2003).

Montane wet meadows in the Sierra Nevada are generally associated with streams or small headwater rivers. However, they also occur along lake and pond margins or springs and seeps at higher elevations (Ratliff 1982; Bombay et al. 2003). Vegetation in the meadows is usually a variety of grasses, forbs, sedges (*Carex* spp.), and rushes (*Juncus* spp.) depending on elevation, hydrology, slope, and management history (Ratliff 1982; Dull 1999; Weixelman et al. 1999; Bombay et al. 2003). Willows (*Salix lemmonii* and *S. geyeriana*) are the dominant riparian shrubs in the open meadows in the area, although other willow species, mountain alder (*Alnus tenuifolia*), creek dogwood (*Cornus sericia*), aspen (*Populus tremuloides*), gooseberries (*Ribes* spp.), and lodgepole pine (*Pinus contorta*) also occur (Storer and Usinger 1963; Ratliff 1982, 1985; Weixelman et al. 1999; Bombay et al. 2003). These shrubs are generally patchy in distribution across the meadows or are restricted to edges of the watercourse (Bombay et al. 2003).

Approach

Our approach to the Sierra Nevada Willow Flycatcher Restoration Project in the central Sierra Nevada, California, is to focus initially on willow flycatcher habitat found near Lake Tahoe. Although willow flycatchers are found throughout the Sierra Nevada, we have narrowed our focus due to feasibility and logistical issues. The USDA Forest Service, Region 5, has been monitoring willow flycatcher populations in the central Sierra Nevada since 1997; therefore, we have chosen the meadows in their study to begin our restoration project (Mathewson et al. 2006b).

Since 1998, 15 meadows have been monitored: Webber Lake/Lacey Valley; Perazzo Meadow; Little Perazzo; Little Truckee 1, 2, and 3; Saddle

Meadow; Independence Lake; Prosser/Carpenter Valley; Upper Truckee Washoe State Park; Uppermost Upper Truckee; Grass Lake; Maxwell Creek; and Red Lake 1 and 2. Six additional sites have been monitored since 2003: Red Lake Peak, Milton Reservoir, Stampede Reservoir, Martis Valley, Cottonwood, and Mabie (Bombay et al. 2003; Mathewson et al. 2006b) (figure 10.1).

Of the 15 meadows used for monitoring, we have chosen three pairs as pilot meadows to test the efficacy of our restoration treatments. Each pair will consist of a meadow slated for restoration and a control meadow of similar size, location, and meadow type (e.g., lake edge or stream). Control sites are necessary to decipher whether changes occurring in the treatment meadows

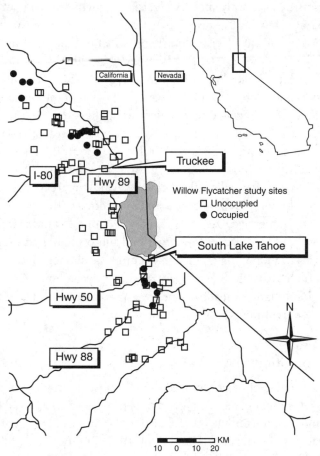

FIGURE 10.1. Meadows incorporated in the willow flycatcher restoration project are a subset of the meadows monitored by USDA Forest Service, Region 5. (From Bombay et al. 2003)

are due to the restoration treatments themselves or are following a similar trend as observed in the unrestored meadows. In addition, we can make stronger comparisons between sites that are relatively close to one another and of similar hydrology and vegetation types than between sites chosen at random. North of Lake Tahoe, we will restore Saddle meadow while using Little Truckee 3 as a control site. South of Lake Tahoe, we will restore Red Lake 1 and use Red Lake 2 as a control site; and also restore Uppermost Upper Truckee using Red Lake Peak as a control site (table 10.1). Although Uppermost Upper Truckee and Red Lake Peak are different types of meadows, they are similar enough in vegetation and hydrology so that reasonable comparisons can be made between the sites.

We chose meadows that are close enough (0.8 km to 9.7 km) geographically to assume that the flycatchers' decisions of which meadows to nest in will not be affected by the distance between the meadows or from other meadows. We based this decision on reported natal dispersal distances of 0.8 km to 19.8 km (average = 6.0 km) (Bombay et al. 2001) and adult dispersal distances of 1 km to 14.5 km in the Sierra Nevada (Stafford and Valentine 1985; Sanders and Flett 1989; Green et al. 2003). All of the meadows chosen for restoration either did not have willow flycatchers detected in 2006, had one territory in 2006 that was not successful (Uppermost Upper Truckee), or had not had a successful nest for several years. Control sites varied in the number of territories and successful nesting attempts in 2006. We chose small meadows with few willow flycatchers because the treatments are less expensive to implement. We also chose small meadows with few birds so that if treatments are ineffective, despite following recommendations from current literature, we will not impact the larger and more populated meadows. Affecting larger, more populated meadows could have a negative effect on the overall population of the Sierra Nevada willow flycatchers. We will extend our restoration to larger meadows once we have assessed the effectiveness of our restoration treatments at the pilot meadows.

Desired Conditions

The main objective of our restoration project is to aid in the recovery of the California state endangered willow flycatchers in the Sierra Nevada. Previous studies found that the willow flycatcher population has been experiencing a continual decline for decades (Harris et al. 1987; Bombay et al. 2003; Stefani et al. 2001; Green et al. 2003). The most recent information shows that the population in this area follows a declining trend and there is concern that the population in the Sierra Nevada will continue this decline (Mathewson et al.

TABLE 10.1

Summary of meadow type, size, location, and treatments included in the meadow restoration project for willow flycatchers in central Sierra Nevada, California

Meadow	Type	Size (ha)	Location (county)	First surveyed	Territories in 2006	Number of successful nests	Treatment
Saddle	river	18	Sierra	1997	0	0	treatment
Little Truckee 3	river	23	Sierra	1997	4	1	control
Red Lake 1	lake margin	9	Alpine	1997	0	0	treatment
Red Lake 2	lake margin	4	Alpine	1997	0	0	control
Uppermost Upper Truckee	river	3	El Dorado	1997	1	0	treatment
Red Lake Peak	spring fed	15	Alpine	2003	1	1	control

2006b). Mathewson et al. (2006b) also found that the population has a very low fecundity rate (number of fledglings per adult female); lower than the minimum thought to be necessary to maintain or increase the current population size. According to Robinson et al. (1993), 2.0–2.5 young/female/season is the minimum fecundity for small passerines to maintain stable populations. Yet, willow flycatchers in central Sierra Nevada showed values from 0.86 to 2.1 (annual mean 1.5) (Mathewson et al. 2006b).

In addition to having low fecundities for small passerines, many willow flycatcher nests are lost to predation. Nest predation is the main factor affecting nesting success of willow flycatchers (Cain 2001; Cain et al. 2003). Studies in the Sierra Nevada estimate that predation accounted for 45% (n = 49) of willow flycatcher nest failure in 1999 and 2000 (Cain et al. 2003), and 70% (n = 30) of willow flycatcher nest failure in 2005 (Mathewson et al. 2006a). Brood parasitism by brown-headed cowbirds is also considered a threat to willow flycatcher nesting success, however, brown-headed cowbird parasitism was responsible for only 6% of nest failures from 1997 to 2005 (Mathewson et al. 2006b). In 2006, 23% (n = 39) of nests were parasitized by brown-headed cowbirds (Mathewson et al 2006b).

In order to increase the abundance of willow flycatchers in the Sierra Nevada, we need to increase nest success. As mentioned, if a nest is unsuccessful, willow flycatchers may renest in the same territory, as close as 50 m from the first nest (Mathewson et al. 2006b). However, later nesting attempts usually produce fewer young than first attempts. Willow flycatcher females usually lay four eggs in their first nests, whereas 2nd and 3rd nesting attempts often produce only two or three eggs (Mathewson et al. 2006b). Furthermore, greater nest success contributes to more juvenile recruitment, which is positively related with productivity, because juveniles come back for the next breeding season (Mathewson et al. 2006b). Increasing the success of first nesting attempts will lead to a higher rate of population growth, termed lambda (λ) (where λ = adult survival + [juvenile recruitment*fecundity], and eventually increase willow flycatcher abundance (Mathewson et al. 2006b).

To increase willow flycatcher nest success, we propose restoring the hydrology and vegetation of our three treatment meadows. Previous studies have found significantly greater water depths in successful (0.60 m) than in unsuccessful (0.50 m) territories (Bombay 1999). Greater water depth may lead to successful territories by increasing the amount of willow flycatcher prey requiring the presence of water in certain stages of their life cycle, and also by impeding some predators from reaching nesting areas (Cain 2001). The abundance of arthropod prey is extremely important during the breeding season, when willow flycatchers need a locally abundant source of insects to maintain energetic requirements related to breeding activities (e.g. territory establishment, mating, nest building, egg laying, and brooding) (Craig and Williams 1998). Moreover, the presence of water or highly saturated soils is needed to support willow clumps and prevent the establishment of lodgepole pine.

Furthermore, inhibiting predator access to the nesting areas may lead to an increase in successful first nesting attempts. Many nest predators have been recorded in this area including mammals, birds, and reptiles (Cain et al. 2003; Mathewson et al. 2006b). The presence of some species has been found to be inversely related to hydrological characteristics. For instance, the activity of Douglas squirrel and chipmunks has been found to decrease in wetter meadows; the presence of these species is negatively related with willow flycatcher nest success (Cain et al. 2003).

The annual fluxes in the water table during the willow flycatcher breeding season begin in the spring when snowmelt swells the streams and rivers and saturates the wet meadows that the birds occupy. Throughout the season, the water table lowers as the water escapes through drainage systems or evaporates. As mentioned before, the hydrology of these riparian areas has been altered, which has led to drier meadows and a lower water table in general. Bombay et al. (2003) found that meadows that were occupied by willow flycatchers had 57% of their area covered by water or saturated soils versus 42% of the area covered in unoccupied habitat. Moreover, Bombay et al. (2003) found that the mean percentage of standing water or highly saturated soils in willow flycatcher territories was 44%. The presence of water (standing or flowing) or saturated soil is a very important factor for the successful reproduction of the willow flycatchers because it may inhibit some predators from reaching nests, create conditions to increase willow cover, and improve habitat for willow flycatcher prey species (e.g., wasps, flies, moths, caterpillar, and bees) (Cain et al. 2003; Green et al. 2003). Based on these findings, we propose restoring and maintaining the hydrology of the meadows so that at least 57% of the treated meadows will be covered by water or saturated soils during the breeding season.

Furthermore, Bombay et al. (2003) found that occupied meadows had a significantly higher percentage of shrub component in the meadow (60%) than unoccupied meadows (40%). Therefore, we propose supplementing the available willow stands with plantings to increase available willow to a foliar density greater than 25% for the lower portion of the shrubs (below 2 m) and 75% for the above portion (as suggested by Bombay et al. 2003). These shrubs will be planted in clumps of approximately 0.01 ha and at least 20%–30% of the meadow, which was suggested to be the minimum for suitable habitat

(Green et al. 2003). However, we will strive for a shrub component of ≥60% in the meadow.

Design

The design and timeline for this restoration project are outlined in table 10.2. As mentioned previously, brood parasitism by brown-headed cowbirds is considered a threat to willow flycatcher nesting success. However, parasitism rates have been much lower than the 30% parasitism rate where conservation concerns begin (Mayfield 1977; Laymon 1987), even in 2006 when the parasitism rate was 23%, the highest rate since monitoring began (Mathewson 2006b). However, this higher parasitism rate was not found in the meadows we chose for restoration, which had one or no parasitism events in 2006. Yet, parasitized nests are unlikely to fledge willow flycatcher nestlings (Mathewson et al. 2006b). Therefore, if we find parasitized willow flycatcher nests, we will addle brown-headed cowbird eggs found in the nests to try to improve the survival of willow flycatcher young. We will also attempt to reduce parasitism rates by removing cattle from the area. Cattle removal will also allow for vegetation and hydrology to recover from being trampled by the cattle.

TABLE 10.2

Projected timeline for the proposed restoration actions and monitoring program for willow flycatchers in central Sierra Nevada, California

Project Year	Event
1st	Vegetation survey
	Predator monitoring
	Cattle removal
	Hydrology treatment (fall)
	Survey for willow flycatchers; if present, map territory and monitor nests.
2nd	Plant willows (spring)
	Reassess vegetation
	Predator monitoring
	Survey for willow flycatchers; if present, map territory and monitor nests.
3rd	Monitoring vegetation. Vegetation sampling to assess if goals were met
	Predator monitoring
	Survey for willow flycatchers; if present, map territory and monitor nests.
4th	Predator monitoring
	Survey for willow flycatchers; if present, map territory and monitor nests.
	If the willow flycatcher is not present, begin conspecific attraction.
5th	Predator monitoring
	Survey for willow flycatchers; if present, map territory and monitor nests.
	If the willow flycatcher is present but not successfully reproducing, consider project unsuccessful and reevaluate factors affecting the population.

If cattle are present in the treatment sites, they will be removed before starting the hydrology and vegetation restoration, which will be during the spring of the first year of the project. To accomplish this, we will offer alternate grazing areas to, or buy out the lease from, lease holders to compensate for the loss of these particular meadows. These alternate grazing areas must be at least 7 to 14 km away from the meadows, because brown-headed cowbirds are known to commute these distances (Rothstein et al. 1984; Curson et al. 2000).

Even though nest predation has been the major cause of nest failure for willow flycatchers in the Sierra Nevada (Cain 2001; Cain et al. 2003), we believe that increasing the wetness of the meadows will effectively reduce predator access to nest sites (Cain 2001; Cain et al. 2003). Monitoring will be conducted to quantify how our treatments affect the use of meadows by predators. This monitoring will begin in the first year of the project before any treatments are applied and continue as long as other monitoring is conducted (see Monitoring in this case study).

We will begin restoring the hydrology of the treatment meadows in the fall, once the willow flycatchers have left the breeding areas. For the river sites, we may change the hydrology by redesigning the road culverts and by mechanically manipulating the streams (e.g., increasing sinuosity, creating oxbows, using a plug-and-pond method). At the lake edge sites, one way to change the hydrology is to adjust the water level by reducing the flow from the dam during the breeding season. However, meadows at Red Lake 1 and 2 receive most of their moisture from springs on the hillsides above the meadows and little water from fluctuating lake levels. The only way to increase the wetness of these meadows permanently is to manage the land above the springs for effective groundwater capture; however, this is beyond the scope of our project. In these sites, we will rely on willow plantings to reach our desired conditions along with raising the lake levels to flood the lower portions of the meadow without drowning the existing willow stands. If the conditions are too dry for willow plantings, then they will be irrigated until their root systems are developed enough to support plant growth.

We propose that 60% of the area of each treated meadow should be covered by riparian shrub, and from this percentage 85% should be willow, as this is considered the minimum for suitable habitat for the willow flycatcher (Green et al. 2003). However, within territories, approximately 44%–49% of the territory should be covered by willow (Green et al. 2003). Bombay (1999) also found successful territories had more shrub cover than unsuccessful ones (52% versus 43%). Also, dense thickets of willow are generally avoided in favor of more patchy willow sites providing considerable edge (Green et al.

2003). The willows should not be spread uniformly across the meadow, but should be clumped into patches such that within the territories there will be 50% willow cover (Bombay et al. 2003).

If necessary, we will actively plant willows to improve the vegetation conditions in the treated meadows. Willow planting will be done in the spring of the second year, following the hydrology restoration actions. Both Lemmon willow and Geyer willow, which grow in widely spaced clumps, are useful in streambed stabilization. These willows are considered to have a rapid growth rate and grow up to 5 m high, which is reached within five to six years. Furthermore, plants degraded by overgrazing recover within five to six years (USDA Plant Database 2007).

Planting will be done during the spring following the hydrology restoration completed the previous fall, when the higher water table will help the roots to grow and establish. Both Lemmon and Geyer willows can be planted as stem cuttings, and both rooted and unrooted cuttings can be used. However, rooted cuttings have a higher survival rate (USDA Plant Database 2007). Moreover, the planting will be done at least 100–150 m from the forest edge, because being close to the edge may expose the nests built in these willows to higher predator activity (Cain 2001).

We will compare the vegetation, hydrology, predation rates, and willow flycatcher territories of the treatment meadows to the control meadows, following the methods outlined later. The control meadows will allow us to interpret whether changes occurring in the treatment meadows are due to the restoration treatments themselves or are following a similar trend as observed in the unrestored meadows. Control meadows will continue to be monitored for willow flycatchers as they have been since 1997 (Mathewson et al. 2006b). We will also take vegetation measurements to make comparisons between treatment and control sites.

Monitoring

Here we outline the three primary sampling procedures used for monitoring restoration effects, including vegetation, predators, and surveys for birds.

Vegetation Sampling

We will survey the existing vegetation in the meadows during the first year previous to implementing the planting. If needed, we will plant willows during the second year and monitor the planted vegetation during the third year of the project. Waiting until the third year to monitor the planted vegetation will allow the willow plantings a full growing season to establish. We will evaluate if we met the desired conditions (e.g., percentage of meadow covered by shrub, percentage of willow in vegetation patches) at this time. If necessary, we will adjust the hydrology or vegetation (using the methodologies explained earlier) to achieve our goals. We will continue the vegetation monitoring after year three to evaluate the development and establishment of the vegetation.

To sample the vegetation we will establish transects within the meadows. We will randomly (using a random compass bearing, taken from the nearest forest edge) establish a baseline transect parallel to the main drainage. Then, individual transects will be randomly established at both sides of the baseline transect (USDA/USDI 1996). The first transect will be randomly established, and the next consecutive transects will be systematically established 20 m apart from each other. The number of transects will depend on the size of the meadow. In each transect, we will implement the point intercept method, using a pole (2 m tall × 1.25 cm diameter). We will take measurements every 5 m, and record every species touching the pole. We will use the contiguous quadrat method (1 m × 1 m) as well, to complement the point intercept (Sherry et al. 2003). We will record height and cover for each individual plant species once a year at the end of the willow flycatcher breeding season.

Predator Monitoring

To measure changes in the presence of predators in the meadows, we will also monitor predators in the meadows before and after treatments in both control and treatment sites. We will monitor for predator abundance using extra large (10.2 × 11.4 × 38.1 cm) Sherman traps (H. B. Sherman Traps, Tallahassee, FL) distributed in a systematic random pattern divided into 50 m buffers from the forest edge. Trapping will be conducted for five nights, two times a year; once at the beginning of the breeding season and once at the end of the breeding season. Using this design, we should be able to determine how frequently and how deeply predators are penetrating the meadows.

Bird Surveys

We will start bird monitoring during the first year of the project to evaluate whether willow flycatchers are present in the project area. Willow flycatchers can use a range of different heights for nesting (0.40 m to 2.50 m; Bombay 1999), so they should already be present in meadows being monitored. Surveys for willow flycatchers will be conducted each year following the protocol outlined by Bombay et al. (2000). If we determine that willow flycatchers are present (i.e., residents of the meadow), we will perform territory mapping and nest monitoring to quantify nesting success.

In years two and three, we will continue to survey the meadows for willow flycatcher presence. We will not take any additional action at this time if flycatchers are not present. Not taking any additional action in years two and three will allow the sites to stabilize after the disturbances associated with the restoration actions, and will also allow the birds to find these restored areas. However, if flycatchers are present, we will monitor the territories to evaluate nest success and predation rates.

If in year four the treatment meadows are considered to have met the hydrology and vegetation requirements, yet the species is still not present, we will attempt to use conspecific attraction to lure breeding willow flycatchers to the meadows using vocalization playback. Conspecific attraction has been shown to be an effective method of attracting birds to apparently suitable but unoccupied locations (Ward and Schlossberg 2004). Again, we will continue to monitor the territories to evaluate nest success and predation rates if willow flycatchers are present.

In year five, if the implementation of vocalization playback did not attract willow flycatchers, or willow flycatchers are present but not successfully reproducing due to high predation rates, we will consider the restoration unsuccessful. We attempted to provide the proper hydrology and vegetation to reduce the predation rates and therefore increase nesting success, yet we were unable to accomplish the goal. If this is the case, we will evaluate other factors that may be affecting the presence and successful reproduction of the willow flycatchers in the Sierra Nevada and attempt other restoration practices that may have a positive effect on these factors. Alternative methods to reduce predation rates (e.g. active control of predators by trapping and relocation), as well as isolation of known nesting areas, are discussed later.

Synthesis

The willow flycatcher is a neotropical migrant whose populations have been in decline in the Sierra Nevada. The two subspecies (*E. t. adastus* and *E. t. brewsteri*) that inhabit the Sierras are listed as endangered by the state of California. These birds inhabit montane wet meadows and nest in shrubby plants (e.g. willow) that are distributed in clumps throughout the meadows. Past and current management activities, including livestock grazing and road development, have led to meadows that are drier than in the past. Drier meadow conditions make it easier for nest predators to enter the meadows and prey upon willow flycatcher nests. High predation rates have been a major factor in the decline of the species, by decreasing the nest success and thereby reducing the number of individuals in the population. This loss of meadow wetness may also have an effect on the abundance of arthropod prey that the flycatchers depend on for their survival. The decline in habitat and increase in predation rates has led us to take action and propose a project that is necessary to restore willow flycatcher habitat, increase habitat quality, and deter nest predators. Restoring willow flycatcher habitat will hopefully increase the nesting success of the willow flycatcher and increase the current population numbers, reversing the current declining trend. If this project is deemed successful, we hope that our techniques may be tested and implemented elsewhere, which will allow for recovery of the willow flycatcher population and eventually lead to delisting these subspecies.

Our restoration project is small in scale, yet we are examining the effects of restoration on different types of meadows (stream or lake edge). Examining small meadows of various types allows us to evaluate the efficacy of our restoration on these meadow types before initiating restoration of larger meadows. The ideal outcome of this restoration project would be for willow flycatchers to begin to utilize the restored habitat in these meadows within a few years of the restoration efforts. It would also be ideal if the birds were successfully reproducing after the first year. If willow flycatchers are successfully reproducing, the restoration would be considered a success, and these restoration techniques could be extended to other, often larger, meadows in the area. If we have done all that we think would be necessary to attract willow flycatchers to breed in the meadows and reduce predation rates, but are unsuccessful in our attempts, we would know before undertaking a larger and more expensive restoration project.

We have planned our restoration project using restoration methods considered to adequately supply the habitat needed for willow flycatcher recovery. However, if these methods prove unsuccessful, it could be rewarding to evaluate the role of prescribed fire as a restoration technique that may be more effective than restoring the hydrology and vegetation of the meadows alone. While subalpine meadows historically have not burned as often as the surrounding lodgepole pine forests, these areas were exposed to fire dynamics in the past (DeBenedetti and Parsons 1979). These fires were known to kill the lodgeplole pine seedlings less than 2 m in height that encroach meadows during dry periods (DeBenedetti and Parsons 1979).

Further research is needed to evaluate the impact of hydrology and vegetation changes in predator activity. There is little research on how these predators utilize eggs, nestlings, or adults as a food source or how this predation rate can be reduced. Studies of the ecology of predators, including distribution and activity patterns, as well as their relationship with different wetness conditions are needed. Experiments could be conducted to evaluate how the predators access the nesting areas and the availability of alternate prey items for the predators.

If research shows that an increase in predator abundance affects nesting success, then some system of predator control should be considered (e.g., trapping and relocation). Moreover, the isolation of known nesting areas can be tested to evaluate the potential use of fences or other barriers to prevent predator access. More research is also needed on other factors that affect willow flycatcher habitat selection.

We have also proposed using conspecific attraction, which has proven to be a useful tool in manipulating other species. However, the effectiveness of conspecific attraction has not been specifically studied in willow flycatchers. There is a possibility that using playbacks of vocalizations could repulse conspecifics and deter the formation of territories in our meadows. It should also be considered that using conspecific attraction to lure birds to an uninhabited meadow could cause them to nest in an area they would not naturally choose. Additional monitoring of predator communities and available food sources in these uninhabited areas is needed to assess why the birds are not present and the potential failure of the birds that are lured to these areas.

Research is currently being conducted in the Sierra Nevada on the effectiveness of conspecific attraction and the ecology of predators of willow flycatchers. When this research is completed, we can use this new information to revise our current restoration plan if this new research indicates that we may not have proposed the most effective methods.

RESTORATION OF HIGHLY DEGRADED WATERSHEDS IN THE LAKE TAHOE BASIN, CALIFORNIA

Julie Groce, *Department of Wildlife and Fisheries Sciences, Texas A&M University.*
Anna Knipps, *Department of Wildlife and Fisheries Sciences, Texas A&M University.*
Stephanie Powers, *Department of Wildlife and Fisheries Sciences, Texas A&M University.*
Yara Sánchez Johnson, *Department of Wildlife and Fisheries Sciences,*
Texas A&M University.

Due to human activities over the past 200 years, the Lake Tahoe Basin has incurred dramatic ecological changes. Extensive logging in some areas opened the land to erosion and shifted the species composition of the forests (Murphy and Knopp 2000). Overgrazing in riparian zones reduced vegetative cover and compacted the soils of once wet meadows (Lindstrom et al. 2000). The construction of roads and houses altered natural water flows and limited ground permeability (Manley et al. 2000). Wetlands were drained for residential developments; small dams and water diversions in and around the basin disrupted the continuity of riparian systems and reduced flows to downstream meadows (Kattelmann and Embury1996). Extirpation of native species and introduction of exotic species altered the wildlife composition (Murphy and Knopp 2000). These activities, along with fire suppression and increasing development and recreation, have left portions of the basin notably different from pre-1800 conditions (Beesley 1996; Murphy and Knopp 2000).

Restoring the diverse ecological communities in the Lake Tahoe Basin to pre-1800 conditions is now the goal of several local and government agencies (e.g., Lake Tahoe Basin Management Unit, Sierra Nevada Alliance). Projects range from species-specific habitat restoration (e.g., Knapp et al. 2007) to reestablishing natural flow regimes of streams within a watershed (e.g., Rood et al. 2003). A watershed approach is beneficial since it acknowledges and incorporates the connections between water resources and the aquatic, semi-aquatic, and terrestrial ecosystems within the region (Roques and Gaman 2002). Goals for such diverse restoration projects are likewise far ranging but have in common the need to be specific, reasonable, and attainable. We make recommendations herein regarding the restoration of two watersheds within the basin, Taylor and Tallac, to benefit wildlife and also to serve as a template for restoring other watersheds in the basin.

Goals and Objectives

The goal of our restoration plan is to provide direction for the restoration of the Taylor and Tallac watersheds in the Lake Tahoe Basin. We aim to promote conditions in the watersheds comparable to pre-1800 by implementing prescribed fire and restoring water flow, for the benefit of a healthy system and persistence of wildlife. Specific objectives include the following:

- Reintroduction of fire to the watershed forests
- Improvement of water flow to the sensitive Taylor and Tallac wetlands
- Reduction of grazing and recreation impacts
- Monitoring and possible control of nuisance species

Our plan outlines steps for the design and implementation of restoration projects related to each objective. Using a multispecies approach (e.g., Lambeck 1997), our plan focuses on restoring appropriate habitat for species of

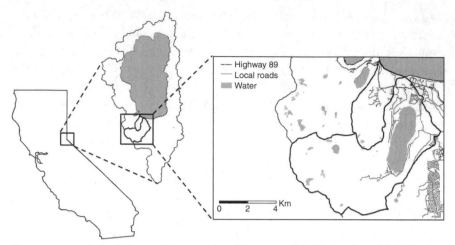

FIGURE 10.2. Lake Tahoe and its associated basin straddle the border of California and Nevada. The Taylor and Tallac watersheds are located in the southwest section of the basin.

interest, monitoring those species, and providing management options based on the species' responses to restoration.

Study Area

The Taylor and Tallac watersheds encompass ~6000 ha in the southwest section of the Lake Tahoe Basin (figure 10.2). Elevations range from 1,900 to 3,040 m, enabling a variety of vegetation types such as wetlands, montane meadows, quaking aspen (*Populus tremuloides*) groves, and coniferous forests. The watersheds are characterized by cold, wet winters and warm, dry summers. Average temperatures range from –1°C in the winter to 18°C in the summer. Mean annual precipitation is 78.4 cm, 86% of which falls as snow between November and April. Taylor and Tallac watersheds are representative of other watersheds in the basin with regard to human disturbances (e.g., timber harvest, road construction, grazing, and recreational use).

Approach

Restoration activities will center primarily on reintroducing fire to the landscape and reinstating a more natural water flow between the upland creek (south of highway 89) and the lowland marsh and wet meadow area (north of highway 89). Within this context, our plan focuses on reestablishing healthy, sustainable habitat capable of maintaining viable populations of an assemblage of vertebrate species, hereafter referred to as focal species. A list of focal species for the Taylor and Tallac watersheds was originally developed in 2004 (Morrison, unpublished report) and included (1) species whose abundance and distribution had potentially declined in the Lake Tahoe Basin, and (2) nuisance species that had detrimental effects on desired species (table 10.3). To understand the effects of restoration activities on focal species, we will monitor their populations. Although determining the population dynamics of a species may be an ideal measure of a population's status (Block et al. 2001;

TABLE 10.3

List of desired condition species (focal species) for the Taylor and Tallac watersheds

Common name	Scientific name
Herptofauna	
Pacific chorus frog	*Hyla regilla*
western toad	*Bufo boreas*
western aquatic garter snake	*Thamnophis couchii*
western terrestrial garter snake	*Thamnophis elegans*
Gilbert skink	*Eumeces gilberti*
Birds	
eared grebe	*Podiceps nigricollis*
pied-billed grebe	*Podilymbus podiceps*
American bittern	*Botaurus lentiginosus*
black-crowned night heron	*Nycticorax nycticorax*
great blue heron	*Ardea herodias*
gadwall	*Anas strepera*
green-winged teal	*Anas crecca*
northern pintail	*Anas acuta*
redhead	*Aythya americana*
common merganser	*Mergus merganser*
osprey	*Pandion haliaetus*
bald eagle	*Haliaeetus leucocephalus*
sora	*Porzana carolina*
spotted sandpiper	*Actitis macularia*
common snipe	*Gallinago gallinago*
Wilson's phalarope	*Phalaropus tricolor*
black tern	*Chlidonias niger*
northern pygmy owl	*Glaucidium gnoma*
common nighthawk (feeding only)	*Chordeiles minor*
calliope hummingbird	*Stellula calliope*
belted kingfisher	*Ceryle alcyon*
white-headed woodpecker	*Picoides albolarvatus*
williamson's sapsucker	*Sphyrapicus thyroideus*
red-breasted sapsucker	*Sphyrapicus ruber*
willow flycatcher	*Empidonax traillii*
pygmy nuthatch	*Sitta pygmaea*
house wren	*Troglodytes aedon*
marsh wren	*Cistothorus palustris*
American dipper	*Cinclus mexicanus*
ruby-crowned kinglet	*Regulus calendula*

TABLE 10.3

Continued

Common name	Scientific name
Swainson's thrush	*Catharus ustulatus*
hermit warbler	*Dendroica occidentalis*
yellow warbler	*Dendroica petechia*
Lincoln's sparrow	*Melospiza lincolnii*
blue grosbeak	*Guiraca caerulea*
yellow-headed blackbird	*Xanthocephalus xanthocephalus*
Mammals	
American marten	*Martes americana*
Belding's ground squirrel	*Spermophilus beldingi*
broad-footed mole	*Scapanus latimanus*
bushy-tailed woodrat	*Neotoma cinerea*
common porcupine	*Erethizon dorsatum*
Douglas' squirrel	*Tamiasciurus douglasii*
long-tailed weasel	*Mustela frenata*
mountain beaver	*Aplodontia rufa*
mule deer	*Odocoileus hemionus*
Trowbridge's shrew	*Sorex trowbridgii*
western red bat	*Lasiurus borealis*
Nuisance species	
American bullfrog	*Rana catesbiana*
European starling	*Sturnus vulgaris*
brown-headed cowbird	*Molothrus ater*
coyote	*Canis latrans*

Smallwood 2001), it is infeasible to do so for all 56 listed species. Thus, we decided to separate the species into three levels of priority (table 10.4). We will determine abundance and reproductive success for the highest ranked species, determine abundance for species of moderate rank, and establish presence for the lowest ranked species. While we understand that species presence and abundance may be a misleading indicator of habitat quality (Van Horne 1983), our intention for long-term monitoring of species presence and abundance will nevertheless provide an indication of population trends, which itself may be related to habitat quality (Sergio and Newton 2003; Padoa-Schioppa et al. 2006).

Although many researchers have offered suggestions on how to choose focal species for monitoring studies (Lambeck 1997; Block et al. 2001; Beazley and Cardinal 2004; Coppolillo et al. 2004), lack of such detailed life history information on our focal species compelled us to develop our own ranking system. In developing the focal species ranking system, we first compiled life history information on all listed species and then divided the species into three vegetation types in which they are most commonly detected:

TABLE 10.4

Priority level for each focal species and justification for their rank

	Priority Level			Justification
	High	Moderate	Low	
Wet meadow				
western aquatic garter snake	X			Listed as potentially vulnerable; prey on amphibians, therefore possible indicator of that taxon; require aquatic habitats therefore may be more responsive to restoration efforts
western terrestrial garter snake		X		Of similar concern as w. aquatic garter snake, but found in a broader range of habitats and possibly not as responsive to restoration
Pacific treefrog		X		Listed as of least concern by IUCN; will determine abundance because amphibians in general are of increasing concern throughout the country
western toad		X		Listed as near-threatened by IUCN; will determine abundance because amph bians in general are of increasing concern throughout the country
Gilbert skink		X		May be difficult to detect
Common nighthawk			X	Listed as of least concern by IUCN; not mentioned by Manley et al. (2000)
Calliope hummingbird			X	Listed as of least concern by IUCN; not mentioned by Manley et al. (2000)
willow flycatcher	X			CA state endangered; population decline throughout range; tends to be found in isolated areas in the basin, so may not be found in Taylor/Tallac; susceptible to cowbird parasitism
yellow warbler	X			CA species of special concern; susceptible to cowbird parasitism
Lincoln's sparrow		X		Listed as of least concern by IUCN, but mentioned as potentially vulnerable by Manley et al. (2000)
blue grosbeak			X	Listed as of least concern by IUCN; not mentioned by Manley et al. (2000); susceptible to cowbird parasitism
Belding's ground squirrel	X			Listed as of least concern by IUCN, but mentioned as potentially vulnerable by Manley et al. (2000)

TABLE 10.4
Continued

	Priority Level			Justification
	High	Moderate	Low	
mule deer		X		Listed as of least concern by IUCN, but mentioned as potentially vulnerable by Manley et al. (2000); harvestable, watchable; agency emphasis
Trowbridge's shrew			X	Listed as potentially vulnerable by Manley et al. (2000), but possibly too rare for estimates of abundance
broad-footed mole			X	Listed as potentially vulnerable by Manley et al. (2000), but possibly too rare for estimates of abundance
mountain beaver		X		CA species of special concern; listed as near threatened by IUCN; large impact on wet meadow conditions
western red bat			X	Apparently very rare, thus need more info on presence/absence before moving on to more in-depth studies
Water/marsh				
eared grebe			X	Listed as of least concern by IUCN; not mentioned by Manley et al. (2000)
pied-billed grebe		X		Listed as least concern by IUCN, but mentioned as vulnerable by Manley et al. (2000) because small or declining population
American bittern			X	Listed as of least concern by IUCN; not mentioned by Manley et al. (2000)
black-crowned night heron		X		Listed as of least concern by IUCN, but mentioned as potentially vulnerable by Manley et al. (2000)
great blue heron		X		Listed as of least concern by IUCN, but mentioned as potentially vulnerable by Manley et al. (2000); of interest to the public
gadwall			X	Migrant, not known to breed in area; not of general concern
green-winged teal			X	Migrant, not known to breed in area; not of general concern
northern pintail		X		Mentioned as potentially vulnerable by Manley et al. (2000); potential general indicator of wetland conditions

Species			Notes
redhead	X		Not of general concern but may be potential general indicator of wetland conditions
common merganser		X	Listed as of least concern by IUCN, but mentioned as potentially vulnerable by Manley et al. (2000); determining breeding status may be indicative of other waterfowl (i.e., needs cavities for nesting); quite common in early 1900s
osprey	X		CA species of special concern; Tahoe Regional Planning Agency (TRPA) special interest species; watchable, but large home range limits ability to quantify changes in population; known to breed in area
bald eagle	X		Listed, watchable, agency emphasis, but large home range limits ability to quantify changes in population; at least one nest near Fallen Leaf lake in late 90s
sora	X		Not of general concern but may be potential general indicator of wetland conditions; perhaps too secretive and difficult to determine breeding status
spotted sandpiper	X		Potential decline in recent surveys (based on 2–3 yrs of data); may be potential general indicator of wetland conditions
common (Wilson's) snipe	X		Potential decline in recent surveys (based on only 2–3 yrs of data); may be potential general indicator of wetland conditions
Wilson's phalarope	X		Listed as of least concern by IUCN; not mentioned by Manley et al. (2000); possibly too rare for good estimates of abundance
black tern	X		CA species of special concern; but likely no longer occurs in Tahoe area; would probably require lots of time and effort to establish them in area again
belted kingfisher	X		Listed as of least concern by IUCN, but numbers appear to have declined in the basin
marsh wren	X		Listed as of least concern by IUCN, but mentioned as potentially vulnerable by Manley et al. (2000); may be potential general indicator of emergent wetland conditions

TABLE 10.4

Continued

	Priority Level			Justification
	High	Moderate	Low	
American dipper		X		May not be able to create the right conditions for the species, but they are potential indicators of healthy streams
yellow-headed blackbird		X		Listed as of least concern by IUCN, but mentioned as potentially vulnerable by Manley et al. (2000); may be potential general indicator of emergent wetland conditions
Forest				
northern pygmy owl			X	Not mentioned by Manley et al. (2000) but of growing interest due to relative lack of data
white-headed woodpecker	X			Mentioned by Manley et al. (2000) as potentially vulnerable due to dependence on old-forest; may be potential general indicator of old-forest conditions; common in early 1900s
Williamson's sapsucker		X		Mentioned by Manley et al. (2000) as potentially vulnerable; numbers may increase with prescribed fire
red-breasted sapsucker		X		Mentioned by Manley et al. (2000) as potentially vulnerable due to small or declining population; numbers may increase with prescribed fire
pygmy nuthatch	X			Mentioned by Manley et al. (2000) as potentially vulnerable due to dependence on old-forest; may be potential general indicator of old-forest conditions
house wren			X	Listed as of least concern by IUCN, although potential decline in recent surveys (based on 2–3 yrs of data)
ruby-crowned kinglet			X	Listed as of least concern by IUCN, although susceptible to cowbird parasitism
Swainson's thrush		X		Listed as of least concern by IUCN, but mentioned as vulnerable by Manley et al. (2000); widespread population declines

Species	Col 1	Col 2	Notes
hermit warbler	X		Mentioned by Manley et al. (2000) as potentially vulnerable due to dependence or old-forest; may be potential general indicator of old-forest conditions; common in early 1900s but not noted in recent years; maybe move to high priority if any found
American marten	X		Mentioned by Manley et al. (2000) as potentially vulnerable due to dependence on old-forest; may be potential general indicator of old-forest conditions
long-tailed weasel		X	Not listed; perhaps too much of a habitat generalist to respond noticeably to restoration efforts
Douglas' squirrel	X		Mentioned by Manley et al. (2000) as potentially vulnerable (and also noted as a nuisance species); potentially important prey of northern goshawks; important role in forest nutrient cycling
bushy-tailed woodrat		X	Not listed; appropriate habitat may not occur in Taylor/Tallac (i.e., needs rocky crevices)
common porcupine	X		Not listed, but considered natural pruners of the forest
Nuisance species:			
American bullfrog	X		Need to better understand their abundance and distribution in the area, since they are not expected to occur in the basin
European starling	X		Need to increase understanding of density and habitat associations in the basin
brown-headed cowbird	X		Manley et al. (2000) recommended increasing understanding of density and habitat associations in the basin, and establishing direct connection between parasitism and host decline before implementation of cowbird management
coyote	X		Top predator in the basin and thus their choice of, and influence on, prey species (possibly desired focal species) is of high interest

marshland/water, wet meadows, and forest. It is in these vegetation types where we will focus our restoration efforts. A fourth category consists of the nuisance species. We then based our ranking system on the factors below. We provided justification specific to each species in table 10.4.

- Whether the species was state or federally listed as endangered, threatened, or of special concern.
- The rate of population decline as suggested in the literature, including early 20th century status descriptions by Orr (1949) and Orr and Moffitt (1971).
- Species for which physical alterations to the habitat (via restoration projects) may be realized in their population dynamics or abundance.
- The ability to adequately detect the species within the study area so that changes in population as result of restoration practices would be discernable. For example, population dynamics of rare species or species with large home ranges would not be an effective response variable due to small sample sizes.
- Seasonality: that is, we will monitor during the breeding season and therefore focus on species that are known to breed in the area.
- The extent of a species influence on its surrounding environment, as indicated in the literature.
- How representative the species is of certain environmental characteristics.
- Public interest.

Design

Restoration efforts will entail establishing a more natural burn cycle in the upland areas of the watersheds and establishing adequate flow and depth of water to the wetlands. To determine the effects of restoration efforts on the focal species, we will conduct multispecies surveys (see Monitoring section in this case study) in the Taylor and Tallac watersheds and in watersheds in which no restoration projects are planned (i.e., control sites) (figure 10.3). Control sites are used to help determine if changes observed on restoration sites are due to management actions. That is, if a difference in the trend of species occurrence or abundance is noted between control sites and a restoration site, a conclusion that management actions were responsible for the trend or changed observation is supported. Comparing restoration sites to multiple controls is ideal to reduce the amount of variation inherent in environmental data. The use of multiple control sites over multiple years allows

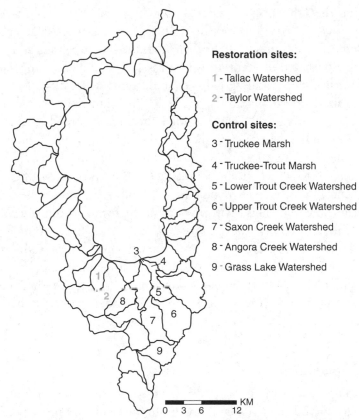

Restoration sites:

1 - Tallac Watershed

2 - Taylor Watershed

Control sites:

3 - Truckee Marsh

4 - Truckee-Trout Marsh

5 - Lower Trout Creek Watershed

6 - Upper Trout Creek Watershed

7 - Saxon Creek Watershed

8 - Angora Creek Watershed

9 - Grass Lake Watershed

KM
0 3 6   12

FIGURE 10.3. Location of restoration and control sites in the Lake Tahoe basin.

researchers to distinguish between natural temporal and spatial variability on the restored site (Block et al. 2001). We chose control sites based on their ecological similarity and relative proximity to the Taylor and Tallac watersheds. Survey data from wetland habitats in the Taylor and Tallac watersheds will be compared with data collected at wetland habitats of Truckee Marsh and Truckee-Trout Marsh. The effects of prescribed fires in the lower elevations (< 2200 m) of the Taylor and Tallac watersheds will be compared with survey data from lower elevations of Saxon Creek, Upper and Lower Trout Creek, and Angora Creek watersheds. The effects of higher elevation fires (~2,200–2,600 m) will be compared with survey data from similar elevations of the Saxon Creek, Upper Trout Creek, and Grass Lake watersheds. We anticipate that restoration projects and monitoring in the watersheds will take place over a 20-year period (see Synthesis section in this case study).

   In addition to the aforementioned restoration efforts, we include options for control or eradication of American bullfrogs (*Rana catesbiana*) and for monitoring the influences of other nuisance species (e.g., brown-headed cowbirds [*Molothrus ater*], European starlings [*Sturnus vulgaris*], and coyotes [*Canis latrans*]). Bullfrogs were introduced to the Sierra Nevada as a source of food in the late 19th century (Jennings 1985) and have since spread throughout the state. Bullfrog tadpoles and adults will eat most anything smaller in size, and can contribute to the decline of myriad wetland species (Hayes and Jennings 1986). Brown-headed cowbirds are brood parasites, laying their eggs in the nests of other bird species and leaving their young to be raised by the hosts (Lowther 1993). Cowbird activity can ultimately reduce host breeding productivity (Robinson et al. 1995); studies suggest parasitism that exceeds 30% may cause population instability (Laymon 1987). European starlings are nonnative cavity-nesters and aggressively compete with native cavity-nesters for nest sites (Cabe 1993). As a relatively abundant top predator in the basin, coyotes may be influencing population levels of smaller desired focal species (Bekoff 1977). Monitoring nuisance species before and after restoration activities is therefore important to understand how restoration may affect their occurrence and to increase understanding of their roles in influencing focal species.

Upland Restoration

In the last 150 years, the forests in the Lake Tahoe Basin have shifted from mostly pine and other fire-tolerant species to fire-intolerant firs. Jeffrey pine (*Pinus jeffreyi*), for example, has declined by about half, whereas white fir (*Abies concolor*) has doubled in frequency of occurrence. Forest composition has also generally shifted from trees of multiple age classes to dense, smaller-diameter trees (Murphy and Knopp 2000). Current species composition is due primarily to intense logging of the late 1800s and fire suppression in the last century. Research indicates that fires historically occurred late in the growing season (i.e., midsummer to early fall) in the Sierra Nevada (Caprio and Swetnam 1995; Taylor 2007). Caprio and Swetnam (1995) estimated that, on average, fires occurred every 23 years (range 4–40 years). We will utilize prescribed fire to reinstate natural fire cycles and intensities similar to pre-1800 with the intent of restoring appropriate forest conditions for the focal wildlife species. We intend to accomplish the following with prescribed burns:

- Maintain adequate densities of snags that are of appropriate size and decay class for utilization by woodpeckers and secondary cavity-nesters (e.g., National Forest guidelines currently recommend retaining 6–10 snags/ha of >25 cm diameter at breast height [Hutto 2006]).

- Decrease the proportion of fir and encourage the establishment of pine to benefit focal species (e.g., woodpeckers and pygmy nuthatches).
- Decrease the encroachment of conifers around wet meadows and aspen groves to benefit wet meadow focal species (e.g., willow flycatchers).
- Maintain a mosaic of percentage canopy closure to meet the diverse requirements of the focal species. For example, white-headed woodpeckers are more common in areas with 40%–70% canopy cover (Garrett et al. 1996), whereas American martens required more closed canopies (Simon 1980).
- Maintain a mosaic of fuel loads to meet the diverse requirements of the focal species and human safety. For example, while many species (e.g., small mammals) require some extent of downed woody debris for nesting, foraging, and avoiding predation (e.g., Converse et al. 2006), fuel loads must be reduced around areas of human development so as to lessen the risk of fire damage to property.
- Promote the growth of shrub layers to benefit focal species (e.g., small mammals).

The timing of the prescribed burns will be an important factor in restoring a natural fire regime to the area. Varner et al. (2005) cited specific unfavorable responses to fire reintroduction in forests in which long-term fire suppression had occurred. Excessive tree mortality can result from severe burns produced by unnaturally high fuel loads that have accumulated in the forest over time (Varner et al. 2005; Monroe and Converse 2006). Early season burns, however, tend to burn less intensely than late season, resulting in patchier burns and less consumption of dead organic matter (Knapp et al. 2005). Therefore, early season fires may be an effective method of initially reintroducing less severe fires to the area and provide the opportunity to evaluate the response of avian, amphibian, and small mammal species to these events (Knapp et al. 2005).

The effects of fire (i.e., snag creation, opening of canopy, clearing of woody debris) have been shown to cause shifts in avian species composition (e.g., Brennan et al. 1998; Kreisel and Stein 1999) and can potentially be harnessed to increase the presence of desired species. Smucker et al. (2005) discovered changes in relative abundance of avian species based on burned versus unburned sites. Although a number of desired avian species (e.g., ruby-crowned kinglet and Swainson's thrush) were shown by Smucker et al. (2005) to decrease in relative abundance after fire, other species,(e.g., white-headed woodpecker) were shown to benefit from burning due to an increase in bark beetles and snag occurrence. In the long term, fire may favor species associated with pine (Dickson 2002), such as white-headed woodpeckers.

Small mammals and many herpetofauna rely on herbaceous vegetation, shrubs, and woody debris on the forest floor, and changes in these components could affect their populations (see literature cited in Converse et al. 2006). Implementation of a fire regime initially results in increased herbaceous layer, a reduction in shrub layer, and a reduction in woody debris. Low-intensity fires may have no noticeable effect on the relative abundance of animal species (Vreeland and Tietje 2002), yet high-intensity fires will reduce the duff layer (Fule et al. 2004) which may reduce amphibians. Too large or too intense a fire may be detrimental to certain species since they require some amount of woody debris for breeding or cover (Ford et al. 1999; Monroe and Converse 2006). Ultimately, there is a need to balance fuel reduction with fuel retention (i.e., woody debris, dead/dying trees) (Knapp et al. 2005).

Conclusions drawn from the aforementioned studies indicate the need to create a mosaic pattern of burn and successional stages in the watersheds to meet the habitat requirements of a variety of species. Furthermore, the introduction of prescribed fire must involve adaptive management. There are still many unknowns regarding the effects of fire because of inconclusive findings (e.g., Ford et al. 1999; Vreeland and Tietje 2002), a lack of long-term or large-scale studies (e.g., Monroe and Converse 2006; Converse et al. 2006), and a lack of studies in general (e.g., Bury et al. 2002; Pilliod et al. 2003; Monroe and Converse 2006). Thus, it is extremely beneficial to initially implement prescribed fire in the Taylor and Tallac watersheds in the form of multiple experiments (both short and long term) and alter future fire management based on the results of the experiments. In addition to the standard fuel reductions near human developments, we will experiment with different size and intensities of fires in these watersheds. We will establish paired treatment (burned) and control (unburned) plots within the Taylor and Tallac watersheds (figure 10.4) and monitor the responses of focal species and vegetation (including snag recruitment and retention) over time. The treated plot within each pair will be randomly selected. We suggest the following four fire experiments:

1. Small (~5–50 ha) compared to large (~100 ha) fires (Carlton et al. [2000] estimated that fires over 400 ha were historically rare). This includes examining the effects of several small fires compared to one to two large fires:
2. Early season (i.e., low-intensity fire) compared to late season (i.e., high-intensity fire).

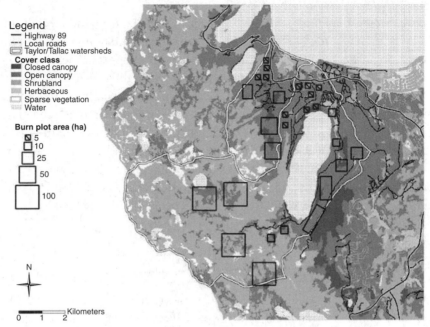

FIGURE 10.4. Potential plots for prescribed fire in the Taylor and Tallac watersheds, Lake Tahoe Basin, California. Boxes indicate size and location of burn and treatment/control pairs. (Note: this plan was developed prior to the Angora Fire that burned western sections of the study area in June 2007).

3. The effects of fire along different slopes or aspects.
4. Maintain early succession stages of vegetation on some plots and allow late succession to occur on other plots; thus, compare short-duration fire cycles (3 to 5 years) with longer-duration fire cycles (10 to20 years).

Plots can be used for more than one comparison at a time. For example, the effects of several small, early-season fires can be compared to the effects of several small, late-season fires and also to the effects of one to two large early-season fire(s). We recommend utilizing the 5 and 10 ha plots for experiments 1–3, the 25 ha plots for experiments 1–4, the 50 ha plots for experiments 1 and 4, and the 100 ha plots for experiments 1, 2 and 4 (figure 10.4). Implementing several concurrent experiments will allow for the creation of a mosaic pattern across the landscape to benefit a larger number of focal species for longer periods of time.

Upland Restoration Monitoring

Focal species monitoring and snag surveys should occur in and around the plot pair each year for 3 to 5 years prior to the implementation of a burn. The extent of posttreatment monitoring will depend on the experiment; in general, however, monitoring should occur each year for three years after a burn and every 2 to 3 years for the following 10 years. Thus, focal species on plots where short-duration fire cycles are implemented would be monitored nearly every year for the duration of the experiment. Sediment deposition in nearby streams and focal species inhabiting nearby riparian areas should also be monitored to verify that the fires are not detrimental to those species and landscape functions. Establishing long-term monitoring is necessary to adequately determine the responses of focal species to the habitat changes caused by different fire prescriptions.

Thresholds and Triggers

It may take several years before the effects of fire treatments are realized in the density and distribution of focal species. If results (e.g., species composition) are similar between plots of low- and high-intensity fires, we recommend continuing the implementation of low-intensity fires to reduce the risk of wildfires. For the same reason, if results are similar between many small fires and one to two large fires, we recommend continuing the implementation of many small fires. For long-duration fire plots, focal species composition is expected to change over time with successional stages of vegetation. However, throughout all experimental plots, we expect the abundance of ≥50% of the focal species to be stable or increasing within five years and ≥75% of the focal species to be stable or increasing within 10 years. Numerous factors may be involved in the lack of meeting these guidelines, and it is currently difficult to predict how large of a role fire may play. If fire alone is found to be inadequate in stabilizing or increasing the focal species' populations, we recommend the following, more management-intensive, options:

- If increased clearing of the understory is deemed necessary, use herbicides or mechanical removal (e.g., Brennan et al. 1998; Fule et al. 2004).
- If snags are not being recruited or retained by fire alone, use mechanical or biological methods to create snags (see Bull 1997).
- Control predators or nuisance species.

Wetland Restoration

The wetlands associated with Taylor and Tallac Creeks are two of the few remaining wetlands in the Tahoe Basin. These wetlands were once joined in an extensive (over 160 ha) marshland (USFS 2006a). Long-term cattle grazing, road building, clogged culverts, and channelization of the creeks have all contributed to the reduction in size and subsequent separation of these wetlands (S. Muskopf, USFS Aquatic Biologist, personal communication) (figure 10.5). There are several obligate wetland species among our list of focal animals (e.g., marsh wren, sora) as well as species that use the wetlands for nesting and/or foraging (e.g., northern pintail, redhead, yellow-headed blackbird, black tern, common snipe, Wilson's phalarope, western aquatic garter snake, Pacific treefrog, western toad). Because wetland areas act as important filters for sediments from upstream erosion (Mitsch and Gosselink 1993), the restoration of this landscape component will benefit the wildlife using the wetland and also improve the water quality flowing into Lake Tahoe. We propose the following restoration steps for at least the first three years:

- Identify and alter areas that appear to restrict water flow to the wetlands (e.g., culverts and roads).
- Discontinue the current grazing permit on the Baldwin allotment.
- Restrict human access to the wetlands via educational signage.
- Begin a bullfrog eradication program.

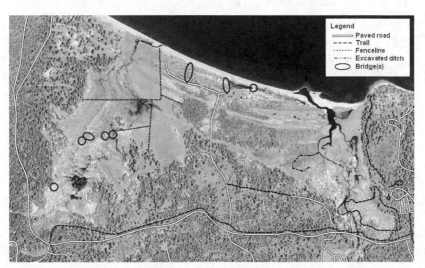

FIGURE 10.5. Wetlands area of the Taylor and Tallac watersheds and the associated human disturbances. (USFS 2005)

WATER FLOW

The first step in restoring the wetland areas is to allow surface water to flow more freely into the marsh. Highway 89 and other access roads currently impede this process. Beach access roads and trails should be either rerouted and rehabilitated, or altered to allow improved water flow. Culverts under roads and parking areas must be checked for effectiveness and cleared of woody debris if necessary. Culverts may need to be replaced with appropriately designed bridges if water is consistently unable to flow freely. Bridges (current and future) must be wide enough to allow high water flows to pass through without scouring the sediments from the banks.

Increased water flow to the wetlands should reduce the encroachment of woody species that are intolerant of standing water or perpetually wet soil. Mechanical removal of undesired trees is not recommended at this point, as this activity has the potential of compacting wetland soils. The original marsh vegetation may automatically respond to the increased water flow, but if it is necessary to replant the wetland vegetation, the seed bank can provide a source for native plant seeds (van der Valk et al. 2004). The increase in area of the wetland vegetation should provide more nesting substrate for spotted sandpipers, black terns, soras, pied-billed grebes, Wilson's snipes, marsh wrens, and yellow-headed blackbirds. Snags created by inundation may provide nesting sites for bald eagles and osprey, though both species may also use artificial nesting platforms (Hunter et al. 1997; Marion et al. 1992).

The USDA Forest Service has monitored groundwater levels in the Taylor and Tallac wetland areas with a series of piezometers since 2002. These instruments measure water table height and groundwater flow, and will be useful in determining if the restoration efforts are indeed improving water flow to the area (USFS 2006b).

GRAZING AND RECREATION

Long-term grazing in and near the Taylor and Tallac wetlands has compacted and eroded the soil, altered stream channels, and altered plant species diversity (USFS 2005). Wetland soils typically have high porosity through which water slowly flows and is filtered (Mitsch and Gosselink 1993). Animal (e.g., horse) hooves greatly compact these soils, leading to a decreased residence time of water flow. Currently there is a grazing permit for pack animals in the Tallac wetland area. We recommend grazing not be allowed in sensitive wetland areas.

Recreation is also a management concern in the wetlands, as excessive foot travel can compact sensitive wetland soils and negatively impact vegetation. Human presence may also disturb nesting bird species and other wetland wildlife. Restricting human access to the marshlands can be beneficial to the restoration and maintenance of this area. Access restriction can initially be encouraged using educational signs.

There are currently no bald eagle nests within the Taylor and Tallac wetlands, though a nest was present near Fallen Leaf Lake in the late 1990's. Snag creation from planned restoration activities may increase nesting and/or roosting sites for bald eagles. If bald eagles again choose this area for nesting, steps must be taken to prohibit human presence. Steidl and Anthony (2000) found that nesting bald eagles in Alaska were agitated and spent less time feeding or caring for their young when humans were 100 m from the nest than they were when humans were 500 m from the nest. Furthermore, Wood et al. (1998) estimated that nestlings were sensitive to disturbance up to 11 weeks postfledging. We suggest the elimination of human presence within 200 m from nest sites for the duration of the breeding season and postfledging period.

Bald eagles also winter in Lake Tahoe Basin, and might use the new wetland snags as winter roosts. We propose to extend the 200 m radius buffer zone for the winter months when eagles are present. This may require rerouting ski trails and possibly blocking access points.

BULLFROG ERADICATION

The control of introduced bullfrogs is necessary for the restoration of the Taylor and Tallac marshlands, as they are voracious predators of a variety of wetland animals. One method of controlling bullfrog populations is to have an annual bullfrog "roundup," in which trained volunteers harvest all detected bullfrogs. The volunteers, as well as the general public, should be educated as to the negative effect introduced bullfrogs have on their surroundings, in addition to proper identification (Orchard 2005). To garner public support for bullfrog control, the public must be educated in a similar way to view these roundups as an important conservation tool rather than cruelty to animals. Govindarajulu et al. (2005) recommend targeting juvenile bullfrogs in the fall to minimize the effects of density-dependent competition release.

Rather than limit eradication efforts to the Taylor and Tallac wetlands, we suggest coordinating efforts to maximize control potential. That is, begin control efforts in the upstream and neighboring wetland areas and work into the Taylor and Tallac wetlands. While it is inevitable that bullfrogs will enter the area from other parts of the basin, a concerted and continued effort may keep bullfrogs at manageable levels.

THRESHOLDS AND TRIGGERS

Monitoring of the focal species will continue throughout the restoration process to determine the results of our restoration activities in the wetlands. We expect the abundance of ≥50% of the focal species to be stable or increasing within 3 years and ≥75% of the focal species to be stable or increasing within 10 years. In the event these thresholds are not met, we suggest the following, more management intensive, steps:

- If water flow is not increasing sufficient to expand the wetland area by at least 25% in the first three years, we recommend allowing water to be released from the dam at Fallen Leaf Lake on an experimental basis.
- Elimination of recreation activities in the wetland area plus a 100 m buffer zone around the wetland area.
- Continued, and possibly increased, bullfrog control efforts.

Forest Changes and Hydrology

The changes in species composition and stand densities resulting from the prescribed fires in the upland areas may also affect the hydrology of the wetlands. Initially, a burned landscape will have greater water repellency and less vegetation, so there will be an increase in water runoff and erosion. Because of this, we propose to initiate fire activities no earlier than the second year after wetland restoration to give the wetland vegetation an opportunity to respond to the culvert and road improvements. The increase in wetland vegetation will assist in the filtration of sediments from postfire upland runoff. If the vegetation does not respond to the increased water flow, or if the flow itself does not increase after the first year of restoration, we may need to postpone the initiation of prescribed fire activities until there is ≥ 25% increase in wetland vegetation area.

With each successive fire plot burned, the forest will gradually change to a more open forest with increased herbaceous groundcover. While more water will be available to flow to the wetlands, the groundcover will keep water from flowing too quickly and eroding soils. Our suggestion of beginning fire restoration with low-intensity fires serves two important factors in relation to water flow:

- Low-intensity fire is less likely to produce the hydrophobicity in soils that high-intensity fires create, thus allowing soil to hold more water.
- Low-intensity fire will leave an adequate amount of the present vegetation to help control erosion (Ice et al. 2004).

Although we suggest experimenting with higher-intensity fires in subsequent years if the response to initial fires is positive, we recommend the following to minimize sedimentation:

- Maintain appropriate no-burn buffer zones around riparian and lentic systems.
- Avoid high-intensity fires on steep slopes.

Monitoring

The methods used to monitor focal species will vary depending on the vertebrate group (i.e. birds, mammals, or herpetofauna) and species' rank (table 10.4). Although we are primarily interested in data pertaining to the focal species, we will record all species (focal and nonfocal) detected during the surveys. Specifically, we want to determine the (1) abundance and reproductive success of the high priority species, (2) abundance of the moderate priority species, and (3) presence of the low priority species.

All surveys discussed below will occur within the Taylor and Tallac watersheds and in the corresponding habitats of the control watersheds. If more than two of the appropriate habitat patches exist in each of the control watersheds, surveyed patches will be randomly determined. We recommend three surveys per site per breeding season.

Birds

Point counts will be used for all medium- and low-priority bird species to determine presence (Bibby et al. 1992; Morrison 2002) while distance sampling, in conjunction with the point counts, will allow for estimates of abundance (e.g., Kissling and Garton 2006). Points will be placed to adequately cover the control and treatment plots in the upland forest and throughout the marsh. Nest searching and monitoring of high-priority species will focus on determining reproductive success, number of nesting attempts, and brood parasitism.

Herpetofauna

To determine abundance of reptile and amphibian species of moderate priority, we recommend using mark-recapture methods (Crump and Scott 1994; Ryan et al. 2001). Capturing individuals can be accomplished by setting drift fences with pitfall traps in the forest (Ryan et al. 2001) and drift fences with minnow traps in wet meadows and marsh habitats (Crump and Scott 1994).

Stratified sampling of the variety of vegetation types that exist within the forest and wetlands will provide a more complete picture of species occurrence (Morrison et al. 2001).

Western aquatic garter snakes (high-priority focal species) tend to breed in late summer (Stebbins 2003). To determine abundance and population dynamics of this species we recommend the aforementioned mark-recapture methods (Crump and Scott 1994). Although capturing reproductive females may indicate reproductive success in the area, it may also be necessary to implant radio transmitters in individual snakes to determine activity budget and reproduction (e.g., Blouin-Demers and Weatherhead 2002).

Mammals

The presence of low-priority mammal species and the abundance of moderate-priority species will be determined via live trapping and mark-recapture methods (Nichols and Conroy 1996). Trapping grids, consisting of both large and small traps, will be established via a stratified sampling scheme within the forest and wetlands (Morrison et al. 2001). Visual encounter surveys and track plate surveys can augment live trapping for low-priority species (Wemmer et al. 1996). The placement of infrared-triggered cameras can aid in detecting medium or large mammals, including mule deer (e.g., Roberts et al. 2006). Bats will be surveyed with ultrasonic bat detectors (Kunz et al. 1996); detectors should be placed in suitable openings, near habitat transition zones, or in likely movement corridors (Shump and Shump 1982). Audio recordings from the bat detectors can be displayed as sonograms through the use of computer software (e.g., SonoBat 2007) and analyzed to identify bat species and calculate relative occurrence of each species. To assess reproductive status of high-priority mammal species we suggest the aforementioned methods along with radio tracking individuals.

Nuisance Species

Nuisance species (i.e., bullfrogs, European starlings, brown-headed cowbirds, and coyotes) should be monitored not only to improve our understanding of their population levels and dynamics in the Lake Tahoe Basin, but also to determine their effects on the desired focal species. The abundance of brown-headed cowbirds, European starlings, and bullfrogs can be determined with the same methods as noted earlier. As with desired mammal species, coyote abundance and breeding success can be estimated with the use of infrared-triggered cameras (Wemmer et al. 1996). Coyote abundance in each area can be estimated by differentiating individuals from the pictures (Nichols and Conroy 1996). Radio tracking may elucidate preferred foraging routes and prey of the coyotes.

Synthesis

The restoration plan for the Lake Tahoe Basin focuses on ecological restoration within the Taylor and Tallac watersheds over a 20-year period (tables 10.5 and 10.6). We will employ a multispecies approach to assess restoration efforts within the two watersheds. After the first three years of project implementation, management options may be altered for the upland and wetland areas based on the responses of the 56 focal species.

Restoration activities for the upland areas within the watersheds include reintroduction of prescribed fire and limitation of human access (table 10.5). We anticipate these activities will increase species presence and abundance

TABLE 10.5

Summary of restoration activities for Taylor and Tallac watershed uplands

Year	Activities	Expected Results	Alternatives
1	• Wetland restoration	• Increase of wetland vegetation	
2–3	• Start annual prescribed fires • Restrict human access	• Creation of nesting and foraging areas • Increase species presence/abundance	• Continue with same efforts
4–5	• Increase number and size of burn plots	• Breeding success of some high priority of species • ≥50% of focal species should be present in the area	• Predator control • Vegetation modification
6, 7, 8	• Continue with prescribed fires	• Continued increase or stabilization of focal species' populations	• Increase intensity of fires • Predator control • Vegetation modification
9–10	• Continue with prescribed fires	• ≥75% of focal species should be present in the area	• Control of nuisance species
11–20	• 2nd cycle of long-duration burn experiment • Continue monitoring (as described in Upland section)	• ≥75% of focal species should remain in the area	• Implement mechanical removal of trees and/or creation of snags (if deemed necessary • Continued control of nuisance species

TABLE 10.6

Summary of restoration activities for Taylor and Tallac watershed wetlands

Year	Activities	Expected Results	Alternatives
1, 2, 3	• Improve hydrology • Restrict human access • Bullfrog roundups	• Expand wetland area by 25% • Increase species presence/abundance • Bullfrog control • By year 3, ≥50% of focal species should be present in the area	• Vegetation modification • Release water from Fallen Leaf Lake dam
4–5	• Monitor effects of prescribed fire experiments on hydrology and animal species • Bullfrog eradication • Reduce human activities	• Limited to no increase in sediments downstream from upland fire areas • Increase species presence/abundance • Breeding success of some species	• Predator control • Vegetation modification • Decrease human activities
6–7	• Continue with same efforts	• Continued increase or stabilization of focal species	• Continue with same efforts
8, 9, 10	• Continue with same efforts	• Control of nuisance species • By year 10, ≥75% of focal species should be present in the area	• Eliminate human activities • Eradication of bullfrogs
13, 16, 19	• Continue to monitor effects of prescribed fire experiments on hydrology and animal species	• ≥75% of focal species should be present in the area • No increase in sediments downstream from upland fire areas	• Limit fire experiments if causing species decline or sediment increase

by creating and protecting nesting and foraging areas within the upland areas. We will monitor species' responses within experimental burn plots and control plots to determine the appropriate number and intensity of prescribed fires applied after the initial project implementation phase.

Restoration activities for the wetland areas within the watersheds include hydrologic alterations, limiting human access, and removal of nuisance species (table 10.6). We anticipate these activities will increase species presence and abundance by enhancing the existing wetland areas and controlling predation of native species. Monitoring in these areas over the initial three-year project-implementation phase will determine the necessity of additional hydrologic adjustments and vegetation modifications.

The projects set forth in this restoration plan endeavor to return, to the extent possible, the Taylor and Tallac watersheds to conditions similar to pre-

1800. Successful execution will be realized when ≥75% of the listed focal species are present and stable within the upland and wetland areas of the watersheds.

RESTORATION OF ENDANGERED SPECIES ON PRIVATE LANDS: THE GOLDEN-CHEEKED WARBLER AND BLACK-CAPPED VIREO IN TEXAS

Krystal Windham, *Department of Wildlife and Fisheries Sciences, Texas A&M University.*
Bryan Ray, *Department of Wildlife and Fisheries Sciences, Texas A&M University.*
Andrew J. Campomizzi, *Department of Wildlife and Fisheries Sciences, Texas A&M University.*
Shannon L. Farrell, *Department of Wildlife and Fisheries Sciences, Texas A&M University.*

The golden-cheeked warbler (*Dendroica chrysoparia*) and black-capped vireo (*Vireo atricapillus*) are federally endangered songbirds that breed in central Texas. The U.S. Fish and Wildlife Service (USFWS) listed habitat fragmentation, habitat loss, and parasitism by brown-headed cowbirds (*Molothrus ater*) as the primary factors leading to the decline of both the golden-cheeked warbler and black-capped vireo (USFWS 1991, 1992). Management for recovery of these species has occurred primarily on public land, including parks and military installations (Wilkins et al. 2006). However, about 90% of land in Texas is privately owned (Texas Environmental Profiles 2000; Texas Parks and Wildlife Department 2007), so much of the potential habitat for the two species is located on private land. Endangered species management on private land presents challenges not encountered on public land, but effective recovery of these species must include restoration plans on private land. In this section we have developed a restoration plan for the golden-cheeked warbler and black-capped vireo that focuses on private land in the Lampasas watershed in central Texas. Although our restoration plan focuses on specific issues in the Lampasas watershed, the plan can be used as a general model for application range-wide for both species.

Our focal region is the Lampasas Watershed that covers 209,000 ha in Mills, Hamilton, Lampasas, Coryell, Burnet, Bell, and Williamson counties in central Texas (figure 10.6). The Lampasas watershed is located in the Edwards Plateau ecoregion, characterized by hills and mesas with elevation range of 200–500 m, and includes Lampasas cut plain, live oak–mesquite savanna, and llano uplift subcoregions (Texas Parks and Wildlife Department 2007). Primary land uses in the ecoregion include ranching, farming, and hunting.

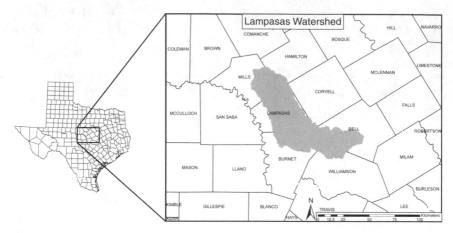

Figure 10.6. Our proposed restoration area is in the Lampasas watershed (209,000 ha) in central Texas, USA covering parts of Mills, Hamilton, Lampasas, Coryell, Burnet, Bell, and Williamson counties.

Vegetation in the Lampasas Watershed

Vegetation in the Lampasas watershed consists primarily of open grassland and stands of Ashe juniper (*Juniperus ahei*) and deciduous trees such as live oak (*Quercus fusiformis*), post oak (*Quercus stellata*), pecan (*Carya illinoinensis*), hackberry (*Celtis occidentalis*), cedar elm (*Ulmus crassifolia*), sycamore (*Platanus occidentalis*), bigtooth maple (*Acer grandidentatum*), and mesquite (*Prosopis* spp.).

Land use, fire suppression, and grazing have altered the vegetation in the Lampasas watershed, changing the natural succession and disturbance regime that historically maintained a patchwork of grassland, shrubland, and woodland. Clearing of woody vegetation by mechanical techniques or fire is a common land management practice on private land in Texas to improve grazing and farming conditions (e.g., Kreuter et al. 2005).

Natural History

The golden-cheeked warble is a neotropical songbird listed as federally endangered in 1991 (USFWS 1992). Golden-cheeked warblers winter in southern Mexico and Central America but breed only in central Texas. Golden-cheeked warbler breeding habitat has been described as mature, closed-canopy stands of mixed oak-juniper woodland (Pulich 1976; Ladd and Gass 1999) with 25 to 50 year-old Ashe juniper in mixed age stands interspersed with hardwoods (Kroll 1980).

Golden-cheeked warblers arrive at their breeding grounds in central Texas in early March. Nests are constructed mainly by females. Females use Ashe juniper bark, spider webs, and other plant materials (e.g. leaves, lichens) to build a cup nest in a fork of Ashe juniper or hardwoods 5–7 m above ground (Ladd and Gass 1999). Golden-cheeked warblers leave breeding grounds for wintering grounds in late July and August. Golden-cheeked warblers have shown strong site fidelity in banding studies on Fort Hood, Texas: 73% of males and 55% of banded females were resighted during subsequent years within 300 m of where they were banded (Ladd and Gass 1999). Golden-cheeked warbler density in breeding habitat is estimated as 9.5 to 20 pairs per 100 ha (USFWS 1992).

Several threats to golden-cheeked warbler populations have been identified, including habitat loss and brood parasitism by brown-headed cowbirds and the underlying causes of these threats (Zwartjes 1999; figure 10.7). Because Ashe juniper is considered a nuisance species (Patoski 1997), removal of Ashe juniper to provide grazing lands, building materials, fence posts, charcoal, and chemical extracts (Smeins et al. 1997; Zwartjes 1999) has reduced the amount and contiguity of mature woodland in central Texas. White-tailed deer browsing of hardwood seedlings may also contribute to golden-cheeked warbler habitat loss by reducing the regeneration of oak-juniper woodlands (Zwartjes 1999).

Golden-cheeked warblers are known brown-headed cowbird hosts (Ladd and Gass 1999) and parasitism leads to nest failure. Brown-headed cowbird abundance was found to be correlated with proximity to agriculture and livestock (Goguen and Mathews 1999). Livestock ranching is common and

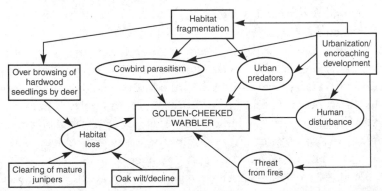

FIGURE 10.7. Visual threat assessment for the golden-cheeked warbler on its breeding grounds. Stresses are enclosed in ellipses, sources of stress in rectangles. (From Zwartjes 1999)

widespread in central Texas leading to the regular presence of brown-headed cowbirds near many golden-cheeked warbler breeding areas. Removal of brown-headed cowbirds by trapping (Eckrich et al. 1999) and shooting (Summers, Kostocke, and Norman 2006) has had positive effects on productivity for golden-cheeked warblers and other species susceptible to brood parasitism including Kirtland's warbler (*Dendroica kirtlandii*; Kelly and De-Capita 1982) southwestern willow flycatcher (*Empidonax traillii extimus*; Whitefield et al. 1999).

Nest depredation is the primary cause of nest failure for many song bird species including golden-cheeked warblers (Nice 1957; Ricklefs 1969; Martin 1993; Cimprich 2005). Stake et al. (2004) found that 28% of nests failed due to nest predators on Fort Hood, Texas. Using video cameras at active golden-cheeked warbler nests, Stake et al. (2004) found that rat snakes (*Elaphe* spp.) were responsible for 63% of depredations; avian predators and fox squirrels (*Sciurus niger*) depredated the other 37% of nests. Golden-cheeked warblers nesting on private land are likely to be exposed to these predators and others associated with human activities including raccoons (*Procyon lotor*), opossums (*Didelphis* spp.), rats (*Neotoma* spp.), mice (*Peromyscus* spp.), domestic cats (*Felis*), and dogs (*Canis*).

The black-capped vireo is a neotropical songbird listed as federally endangered in 1987 due to habitat loss and parasitism by brown-headed cowbirds (USFWS 1992). Black-capped vireos winter on the Pacific slope of Mexico, but little else is known about their wintering ecology. Their historical breeding range included northern Mexico, central Texas, southern Oklahoma, Kansas, and possibly Nebraska (Graber 1961). Their current breeding range includes northern Mexico, central Texas, and a few isolated patches in southern Oklahoma (Grzybowski 1995). Black-capped vireo breeding habitat is described as midsuccessional, structurally heterogeneous shrub land (Graber 1961; Grzybowski et al. 1994).

Black-capped vireos arrive on breeding grounds in mid- to late April with adult males arriving first, followed one to two weeks later by females and second-year males (Grzybowski 1995). Males establish and defend territories of 1–10 ha (Grzybowski 1995). Pairs use spider and worm webs, lichens, grasses, and other fibrous materials to build a deep cup nest in a forked branch about 1–2 m high in a shrub or tree. Black-capped vireos migrate to their wintering grounds in western Mexico in August–September.

Several threats to black-capped vireo populations have been identified, including habitat loss and brood parasitism by brown-headed cowbirds and the underlying causes of these threats (Zwartjes 1999; figure 10.8). Black-capped vireo habitat loss is thought to be due to fire suppression and clearing of

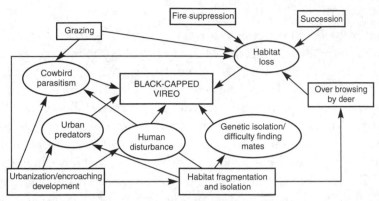

FIGURE 10.8. Visual threat assessment for the black-capped vireo on its breeding grounds at the Barton Creek Habitat Preserve. Stresses are enclosed in ellipses, sources of stress in rectangles. (From Zwartjes 1999)

woody vegetation for agriculture and grazing use. Fire historically created periodic disturbance that helped maintain structurally heterogeneous, midsuccessional vegetation, so some black-capped vireo habitat loss may be due to succession of suitable areas into more mature woodland due to fire suppression (Grzybowski 1995). Overgrazing by livestock may alter vegetation structure and composition and limit postfire regeneration of suitable vegetation, limiting shrub cover and available nest sites for black-capped vireos.

Black-capped vireos are highly susceptible to brood parasitism by brownheaded cowbirds (Grzybowski 1995) and parasitism leads to nest failure for the vireos. Increased risk of parasitism with proximity to cattle ranching and agricultural areas (Goguen and Mathews 1999) coupled with loss of habitat for agriculture and habitat fragmentation may contribute to the low reproductive success for the vireo observed in some areas of central Texas (Farrell 2007).

Nest depredation is the primary cause of nest failure for many song bird species including black-capped vireos (Nice 1957; Ricklefs 1969; Martin 1993; Cimprich 2005). Cimprich (2005) found that up to 75% of nests failed to produce young and 80% of failures were accounted for by nest predators in breeding areas where brown-headed cowbirds were removed on Fort Hood, Texas. Using video cameras at active black-capped vireo nests on Fort Hood, Texas, Stake and Cimprich (2003) found that snakes were responsible for 38% of depredations, red imported fire ants (*Solenopsis invicta*) for 31%, avian predators for 19%, and mammalian predators for 11%. Black-capped vireos nesting on private land are likely to be exposed to these predators and others associated with human activities, including raccoons, opossums, rats, mice, domestic cats, and dogs.

Management on Private Lands in Central Texas

Management for restoration and species recovery has been conducted in central Texas since the listing of these two species. Efforts to manage for endangered species on private land may encounter several challenges. Some landowners express fear and discomfort with governmental agency intervention in land uses on private property (Brook et al. 2003). Landowners may feel threatened by potential land-use restrictions in endangered species habitat, leading to what is sometimes called the *shoot, shovel, and shut-up* approach, extirpating any endangered species habitat on their land to avoid restrictions (Smith 1999). In 1995, the U.S. Fish and Wildlife Service plans to designate critical habitat for the golden-cheeked warbler in Texas led to landowner concern and protest (Hutchinson 1995). Conservation goals are more likely to be achieved by providing landowners with incentives rather than regulatory restrictions and penalties (e.g., land-use restrictions, fines). Landowners may have a variety of economic needs as well as personal and cultural concerns and priorities, so providing flexible and varied incentives (e.g., Langpap 2006) specifically designed for each landowner's unique needs and attitudes (Langpap 2004) may be necessary. Previous research conducted in central Texas classified most landowners as one of three types with unique concerns and priorities: born to the land, reborn to the land, and agricultural business landowners (Sanders 2005).

There have been several previous efforts to incorporate landowner concerns into management efforts in Texas, including the Landowner Incentive Program. The Landowner Incentive Program is open to all private landowners and designed to assist private landowners in protecting, managing, and conserving rare species by providing technical guidance. The Landowner Incentive Program was developed to support the Texas State Wildlife Action Plan, a plan to improve management of both endangered and non-endangered species in Texas and to meet the requirements of the U.S. Fish and Wildlife Service's State Wildlife Grant program (TPWD 2007). Under the Landowner Incentive Program, landowners receive financial, technical, and logistical guidance and support to meet management plan objectives for threatened or endangered species on their property. Funds can be used for a variety of management procedures such as restoring native vegetation, prescribed burns, selective brush management, rotational grazing, and fencing ecologically sensitive areas (TPWD 2007).

The state of Texas also offers other legal incentives to landowners for participation in wildlife and endangered species management. Landowners are protected from power line, gas line, highway, and railroad developments if their land is occupied by endangered species (TPWD 2007). The state also offers a wildlife tax exemption to landowners who manage their properties for the benefit of native wildlife species. According to Texas Parks and Wildlife, the Texas tax code (section 23.51(2)) includes wildlife management in the definition of agricultural uses of land (Stevens 1999). Wildlife management is broadly defined and includes habitat control, erosion control, predator control, providing supplemental supplies of water or food, providing shelters, and making census counts to determine population. This tax exemption is similar to the state agricultural exemption and requires documented proof of all improvements on that land for wildlife practices.

Approach

Our goals to restore populations of golden-cheeked warblers and black-capped vireos on restoration sites in the Lampasas watershed include (1) Creating new habitat in areas where it does not currently exist, (2) modifying currently occupied habitat to improve habitat quality (increasing productivity), and (3) meeting landowner economic and land-use needs.

The specific objectives of this restoration effort include the following:

- Preserving existing mature oak-juniper woodland for golden-cheeked warbler habitat
- Maintaining existing vireo habitat by mechanical techniques and fire to maintain mid-successional vegetation structure
- Improving existing warbler habitat by connecting patches to create larger woodland patches
- Creating ≥50 ha patches of shrub land for black-capped vireo habitat
- Reducing the number of brood parasites in occupied areas
- Reducing the number of nest predators in occupied areas
- Meeting specific landowner objectives as determined by premanagement interviews with each landowner (e.g., increasing cattle and deer forage, increasing land value, increasing aesthetic value, increasing stocking rates, improving hunting conditions)

The restoration plan will be carried out over 10 years. Management actions will be implemented and evaluated for effectiveness in meeting the stated objectives throughout the 10-year period. We will adjust or change management actions based on the results of monitoring at the restoration and control sites. Table 10.7 indicates specific management action, monitoring, desired condition, and adaptive management plans for each year of the restoration plan. If desired conditions are not met for a particular year, then specified

TABLE 10.7

Specific management action, monitoring, desired condition, and adaptive management plans for each year of the restoration plan. If desired conditions are not met for a particular year, then specified adaptive management plans will be implemented to reach our desired condition for the following year. We will adjust or change management actions based on the results of monitoring at the restoration and control sites.

Year	Management Action	Monitoring	Desired conditions Existing habitat	Desired conditions Created habitat	Adaptive management Actions
1	Identify sites	Pretreatment data	—		
2	Vegetation and fire Cowbird control Fire ant management	Territory mapping and Productivity monitoring	GCWA and BCVI present and successful	GCWA and BCVI present If present, successful	
3	Vegetation and fire Cowbird control Fire ant management	Territory mapping and Productivity monitoring	GCWA and BCVI present and successful	GCWA and BCVI present If present, successful	• Cowbird trap relocation if needed • If not present in created habitat add conspecific attract • If not successful implement predator removal
4	Vegetation and fire Cowbird control Fire ant management • Predator removal • Conspecific attraction	Territory mapping and Productivity monitoring	GCWA and BCVI present and successful	GCWA and BCVI present If present, successful	• Cowbird trap relocation if needed • If still not present in created habitat continue conspecific attract • If implemented, continue predator removal
5	Vegetation and fire Cowbird control Fire ant management • Predator removal • Conspecific attraction	Territory mapping and Productivity monitoring	GCWA and BCVI present and successful	GCWA and BCVI present If present, successful	• Cowbird trap relocation if needed • If still not present in created habitat continue conspecific attract • If implemented, continue predator removal

6	Cowbird control
	Fire ant management
	• Predator removal
	• Conspecific attraction
		Territory mapping and
		Productivity monitoring
			GCWA and BCVI
			present and successful
				GCWA and BCVI present
				If present, successful
					• Cowbird trap relocation if needed
					• If still not present in created habitat
					  continue conspecific attract
					• If implemented, continue predator
					  removal

7	Cowbird control
	Fire ant management
	• Predator removal
	• Conspecific attraction
		Territory mapping and
		Productivity monitoring
			GCWA and BCVI
			present and successful
				GCWA and BCVI present
				If present, successful
					• Cowbird trap relocation if needed
					• If still not present in created habitat
					  continue conspecific attract
					• If implemented, continue predator
					  removal

8	Cowbird control
	Fire ant management
	• Predator removal
	• Conspecific attraction
		Territory mapping and
		Productivity monitoring
			GCWA and BCVI
			present and successful
				GCWA and BCVI present
				If present, successful
					• Cowbird trap relocation if needed
					• If still not present in created habitat
					  continue conspecific attract
					• If implemented, continue predator
					  removal

9	Cowbird control
	Fire ant management
	• Predator removal
	• Conspecific attraction
		Territory mapping and
		Productivity monitoring
			GCWA and BCVI
			present and successful
				GCWA and BCVI present
				If present, successful
					• Cowbird trap relocation if needed
					• If still not present in created habitat
					  continue conspecific attract
					• If implemented, continue predator
					  removal

10	Cowbird control
	Fire ant management
	• Predator removal
	• Conspecific attraction
		Territory mapping and
		Productivity monitoring
			GCWA and BCVI
			present and successful
				GCWA and BCVI present
				If present, successful
					• Cowbird trap relocation if needed
					• If still not present in created habitat
					  continue conspecific attract
					• If implemented, continue predator
					  removal

adaptive management plans will be implemented to reach our desired condition for the following year.

Site Selection

We will locate potential golden-cheeked warbler and black-capped vireo habitat in the study region using aerial photographs and remote sensing following the findings of DeBoer and Diamond (2006) and Magness et al. (2006). After potential habitat is located, we will work with county extension agents in the Texas Cooperative Extension (part of the Texas A&M University system) to identify landowners with warbler or vireo habitat, as identified by remote sensing, that may be interested in an incentive-based restoration program. We will acquire access to enough properties to provide five restoration sites and five control sites for each species. Restoration or control sites may span more than one property if patches of suitable habitat extend beyond property boundaries. Each site will be 20–40 ha to provide enough area for several bird territories. Sites will be located only on properties to which we obtain permission to access for ≥ 10 years to allow adequate time for management implementation and evaluation of restoration efforts.

Effective communication with potentially willing landowners is critical to fostering working relationships for effective conservation (McDowell et al. 1989). We will initially meet with landowners and a county extension agent or another trusted, local intermediary who has a working relationship with the landowner. Once we have secured access to properties with potential golden-cheeked warbler and black-capped vireo habitat, we will identify occupied areas for management to maintain or improve existing habitat and unoccupied areas for management to create new habitat.

Landowner Incentives

We will encourage private landowner participation in restoration of black-capped vireos and golden-cheeked warblers by providing financial incentives and technical guidance to meet the needs of landowners and endangered species. We will allow landowners to select incentives that meet their needs such as annual payments and fence or stock tank building/repair. We will provide each landowner with monetary incentives totaling up to $50,000 during the 10-year duration of the restoration project. Landowners can select annual payments ranging from $1,000 to $5,000 per year. An independent third party (e.g., nonprofit organization) will be selected to handle all contracting and accounting. Fence and stock tank building or repair will be carried out as a cost share between the landowner (33%) and by the restoration project (67%). Landowners must follow the requirements of the restoration program for endangered species on their property to continue receiving incentives.

Maintaining Existing Habitat

We will preserve existing golden-cheeked warbler habitat by designating areas on properties that will not be altered by landowners. These designated areas can not be altered by landowners for grazing, farming, or other uses.

We will maintain black-capped vireo habitat by using mechanical vegetation manipulation and prescribed fire to keep vegetation in a midsuccessional stage. We will use mechanical shearing to remove individual juniper trees to maintain open corridors among patches of shrubs and trees. Selective shearing is as efficient and effective as bulldozing but causes fewer impacts to soil and desired plant species (Hamilton 2004). We will use low-intensity, targeted fire to stimulate regrowth of desirable grasses, forbs, and deciduous woody vegetation (White and Hanselka 1991) and to burn juniper removed by shearing. We will conduct prescribed fire in the fall and winter when black-capped vireos are not present in breeding areas. We will limit grazing immediately following prescribed fire to allow regrowth of desired plants. Grazing limitation time will vary among sites with differences in plant regeneration due to factors such as local soil characteristics or rainfall, and will be determined on a case-by-case basis. Moderate grazing will be allowed in subsequent years to maintain open areas among patches of woody vegetation and to meet landowner needs.

Creating New Habitat

We will connect existing golden-cheeked warbler habitat patches by encouraging mature oak-juniper woodland growth in gaps between patches. Different preexisting conditions in these gaps between warbler habitat patches will require different management approaches. Management actions may include juniper thinning, planting deciduous trees, and prescribed fire to induce growth of oak-juniper woodland between existing patches.

In areas that are not currently occupied by vireos, but that may be suitable with some modification, we will create black-capped vireo habitat by using mechanical means and prescribed fire to create midsuccessional stage vegetation. We will use mechanical shearing to remove individual juniper trees to create open corridors among patches of shrubs and trees. We will use prescribed fire on both small and large areas to reduce woody vegetation cover and stimulate growth of desirable grasses, forbs, and deciduous woody vegetation (White and Hanselka 1991) and to burn juniper removed by mechanical means. As in areas where prescribed fire will be used for maintenance, we will limit grazing immediately following fire and moderate grazing will be allowed in subsequent years.

Reducing Brood Parasites

We will reduce brown-headed cowbird abundance in both currently occupied patches and newly created habitat to reduce their effect on black-capped vireo and golden-cheeked warbler productivity. Trapping brown-headed cowbirds during the breeding season (March–July) is likely to be the most effective method for removal (Eckrich et al. 1999). There is also evidence to suggest that shooting brown-headed cowbirds in restoration areas may be a feasible and cost-effective option during certain parts of the breeding season (Summers, Kostocke, and Norman 2006; Summers, Stake, et al. 2006). We will use a combination of trapping and shooting to maintain parasitism frequency below ~30% to minimize effects on productivity (USFWS 1991). Brown-headed cowbird management efforts will be assessed and adjusted based on monitoring of productivity for both species. We will place traps in grassland areas adjacent to warbler and vireo breeding habitat where brown-headed cowbirds congregate to maximize trapping effectiveness. If parasitism frequency on restoration sites remains above 30% after three years, then cowbird removal efforts will be adjusted based on trapping and parasitism data. The number of traps may need to be increased and placement of cowbird traps may need to be altered.

Reducing Nest Predators

We will reduce nest predator abundance in both currently occupied patches and newly created habitat to reduce their effect on black-capped vireo and golden-cheeked warbler productivity. Predator control will be implemented in phases based on productivity monitoring data. Previous research found that fire ants are a primary predator of black-capped vireo nests (Stake and Cimprich 2003) but do not negatively impact golden-cheeked warbler nests (Stake et al 2004). We will reduce red imported fire ant (*Solenopsis invicta*) abundance in black-capped vireo restoration areas using broadcast baits. If we find that >30% of pairs successfully fledge ≥1 young then management for *S. invicta* will be continued. If monitoring indicates that <30% of pairs successfully fledge ≥1 young then we will implement the second phase of predator control. Stake and Cimprich (2003) and Stake et al (2004) found that several snake species are primary predators for vireos and warblers. We will trap and remove rat snakes (*Elaphe* spp.) and western coachwhips (*Masticophis flagellumtestaceus*) from restoration areas using funnel traps (Greenberg et al. 1994). If productivity reaches desired levels following predator removal then we will continue predator removal. If productivity does not reach desired levels we will implement monitoring to identify additional nest predators that may be reducing productivity.

Conspecific Attraction

If black-capped vireos are not present in all restoration sites after four years then we will attempt to attract individuals to unoccupied sites using conspecific attraction. Ward and Schlossberg (2004) found that black-capped vireos settled and successfully nested in previous unoccupied areas after being attracted by broadcasts of black-capped vireo songs. We think it is appropriate and beneficial to attempt to attract black-capped vireos to our restoration areas because management efforts should provide areas where birds can successfully breed. It is not know if golden-cheeked warblers use the presence of conspecifics when selecting breeding habitat, however many species use conspecific cues during habitat selection (reviewed by Ahlering and Faaborg 2006). If warblers are not present in all restoration sites after four years then we will attempt to attract them to unoccupied sites using conspecific cues (e.g., broadcasting golden-cheeked warbler songs during spring migration).

Landowners

We will conduct verbal interviews and written surveys with participating landowners three times during the 10-year restoration management term. Landowners will be consulted prior to the drafting of a management plan to determine landowner preferences and objectives. Landowners will then be surveyed at year 3, year 6, and year 10 of the management term to determine their level of satisfaction with the management process and if their specific predetermined objectives are met such as increases in property value, improved cattle production, or improved aesthetic value of the land.

Desired Conditions

Our management approach is designed to allow for adaptive management. We have specified desired conditions for bird presence and reproductive success for the two birds species and for landowner satisfaction with outcomes on their property. We will monitor bird presence and reproductive success during all years to determine if desired conditions are being met so that management techniques can be adjusted if desired conditions are not being met. Additionally, we will monitor landowner satisfaction at years 3, 6, and 10 to allow us to determine if desired conditions for landowner satisfaction arc being met and to alter management techniques if needed.

Landowners

We expect to have 20 to 30 participating landowners in this restoration project. We expect landowners to be pleased with improvements to their property by cost share projects. Landowners will be consulted prior to the drafting of a management plan to determine landowner preferences and objectives. For each landowner, their stated objectives will be used to set desired outcomes for their property. If surveys conducted during year 3 and year 6 indicate that landowner objectives are not being met, we will hold a second consultation with landowners to discuss potential changes to the management plan to better meet landowner objectives.

Maintaining Existing Habitat and Creating New Habitat

We expect preserved golden-cheeked warbler and black-capped vireo habitat to remain occupied and productive

We expect that new golden-cheeked warbler habitat will be mature oak-juniper woodland that visually resembles those patches that are currently occupied and meets Texas Parks and Wildlife guidelines for warbler canopy cover requirements. We expect new golden-cheeked warbler habitat will be occupied and productive. We expect warblers to use connected habitat patches as one continuous patch. If warblers are not present in all restoration sites after four years then we will attempt to attract them to unoccupied sites using conspecific cues.

We expect that new black-capped vireo habitat will be a patchwork of midsuccesional shrubs and trees that visually resembles currently occupied habitat and meets the descriptions of suitable habitat (Grzybowski 1995). We expect that new black-capped vireo habitat will be occupied and productive. If black-capped vireos are not present in all restoration sites after four years then we will attempt to attract individuals to unoccupied sites using conspecific attraction.

Reducing Brood Parasites

We expect that brown-headed cowbird removal will maintain parasitism frequency <30%. If parasitism frequency on maintained or created habitat sites remains above 30% after three years, then cowbird removal efforts will be adjusted based on trapping and parasitism data.

Reducing Nest Predators

We expect that nest predator removal will maintain productivity for both black-capped vireos and golden-cheeked warblers above 30%. For black-capped vireo areas, if we find that >30% of black-capped vireo pairs successfully fledge ≥1 young then management for *S. invicta* will be continued. If monitoring indicates that <30% of pairs successfully fledge ≥1 young then we will add implementation of snake control measures. If productivity does not reach desired levels we will implement monitoring to identify additional nest predators that may be reducing productivity.

For golden-cheeked warbler areas, if monitoring indicates that <30% of pairs successfully fledge ≥1 young then we will implement snake control measures. If productivity does not reach desired levels we will implement monitoring to identify additional nest predators that may be reducing productivity.

Monitoring

We will conduct monitoring to determine whether the desired conditions are being met for bird presence and reproductive success for the two bird species and for landowner satisfaction with outcomes on their property. If desired conditions are not being met, management will be adjusted and further monitoring will allow us to observe the effect of the management changes. We expect that this adaptive process, using feedback provided by monitoring to make management adjustments, will increase our success in meeting restoration objectives.

Pretreatment Surveys

We will monitor the number of males defending territories and estimate productivity for black-capped vireos and golden-cheeked warblers in restoration and control/reference sites to determine the success of restoration efforts. We will use spot-mapping (Kendeigh 1944) to determine the number of singing males defending territories. Mapping the location of singing males will allow us to identify areas occupied and not occupied by golden-cheeked warblers and black-capped vireo to guide restoration efforts.

We will locate and monitor nests for black-capped vireos to evaluate nesting success and productivity (Martin and Geupel 1993). Nest monitoring results will guide brown-headed cowbird and nest predator removal on restoration sites.

We will monitor golden-cheeked warbler productivity using an index of breeding success (Vickery et al. 1992). The Vickery index will provide the necessary information on productivity by locating breeding pairs and fledglings, and is less time consuming than nest searching (Kristoferson and Morrison 2001) because golden-cheeked warbler nests are difficult to find relative to many songbirds.

Maintaining Existing Habitat

We will monitor existing habitat areas each year following the same procedures as described for the pretreatment surveys, to determine the presence, abundance, and productivity for black-capped vireos and golden-cheeked warblers in restoration and control/reference sites to determine the success of restoration efforts. Monitoring the number of singing males each year will allow us to determine if the number of individuals is increasing on restoration sites versus control/reference sites over time. Identifying the location of singing males will allow us to determine if birds are occupying specific areas where restoration efforts have been focused.

In addition to the methods used in pretreatment surveys, we will compare nest success on restoration sites to control sites to determine the effect of restoration efforts. Nest monitoring will allow us to determine if brown-headed cowbird, fire ant, or snake management needs adjustment to increase nesting success to desired levels of productivity or if additional investigation is necessary to identify other nest predators.

Creating New Habitat

We will use remote sensing and line transect measures to quantify vegetation cover. We will evaluate if new habitat patches meet habitat requirements for black-capped vireos and golden-cheeked warblers.

We will monitor new habitat areas each year following the same procedures as described for the maintained habitat areas, to determine the presence, abundance, and productivity for black-capped vireos and golden-cheeked warblers in restoration and control/reference sites to determine the success of restoration efforts. We will determine if the number of singing males is increasing on restoration sites versus control/reference sites over time and if singing males are occupying specific areas where restoration efforts have been focused. We will also compare nest success on restoration sites to control sites to determine if predator management needs to be adjusted to increase nesting success to desired levels of productivity or if additional investigation is necessary to identify other nest predators.

Synthesis

Recovery of the black-capped vireo and golden-cheeked warbler in Texas requires management conducted on private land. Our restoration plan targets factors known to limit the population and productivity of both endangered species including vegetation characteristics, nest depredation, and brood parasitism. By both improving existing habitat and creating new patches of habitat, we can increase the number of occupied habitat areas and improve the productivity of the birds in these areas. This effort may lead to increases in the local populations of these two endangered species. Additionally, if we can conduct successful restoration while also meeting landowner needs, we can improve landowner perception of endangered species management efforts, and improve future receptivity to such efforts. These impacts may help facilitate future endangered species restoration efforts on private land in central Texas and elsewhere.

RESTORING SMALL AND ISOLATED POPULATIONS: THE SAN JOAQUIN KANGAROO RAT AT LEMOORE NAVAL AIR STATION

Christopher Lituma, *Department of Wildlife and Fisheries Sciences, Texas A&M University*
Shannon Farrell, *Department of Wildlife and Fisheries Sciences, Texas A&M University*
Alejandro Calixto, *Department of Entomology, Texas A&M University*
Justin Cannon, *Department of Wildlife and Fisheries Sciences, Texas A&M University*

Naval Air Station Lemoore (NAS Lemoore) was established in 1961 in Kings County, California. The Fresno kangaroo rat (*Dipodomys nitratoides exilis*) was first identified on NAS Lemoore in 1982 near the runway. The population at our study area, Resource Management Area 5 (RMA5), was unknown prior to 1989 (Gorman and Rosenberg 2000). In 1965, RMA5 was converted from agricultural use to an off-road vehicle (ORV) recreation area. The agricultural intensity prior to 1965 is unknown, and the area may have been used as a drainage basin. It is possible that kangaroo rats colonized RMA5 after the cultivation activities ceased or after drainage use. Fresno kangaroo rats shelter in ground burrows usually found in sandy soils like those typical of the motocross track on NAS Lemoore. RMA5 is no longer used for recreational vehicles and has been managed for kangaroo rats since 1992 (Smallwood and Morrison 2006). Management includes burning, mowing, and grazing treatments as part of an Endangered Species Recovery Program (ESRP) intended to create kangaroo rat habitat (Smallwood and Morrison 2006).

Kangaroo rats weigh up to 170 grams and can reach lengths of 30 centimeters. The gestation period is about 32 days with an average of three young per litter. The Fresno kangaroo rat eats mostly seeds and some vegetation. The seeds are cached for consumption at a later time. Despite a patchy distribution, the kangaroo rat has seemingly avoided restricted gene flow and reduced genetic variation (Good 1997). However, the potential for genetic problems and local extinction continues to exist for this small isolated population. We plan to increase the population size and spatial extent of the Fresno kangaroo rat on NAS Lemoore to reduce some of the risks of genetic problems and extinction. Increasing the population size and spatial extent of kangaroo rats on NAS Lemoore may improve the long-term probability of survival of kangaroo rats on the base. We will continue many of the management practices that have been implemented on the base since 2001. Research conducted on the base has shown that the current management practices have successfully promoted the expansion of the kangaroo rats beyond a concentrated area near the motocross track to other parts of RMA5 (Smallwood and Morrison 2006). We will take the next step in expanding the spatial extent of the population by creating a new habitat patch, establishing a population of kangaroo rats in the new patch, and developing a corridor linking RMA5 with the new patch.

Approach

Our goal is to increase the population size and spatial extent of the kangaroo rat population on NAS Lemoore. Specifically, we will (1) increase the abundance of kangaroo rats in the existing occupied patch, RMA5, by improving the habitat quality; (2) create a new patch of habitat; (3) establish a kangaroo rat population in the newly created patch; and (4) create a connective habitat corridor between RMA5 and the new patch using adaptive management techniques. We will implement the restoration plan in two phases (table 10.8); after each phase, we will conduct an assessment to determine whether

TABLE 10.8

Fresno kangaroo rat restoration timeline on NAS Lemoore

Management stage	Years	Actions	Alternative action if desired conditions are not met
Phase 1 Management actions	1–3	• Habitat improvement of RMA5 • Habitat creation in P1 • Start incremental corridor extension	
Assessment 1 Monitoring conditions	4	• Assess vegetation conditions in RMA5, P1, and corridor extension • Assess population in RMA5 • Assess use of corridor extension	
Phase 2 Management actions	5–7	If desired conditions are met: • Translocate individuals from RMA5 to P1 • Continue corridor extension	If vegetation desired conditions not met: • Continue habitat manipulations until desired conditions are met If population in RMA5 is not large enough: • Do not translocate individuals
Assessment 2 Monitoring condition	8	• Assess vegetation conditions in RMA5, P1, and corridor extension • Assess population in RMA5 • Assess population in P1 • Assess use of continued corridor extension	If alternative action was taken in Phase 2: • Assess vegetation conditions in RMA5, P1, and corridor extension • Assess population in RMA5 • Assess use of corridor extension
Postassessment 2 Management actions	9 +	If desired conditions are met: • Continued habitat improvement of RMA5, P1, and corridor to maintain conditions • Continue corridor extension to connect RMA5 and P1	If vegetation desired conditions not met: • Employ alternative habitat manipulations If population in RMA5 is not large enough: • Do not translocate individuals If corridor is not being used, and population in RMA5 is large enough: • Periodic translocation from RMA5 to P1

the desired conditions are met and make necessary changes in management based on the results of each assessment. We developed a decision tree for each assessment a priori to follow proper adaptive management guidelines (figures 10.9–10.11).

In phase 1 we will convert a 36 ha (86 ac) patch of farmland to suitable kangaroo rat habitat using management techniques that have been previously effective on RMA5. Patch 1 (P1) will be located approximately 800 m from RMA5. We will also begin an extension of the suitable habitat off of RMA5 with the intention of further extending this area as a corridor to connect RMA5 to P1. Research has found that kangaroo rats can move long distances when provided with suitable condition for safe movement (Zeng and

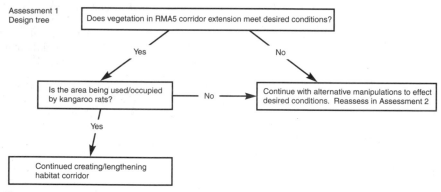

FIGURE 10.9. Decision tree for management action following phase 1 and assessment 1 to determine whether corridor expansion should be continued.

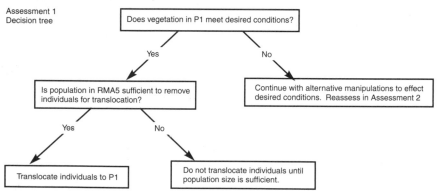

FIGURE 10.10. Decision tree for management action following phase 1 and assessment 1 to determine whether translocation should be conducted.

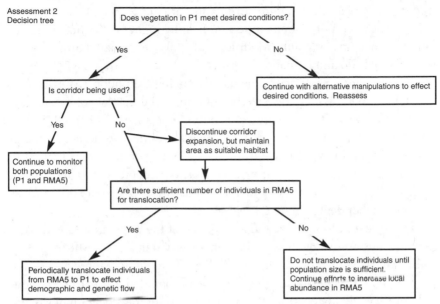

Figure 10.11. Decision tree for management action following phase 2 and assessment 2 to determine what management actions should be taken following phase 2.

Brown 1987; Gorman and Rosenberg 2000). However, we will create the corridor connecting RMA5 and P1 in stages, to determine whether the kangaroo rats are using each extension. The corridor creation will begin with an experimental expansion of RMA5 in phase 1. Assessment 1 (figures 10.9–10.10) will determine the actions of phase 2.

In phase 2, we will translocate kangaroo rats from RMA5 to P1 if a sufficient number of kangaroo rats are present on RMA5. We will continue expansion of the corridor from RMA5 as well as the beginning of an expansion from P1 if we determine that the desired conditions for the corridor have been met in assessment 1. Assessment 2 (figure 10.11) will follow phase 2 to determine if management actions are successful.

Increasing Local Abundance

Several ecological and genetic risks face small populations (Brito and Fernandez 2000a). Extinction is defined as a probabilistic event in which stochastic demographic, environmental, and genetic processes can lead to loss of adaptive flexibility, and is considered a greater threat to small populations due to this probabilistic nature (Brito and Fernandez 2000a). Research has shown that small populations have significantly higher extinction risks than larger populations for plants (Matthies et al. 2004) and small mammals (Bennett 1990b; Brito and Fernandez 2000b).

Genetics are an important factor influencing extinction risk (Brito and Fernandez 2000a). Sheremet'ev (2004) summarized that loss of heterozygosity is the most significant factor contributing to extinction of small, isolated populations. Loss of heterozygosity through the random process of gene frequency distribution is significantly higher for small populations (Li 1978). An increase in the population of kangaroo rats should decrease the likelihood of localized extinction or loss due to deleterious effects of genetic drift and also sustain a higher degree of heterozygosity.

Thus, we aim to increase the abundance of the population on RMA5 to increase the ability of this population to withstand stochastic ecological events and avoid loss of adaptive flexibility due to demographic and genetic processes. Increasing the size of the existing population on RMA5 should help provide adequate numbers of kangaroo rats for translocation of some individuals to the newly created P1. Increasing abundance on RMA5 may also facilitate dispersal of the population into new habitat created on RMA5 as a first step in extending a corridor between patches. Small mammals may exhibit several types of movement through exploratory movement, activities within a home range, and dispersal behavior (Wolton and Flowerdew 1985); dispersal movement is important for facilitating colonization of new locations and gene flow (Hansson 1991).

We will increase population size by improving the quality of existing habitat. Previous studies have suggested that availability of high-quality food resources and susceptibility to predators are often major drivers of populations of small mammals (Batzli 1992). Improved habitat will provide greater resource availability including food resources, burrow sites, shelter from thermal impacts, and shelter from predators.

We will implement a variety of treatments to improve habitat quality during phase 1, including: fire, selective cutting, and other mechanical manipulations. Additionally, we will conduct reseeding of cover plants and application of wooden palates to provide additional cover. We will monitor the effects of phase 1 treatments to improve existing habitat quality during assessment 1 to determine if the treated area (RMA5) meets the desired conditions for vegetation and soil characteristics, shows improved resource availability, and increased kangaroo rat abundance. We will continue the treatments in phase 2 pending the results of assessment 1 (figures 10.9–10.10)

Expanding Spatial Extent of Population

We will expand the spatial extent of the population in two ways: (1) creating new habitat patch P1, and (2) developing a corridor connecting RMA5 to P1. Increasing the spatial extent is likely to reduce the species extinction risk and improve long-term population viability. Sheremet'ev (2004) found that for space limited and isolated populations, populations distributed across many islands, even small islands, were more viable than those distributed across fewer, larger islands. Similarly, we expect that increasing the spatial extent of the species by a new habitat patch will facilitate lower extinction risk for the kangaroo rat population.

Corridors have been found to lead to decreased extinction rate and increased rate of colonization of new patches in several empirical studies (Merriam et al. 1989; Hansson 1991; LaPolla and Barrett 1991). The distribution of gene frequencies depends on specific features of population dynamics (Nei et al. 1975) and may be constrained not only by population size, but isolation. Corridors were found to provide important connections for small mammals in otherwise isolated forest fragments, by facilitating both individual animal movement and gene flow between populations (Bennett 1990a, b; Mech and Hallett 2001). Additionally, researchers have found that even small amounts of movement between connected patches can reduce extinction risk and improve population gene flow (Gauss 1934; Levins 1969, 1970; Newman and Tallmon 2001)

Using adaptively managed steps, we will create a corridor over time with the objective of creating a strip of habitat that connects two larger patches (Forman 1995), and that acts as both a conduit between patches and also as suitable habitat itself (Simberloff et al. 1992). We will create the corridor in small increments between the two patches, beginning with a small extension of habitat off the side of RMA5 (figure 10.12). If kangaroo rats do not appear to be dispersing through the entire length of the incremental extension of the corridor, we will discontinue further extension. If corridor extension is discontinued, we will continue to manage the existing extent of the corridor to act as a habitat area. Many small mammal species have been observed performing relatively long range movements of greater than 1000 m (Szacki et al. 1993). Additionally, previous data collected on RMA5 show kangaroo rats dispersing >300 m during a three-year period. Thus, previous data suggest that a corridor may be used by kangaroo rats, and useful for management of the kangaroo rat population.

We will increase the spatial extent of the population in both phase 1 and phase 2. In phase 1, we will convert a patch of farmland to suitable kangaroo

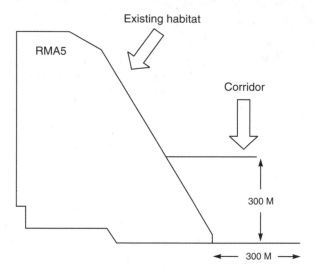

FIGURE 10.12. Current restored area RMA5 and proposed extension area (corridor).

rat habitat (P1). In exchange for the conversion of agricultural land to kangaroo rat habitat, we will provide an equivalent area of land in another part of NAS Lemoore for agricultural use. We will also apply treatment to create suitable habitat extending from the southeastern portion of patch RMA5, beginning the intended corridor between the two habitat patches (figure 10.12). We will implement fire, selective cutting, and other treatments to simulate natural disturbance, and we will conduct reseeding of cover plants and application of wooden palates to provide additional cover. The effects of phase 1 treatments will be monitored in assessment 1 to determine if patch P1 and the extension of RMA5 meet the desired conditions for vegetation, soil characteristics, and resource availability, and if the RMA5 extension area is being used or occupied by dispersing kangaroo rats.

The actions in phase 2 will be contingent on the results of assessment 1 (figure 10.9). If P1 meets the desired conditions for suitable kangaroo rat habitat, and if an adequate population of kangaroo rats has persisted on RMA5, we will translocate kangaroo rats from RMA5 to P1 (see figure 10.10). Additionally, if the RMA5 extension appears to meet the habitat needs of the kangaroo rats and their burrows have been located in the extension area, we will apply treatment to further extend the habitat corridor toward P1.

If the desired conditions are not met in P1, we will continue the effort to create suitable habitat in P1 using other manipulative approaches, but will not translocate individuals until the conditions in P1 are determined to be suitable. If there are not an adequate number of individuals on RMA5 at the time of assessment 1, we will not remove or relocate individuals from the population until an adequate number of individuals are present to allow removal. If kangaroo rats are not using or occupying the corridor extension, the habitat manipulation will be modified. If this extension continues to be unused, we will discontinue corridor extension, but we will continue to manage the existing extension area to maintain suitable kangaroo rat habitat. If we discontinue extension of the corridor but desired conditions are met in P1 and a sufficient number of kangaroo rats are present on RMA5, we will periodically relocate some individuals from RMA5 to P1 to facilitate demographic and genetic flow between the two habitat patches. Previous research suggests that for isolated populations, one migrant per generation can lead to sufficient gene flow to limit the negative effects of isolation and decrease differentiation between populations (Monty et al 2003). Thus, we would suggest that periodic translocations conducted every two to three years should be sufficient to create gene flow between the populations.

Captive breeding will provide an alternative in the event of severe population decline and loss of genetic variability or local extinction for the kangaroo rats on NAS Lemoore. Previous findings suggest that the threat of loss of heterozygosity or local extinction is substantial in the small isolated population like the kangaroo population on RMA5 (Nei et al. 1975; Li 1978, Sheremet'ev 2004); captive breeding can maintain a genetic pool and population of individuals under controlled conditions that can be used to restock a patch that has experienced a genetic bottleneck or local extinction. We will begin captive breeding if population size on RMA5 and the new population in patch P1 drops below 40 individuals, half the lowest population size detected since the year 2000 (Smallwood and Morrison 2006).

We will manage the population of other neighboring species, such as the California ground squirrel (*Spermophilus beecheyi*), if we observe that they are increasing to level at which they are competing with or impeding the movement of the kangaroo rats. Ground squirrels are considered nuisance species by some agriculturalists and may pose problems for agriculture on NAS Lemoore; if our treatments appear to increase squirrel populations we will implement management to mitigate these effects. We will use baited traps to remove and relocate squirrels or other species that we observe that negatively impact the kangaroo rats (Whisson 1999).

Desired Conditions

We expect to see an increase of at least 25 burrows within the first three years of phase 1 treatments in the new area at RMA5, based on previous data on kangaroo rat expansion patterns on RMA5 in response to similar treatments. We expect that treatment effects may not be observed immediately; previous management on RMA5 resulted in the expansion of the kangaroo rat population from a limited area along the motocross track to the entire RMA5 area within four years. Thus, we expect to observe a treatment effect during our assessment after three years.

The corridor extension should be approximately 300 m long and 300 m wide (figure 10.12), sufficient for kangaroo rat burrowing habitat. The appearance of the habitat in the corridor extension should be consistent with habitat in RMA5 and descriptions of suitable habitat (Smallwood and Morrison 2006).

Conditions in P1 should also meet the decriptions of suitable habitat for kangaroo rats and match the appearance of occupied habitat areas on RMA5 (Smallwood and Morrison 2006). We expect to observe a reduction in the thatch layer present, reduced exotic grasses that may impede kangaroo rat movement, and an increase in forbs that provide cover and food for the kangaroo rats.

Translocation

We will conduct translocation in phase 2 if the desired conditions are met in phase 1. Five years after translocation of 20 individuals to P1, we expect a population of >80 burrows in P1 in the fall, based on the lowest burrow abundance recorded on RMA5.

Squirrels and Predators

We will monitor the population of ground squirrels. We expect that the ground squirrel population will not increase in response to our management. Additionally, we expect that the presence and abundance of other neighboring species, including predators of kangaroo rats, will not increase in response to our management (Daly et al. 1990). If we observe an increase in ground squirrels, kangaroo rat predators, or competitors, we will implement trapping and relocation of those species.

Design

We have designed an adaptive approach to implementation of management techniques for creating suitable kangaroo rat habitat. We have also designed monitoring to assess whether the management is achieving desired conditions so that changes in management may be implemented if needed based on the results of monitoring.

Extension of Current Restored Area—Corridor

Using an adaptive management approach (Walters 1986) we will construct a small section of corridor (southeast front of RMA5) to test the different approaches for habitat restoration for kangaroo rat during phase 1 (figures 10.12–10.13). Results obtained during this phase will determine future management of this area. Management techniques that produce areas frequently used or occupied by kangaroo rats will be used for further corridor extension and habitat management.

We will measure kangaroo rat movement to determine if the corridor is being used. We will use a BACI-P (before/after–control/impact Paired) design (Stewart-Oaten et al. 1992). We will monitor kangaroo rat movements during the spring and fall following previous monitoring procedures (Smallwood and Morrison 1999). We will use Sherman traps and burrow surveys to determine kangaroo rat movement and use within each experimental plot.

Experimental Units

Experimental units will consist of randomly-placed 10 × 30 m parallel transects across the corridor area to intercept and detect kangaroo rats (figure 10.13). We will pair units with rows of a single treatment, replicated four times.

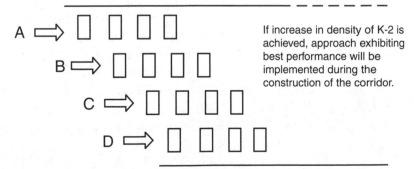

If increase in density of K-2 is achieved, approach exhibiting best performance will be implemented during the construction of the corridor.

Figure 10.13. Proposed restored area (corridor) and management practices.

TREATMENTS

We will apply treatments during phase 1 to evaluate the effectiveness of each treatment on facilitating movement and use by kangaroo rats in each treated area. Treatments consist of (A) prescribed burning, (B) shrub planting, (C) cutting, and (D) untreated control (figure 10.13). We will evaluate success of these treatments by estimating the increase in densities (burrows and individuals caught in traps) after the treatments. Based on preliminary data obtained in RMA5, we expect that an increase of at least five burrows will indicate individuals are positively responding to treatment regime. We will use the results of these experimental applications of treatment to determine what management treatments should be in the continued extension of the corridor in phase 1 and 2.

Monitoring

The kangaroo rats in RMA5 have previously been monitored using *burrow counts* (Smallwood and Morrison 2006). A burrow counts approach assumes one kangaroo rat per burrow (Smallwood and Morrison 1999; Smallwood and Morrison 2006). Some burrows may be home to several kangaroo rats and some kangaroo rats may have several burrows; the relationship between burrow numbers and individuals per burrow is not well understood. Although the number of burrows may not provide a clear estimate of number of kangaroo rats, the number of active burrows can provide information about relative changes in kangaroo rat abundance and activity in an area. We will continue to use burrow counts as an indicator of abundance and activity.

We will systematically survey burrows in the spring and fall each year in RMA5 and all newly created habitat areas (P1). We will survey burrows by walking straight parallel lines east to west, then north to south following the methods of Smallwood and Morrison (1999), counting and recording burrow locations. We will use Global Positioning System (GPS) units to record burrow locations to determine the distribution and spatial extent of burrows within our study areas. We will use burrow counts to assess increases or decreases in the population and distribution of the kangaroo rats within our study area and to determine whether desired conditions have been met in the treated areas. We will also trap kangaroo rats within our study area, to determine whether there is any quantifiable relationship between burrow numbers and kangaroo rat abundance.

We will use the monitoring techniques used in previous studies (Good 1997; Smallwood and Morrison 2006) to determine whether the desired conditions for thatch, shrub, grass cover, and other vegetation condition have been met. We will use photo points and walking surveys to document the characteristics of the vegetation cover layers. We will also conduct systematic walking surveys to look for vegetation types that are common food or cover species used by the kangaroo rats, such as Atriplex bush (*Atriplex spinifera*) and Goldenbush (*Isocoma acradenia*).

We will monitor the populations of California ground squirrels and Botta's pocket gophers (*Thomomys bottae*) following the methods used in previous years. According to recent data, ground squirrel and pocket gopher burrows have been increasing with the kangaroo rat management practices. Although there seems to be no competition between these species, continued monitoring will provide information to determine whether management for these species may be needed. We will also monitor predator species in the area including snakes, feral dogs, and hawks.

Synthesis

The Fresno kangaroo rat has been in decline for the past century and is listed as endangered under both the California and federal Endangered Species Acts (Smallwood and Morrison 2006). One of the few remaining populations of the Fresno kangaroo rat is located on NAS Lemoore. Ongoing research has found that management can successfully increase the abundance and distribution of the kangaroo rat population on NAS Lemoore. We have developed a plan that incorporates effective management techniques to manage existing populations and to expand the abundance and spatial extent of the kangaroo rat on NAS Lemoore. Our plan should increase kangaroo rat population in the existing habitat, create an additional habitat patch, and ultimately populate the new habitat patch, reducing the risks of genetic problems for the small population and the risk of local extinction.

A possible criticism of our plan is that our newly created habitat patch P1 will only be 800 m from the initial patch RMA5. It is true that these patches are still relatively close to one another and a catastrophic event such as a fire, or drought could very well threaten the population at both patches. We initially suggested the restoration of an RMA that was farther away for this reason, but we quickly discovered that the land area needed to connect the patches would then far exceed the area of the patches themselves. Although our plan proposes to double the size of kangaroo rat habitat, it in no way compares to the amount of habitat there once was. The results of our restoration effort should provide information for future management efforts

Closing Comments by Author

As I noted in my introductory comments to this chapter, I hope that these case studies have caused you to consider the many interrelated factors that must be evaluated when crafting a wildlife restoration plan. At a minimum, the plans outlined in this chapter should be thought provoking and serve as a valuable template for your own initial planning efforts. Because all of the plans presented here are works in progress and incorporated aspects of adaptive management, they would in reality be considered working documents. I hope you will use these case studies as a springboard to discuss how they could be modified and improved.

Chapter 11

Wildlife Restoration: Synthesis

This book has focused on integrating concepts of animal ecology for applica-tion to restoration planning. Thus, the foundation of the science of restora-tion is broad and includes such ecological concepts as population ecology and genetics, ecophysiology, demography, community ecology, evolutionary biology, food webs, ecosystem dynamics, habitat and niche ecology, and other related topics.

It should be evident from even a casual reading of this book that I think wildlife studies should not be initiated without careful planning, and must be conducted with statistical rigor. Doing wildlife-habitat restoration requires that the conceptual framework driving our sampling design and methods be clearly stated. We are all operating under some conceptual framework, but this is often not recognized and seldom stated. That is, are we basing our goals and objectives on the adage that "some data are better than no data at all?" Do we restore portions of the historic vegetative community and assume this will be acceptable to the historic animal assemblage? Do we think that competition determines the species composition and abundance? Are we as-suming that prey abundance is a satisfactory surrogate for prey availability?

Regardless of one's conceptual framework, the rigor used to implement a study will in large part determine the results. As outlined in various sections herein, it should be clear that sample size is not a trivial matter. Further, I presented extensive material showing that no single sampling method is ade-quate for quantifying the presence, abundance, and activity of the animals in a project area.

Thus, perhaps the single most important take home message from this book is that no study should be attempted unless it can be done within a clear conceptual framework and with rigor sufficient to convince the reader or user of the results' validity. Further, each restoration project can be thought of as yet another piece of the puzzle that we hope will advance both the preservation of wildlife and our knowledge of basic animal ecology. Thus, monitoring of success should be an essential component of the design and implementation of each project.

Major Messages from Conceptual Development

A central focus of this book is the use of habitat by animals. The term *habitat*, however, is a concept, and as such, cannot be tested per se. It is understood, even by the lay public, to indicate a place where an animal resides. This traditional definition of habitat is, however, useless as a predictive tool because it has numerous similar, but not identical, definitions. Habitat serves as an umbrella under which specific relationships between an animal and its surroundings can be quantified.

Habitat has a spatial extent that is determined during a stated time period. Thus, the physical area occupied by an animal can be described by the observer. The various factors we commonly recognize as components of habitat—cover, food, water, and such—are contained within this area. A functional relationship between resources and animal performance is assumed; the observer often does not define a specific area, or produce a user-defined area that is perhaps based on animal activity. Thus, habitat is a convenient boundary for measurement of vegetation, various other resources, and the environment. A priori, user-defined boundaries are convenient but are likely artificial, because the resources contained within those spatial areas will, of course, vary over time, both in response to abiotic factors and as a result of use of resources by animals. The spatial extent of the habitat, and the precision at which measurements are taken, should be defined to increase communication among workers. Descriptions of habitat should consider the dynamic nature of the components. Terms such as *micro-*, *meso-*, and *macrohabitat* are usually used to characterize the continuous nature of the factors that can be measured within an animal's habitat. Here again, technical limitations prevent microhabitat from being described and measured over a large extent at large mapping scales. Macrohabitat usually includes measures such as canopy cover and tree density, whereas microhabitat will include shrub stem density and pebble cover.

To quantify habitat quality requires discovering which relationships determine individual fecundity/survivorship at the appropriate scale. The strength and frequency of interactions between the individual and its environment define the performance of the animal (e.g., survival, fecundity), and are considered niche relationships (see Morrison and Hall 2002). Thus, we can draw boundaries around where the animal performs activities and interacts with the biotic and abiotic characteristics, which can then be called the spatial extent of the habitat. Within the spatial extent we can then define the spatial and temporal resolution of our observations.

Habitat can certainly be used to develop general descriptors of the distribution of animals. However, we fail repeatedly to find commonalties in habitat for most populations across space because we are usually missing the underlying mechanisms (e.g., size distribution of prey, forage nutrients, competitive factors) determining occupancy, survival, and fecundity. Habitat per se can only provide a limited explanation of the ecology of an animal. Other concepts, including especially the niche, must be invoked to more fully understand the mechanisms responsible for animal survival and fitness. As we have seen in many wildlife studies, *high quality habitat* varies in physical attributes for a species across its range because it is not measuring mechanisms. Rather, our statistical models of habitat are, at best, analyzing surrogates of these mechanisms. Habitat is a useful concept for describing the physical area used by an animal, and should probably retain its simplicity for ease of communication among scientists, managers, and the public.

Information Needs

Integrated in each chapter were discussions of the weaknesses in information available on animal ecology. These weaknesses in basic knowledge inhibit our ability to develop rigorous restoration plans. In the first edition of this book (Morrison 2002, 193–93) I listed what I considered were topics that deserved special attention because of their importance in developing restoration plans that centered on wildlife habitat. Although many researchers have been addressing each of the topics I listed, our knowledge of each topic remains well short of what we need to apply to restoration plans with much confidence. Thus, we need to gather substantially more information on the following:

- Identification of the actual population boundaries of species of interest. This information has clear implications for understanding how individuals occupy a restored site. Thus, studies of immigration, emigration, and dispersal are needed.
- Related to the previous item is the determination of the metapopulation structure of a species of interest. Thus, the success of a project will be enhanced if the structure of animal populations throughout the region in which the project occurs is known.

- The role that competitors and predators have on occupancy of a site is poorly understood. There has been extensive theoretical development on how competitors and predators influence other species, although little practical data are available on how they influence occupancy of specific locations.
- Further work is needed to identify factors that limit the occupancy, survival, and reproductive success of animals. Such critical limiting factors can be used to build restoration plans. These factors will likely include microhabitat variables and niche relationships.
- Little work on the historic distribution and abundance of animals in a project area has been conducted. Field notes and journals, museum records, and fossils and subfossils have been little utilized to design restoration plans.
- The influence of sampling effort and sampling methods on results of field surveys needs much greater attention. Rare and abundant species usually require different sampling methods and sampling intensities.
- The influence of the size, shape, and location of preserves on wildlife occupancy is poorly known. Little empirical data are available to either support or refute the numerous concepts on preserve design that have been offered.
- Recent studies of corridors indicate that they function only in very a species-specific and location-specific manner. Further empirical evidence is needed before corridors should be incorporated into restoration plans.

Working with Wildlife Scientists and Managers

It is simple common sense that people that seek to work together speak the same language. It thus behooves both restoration (plant) ecologists and wildlife scientists to be familiar with both the historic and current thoughts on concepts, theory, and terminology. A simple example—developed in chapter 3—was the contrasting meaning of the term *habitat type* in plant and wildlife ecology. Additionally, there are multiple concepts of the niche, community, the process of plant succession, and other terms and concepts. The message here is that restoration is a multifaceted endeavor that requires all workers to become knowledgeable in a host of ecological fields. We must be critical in our approaches to science and subsequent applications to restoration.

I suspect that many restorationists, especially trained plant ecologists, will become frustrated by the myriad of opinions among wildlife scientists regarding the appropriate variables to measure, and the spatial and temporal scales at which the measurements should take place. Even concepts with a relatively long history in animal ecology are steeped in current controversy. The advent of remote sensing technology and geographic information system (GIS) software has resulted in a substantial increase in large-area analyses of the distribution of wildlife. In fact, some workers have stated that the *landscape* is the most effective scale for conserving wildlife. As developed in this book, I fundamentally disagree with this perspective. This, of course, is an example of the point I made earlier—that wildlife scientists often disagree on rather fundamental issues. But we have substantially enhanced our understanding of how animals select habitat, and most now accept the concept of hierarchical approaches to the study of animal habitat.

Such differences in perspectives and approaches mean that restorationists must be current on fundamental issues of wildlife ecology. These issues are continually being discussed at professional meetings, and debated in the pages of leading journals. It thus behooves restorationists to participate in these discussions and debates through membership in the organizations and participation at annual meetings (e.g., The Wildlife Society [TWS], Society for Conservation Biology [SCB], Ecological Society of America [ESA]). In recognition of the central role that restoration can play in wildlife conservation, The Wildlife Society developed a working group on restoration. Members of the working group participate in focused discussions on issues pertinent to wildlife and wildlife-habitat restoration. It seems reasonable to assume that greater participation by restorationists in TWS, SCB, ESA, and related societies would only enhance the restoration of wildlife and their habitat.

Synthesis

In this closing section I present a collation of the synthesis sections of each preceding chapter. I do this to provide a quick reference reminder of the central points raised in this book.

Chapter 1

The wildlife profession has moved increasingly toward the gathering of rigorous scientific data, followed by development of applications for management of animals based on those data. Likewise, the restoration profession is moving increasingly toward the gathering of sound data as a foundation for development of practical applications of those data. Much of restoration involves, directly or indirectly, improving conditions for native species of wildlife. To be ultimately successful, then, restoration plans must be guided in large part by the needs of current or desired wildlife species in the project area. Thus, proper consideration of wildlife—their habitat needs and numbers—is a complicated process that requires careful consideration during all stages of restoration. This book provides a thorough understanding of the conceptual and practical problems involved in sampling wildlife populations. It is critical that restorationists understand what wildlife biologists can and cannot provide within certain time and monetary limitations.

Chapter 2

A continuing difficulty in the terminology of restoration is determining the condition to which something will be restored. Additionally, there is debate over what is considered natural or native: how do we treat plants and animals introduced to new areas by humans now considered indigenous relative to those introduced by more recent human populations? And, what exactly is considered wildlife, and over what geographic area must we evaluate to evaluate ecosystem processes? There are no established answers to these and related questions, but there is a process being established within the community of restorationists by which a structured methodology can be used to make informed decisions based on available data. Thus, it is important that the restoration of wildlife and their habitats be approached in a holistic approach. In this manner we have a chance of understanding at least some of the factors responsible for determining the distribution, abundance, and success of animals.

Chapter 3

The structure of animal populations has a direct influence on the success of a restoration project. No restoration project should be designed and implemented without some type of assessment of local and regional distribution of the animal species of interest. If time and or funding restrictions prevent a field assessment of the population(s), then a thorough literature review, combined with the opinions of several experts on the species and population ecology in general, might prevent the pursuit of a project whose goals are unattainable. At a minimum, such knowledge will allow all project stakeholders to be informed of the likelihood of colonization of the site by the target animals. Additionally, a clear understanding of the difference between simple occupancy and successful survival and reproduction within the project area can be imparted.

Reintroduction is gaining increasing attention throughout the world as a means of restoring and managing animals, as indicated by the increasing use of the technique. Reintroductions involve all vertebrate groups. As the literature on evaluations and improvements in reintroduction grows, these techniques will likely increase as a valuable restoration tool. Lastly, we need to continue to develop and evaluate techniques for genetic management, especially those applied to metapopulations composed of small subpopulations.

Chapter 4

Advancing our understanding of habitat relationships will require increased cooperation among wildlife scientists, conservation biologists, and restorationists. Standardization of terminology will assist with such cooperation by promoting use of a common language. Wildlife scientists have also expended considerable effort studying efficient means of quantifying animal habitat. Restorationists will be able to accelerate achievement of project goals (for wildlife) by closely studying the strengths and weaknesses of previous wildlife-habitat studies. That is, there is no reason to reinvent the wheel, or repeat past mistakes. Analyses that are relevant to specific spatial scales are critical in restoration planning because we are able to relate the way animals use their environment to the size of the actual area being restored. By placing our studies and subsequent restoration plans into a spatial context allows us to identify what animals appear to require to occupy an area, then to survive and mate in the area, and finally to produce offspring in the area.

Chapter 5

The study of how to assemble species encompasses various approaches to finding rules that govern how ecological communities develop. Contrary to common practice, documenting a pattern is not the study of community assembly. Rules must be explicit and quantitative in nature. My review indicates that if assembly rules do, indeed, exist, they would be very local and not necessarily very simple. As such, it is unlikely that we will be able to identify rules that will be generally applicable except at the larger spatial extents. Restorationists need to pursue or otherwise promote studies that gather far more information than is usually done on the biology of individual species, including a focus on behavioral interactions. We need to be quantifying the often subtle responses of species to one another and to small changes in the physical environment.

Restorationists can design projects that try to modify the effects of filters to allow desired species in and prevent the establishment of undesired species, including managing recolonization, modifying abiotic and biotic filters, and overcoming dispersal barriers. Thus, restoration includes designed efforts to manipulate filters to arrive at a desired species composition. We might be able to develop general guidelines for application to large spatial scales and for predicting gross measures of animal performance (e.g., presence-absence). But, it becomes necessary to identify the mechanisms underlying animal performance as we work in smaller and smaller spatial areas and wish to restore a certain abundance and productivity of target animal species.

Chapter 6

Reconstructing the likely animal species assemblages of a proposed restoration site is a valuable means of developing project goals. Unfortunately, the historic presence of a species does not indicate that it could reoccupy the site even if all necessary habitat conditions were restored. Many species have certain minimum area requirements that cannot be met on small restoration sites. Further, conditions on lands surrounding the restoration site may be unsuitable to allow recolonization. Thus, restoration goals for wildlife should be developed in light of both historic possibilities and current realities.

Many attempts have been made to develop guidelines for selecting species as a focus for restoration. In addition to ecological criteria, factors to consider often include economic, cultural, ethical, and esthetic dimensions. There is no single approach to developing a list of focal species for emphasis during project planning; there will be a multitude of reasons for selecting species. Where we run into trouble is when we try to use one or a few species as surrogates (indicators) of how other species will respond to threatening process and management actions. The key element of any focal species approach is to clearly document the rationale used for all selection criteria.

Chapter 7

Conservation biologists have expended considerable time and effort in developing concepts that might promote species persistence. Although the concepts make intuitive sense, it is important to remember that they are *concepts*, and as such, usually have minimal empirical support. In large part this is because of the difficulty in replicating even small reserves, and the multitude of interacting environmental factors present. Thus, restoration projects that implement ecological and conservation concepts should be viewed as tests of these concepts. If designed properly, such projects can provide invaluable guidance for future projects. This makes publication of projects results—regardless of the success in attaining project goals—an essential part of the growth of the field of wildlife-habitat restoration.

Habitat heterogeneity is the degree of discontinuity in environmental conditions across a landscape for a particular species. Because habitat is a species- or organism-centric term, a particular environmental condition or gradient may constitute habitat for one species and a barrier for another. The species-specific concept of habitat selection, which occurs across spatial scales, must be the focus of restoration with regard to fragmentation (and the broader issues of habitat heterogeneity). Likewise, the concept of species assembly rules is directly relevant here because fragment size, shape, and the quality of the habitat therein will serve as a filter determining in part if a species can exist in a target area. Thus, fragmentation is a crude concept unless the response of animals is placed into an appropriate spatial hierarchy. Thus, the changes that take place in the environment at several scales of resolution: (1) fragmentation in a broad sense that takes place on the landscape scale; (2) fragmentation of different plant associations that take place within a vegetation type; and (3) alteration of habitat quality within small areas.

To best understand effects on wildlife, disturbances should be depicted according to their frequency, intensity, duration, location, and geographic extent. In this way, disturbances to habitats can be incorporated into spatially explicit models of population demography to project effects on species distribution and viability. As vegetation patches change in time so do responses by animals as they use patches for breeding, foraging, refuge from predators, resting, and dispersal. Some wildlife species likely evolved in concert with native disturbance regimes and take optimal advantage of resources distributed through space and time in shifting patches.

The intuitive appeal of corridors resulted in the widespread recommendation for their use in conservation planning and landscape design. However, corridors must be discussed and then described on a species-specific basis. The utility of corridors only makes sense in terms of a particular focal species within a particular landscape. This conclusion follows naturally from our development of the habitat concept as a species-specific phenomenon. As such, generalizations about the utility of corridors are difficult to make.

Perhaps the most prudent approach to planning for wildlife habitat is to consider the landscape as a whole and then decide which elements within it best contribute to specific conservation goals. The first challenge, of course, is to articulate those goals unambiguously and precisely, so it is clear which species, communities, habitats, and vegetation and environmental conditions are of conservation value and interest, which usually include those that are scarce, isolated, disjunct, or declining. A true landscape planning approach then entails integrating management goals across ownerships and land allocation categories.

Empirical studies and models of populations in patchy environments can help provide realistic expectations for occupancy rates within and among habitat patches. This has important implications for designing inventory and monitoring studies and interpreting results of such studies. Configuration of habitat patches may be directly manipulated by management activities, an indirect result of other human land-altering activities such as growth of towns and roads, or a result of stochastic and unpredictable natural disturbance events. We can attempt to guide direct management activities such that habitats are configured so as to afford rescue effects of habitat colonization by organisms over time.

Chapter 8

Knowledge is discovered through the application of scientific methods. There is not, however, any single best scientific method. I view the scientific method as a series of steps that build upon each other in a hierarchical fashion, from observation to induction to retroduction to development of testable hypotheses.

A common misconception among resource managers is that monitoring is something different than research, and as such, requires less rigor in its design and implementation. This misconception is widespread and has lead to many poorly designed studies and subsequent management decisions.

Experiments can be placed into two major classes: mensurative and manipulative. Neither type is necessarily more robust than the other; it all depends upon the goal of the study. Manipulative experiments have the advantage of being able to test treatments and evaluate the response of something to varying degrees of perturbation; and, do so in a manner usually within your control. Critical in the development of any experiment is proper consideration of replication, the use of controls, randomization, and independence. Project directors should avoid the temptation of rushing into study initiation without first developing all phases of the study, including likely statistical analyses. Even crude estimates of necessary sample sizes are essential if field efforts are to be maximized—there is no reason to under- or oversample.

The development of the study goals flow from initial statement of the problem, to the stating of the specific null hypothesis. It is critical that this step be well developed, or all that follows might be for naught. In establishing the goals for a study, it is critically important to establish the spatial and temporal applicability of the results. Such decisions will determine the spatial distribution of sampling areas and the temporal nature of sampling. The statistical literature, and an increasing amount of the wildlife literature, provides in-depth guidance on designing, conducting, analyzing, and interpreting field studies. I highly recommend that restorationists and wildlife scientists increase direct communication on wildlife studies, especially during the design phase.

The terms *significant* and *significance* have caused much confusion in ecology. We typically use the term significance in terms of biological, statistical, and social significance. Unfortunately, most authors do not separate statistical from biological significance, which leads to results of minimal absolute difference often being given ecological importance based on a statistical test. When ecologists say that something is *biologically significant*, we mean that there is enough difference or a strong enough relationship exists to make us believe that it matters biologically. Because we would seldom expect exactly no difference between two biological entities, we need to specify how much difference we think would matter to the entities of interest.

Many, if not most, restoration projects fall into the design category of *suboptimal* because (1) the treated (restored) area is not randomly selected, and (2) pretreatment data are often scanty or altogether unavailable. Thus, the field of impact assessment offers guidance on how to quantify treatment effects in as rigorous a manner as possible, given the suboptimal situation.

Chapter 9

Inventories associated with restoration projects are typically done to establish a baseline data set on existing conditions; that is, before initiating habitat altering activities. Monitoring is usually defined as a repeated assessment of status of some quantity, attribute, or task within a defined area over a specified time period. Inventory and monitoring projects require an adequate sampling design to ensure accurate and precise measurements. Thus, adequate knowledge is required of the behavior, general distribution, biology and ecology, and abundance patterns of the resource of interest. Relationships between animals and the environment are not as strong for animals as they are for plants because of the mobility of most animals. Thus, the use of indicators in monitoring is controversial and should be attempted only after careful evaluation.

A restoration project must begin with establishment of the sampling universe from which samples can be drawn and the extent to which inferences can be extrapolated to areas outside your immediate study area. In many restoration projects, however, the physical size of the area is predetermined by management or regulatory issues. Relatively small areas are unlikely to support viable populations of many vertebrates unless the surrounding area is also suitable for inhabitation or at least provides passage routes. If the study goal is to maintain population viability over the long term, then the sampling area may include locations outside the immediate restoration site. I emphasize that monitoring studies are simply a special type of research project and as such must employ a rigorous study design.

The duration of a monitoring study depends on study objectives, field methods, ecosystem processes, biology of the species under study, and logistical feasibility. A primary consideration for inventory and monitoring studies is the temporal qualities of the ecological state or process being measured. The duration of a monitoring study depends on study objectives, field methods, ecosystem processes, biology of the species under study, and logistical feasibility. There are several main alternatives to long-term studies: retrospective studies, substituting space for time, and modeling how an ecological process might behave under various scenarios.

The concept of adaptive management is centered primarily on monitoring the effects of land-use activities on key resources, and then using the monitoring results as a basis for modifying those activities to achieve project goals. It is an iterative process whereby management practices are carefully planned, implemented, and monitored at predetermined intervals. Adaptive management is not synonymous with trial-and-error approaches to restoration. Attempting to fix a problem after implementation is quite different than developing an action plan prior to initiation of a project. Thus, thresholds and triggers represent a predetermined level that when violated will lead to specific responses.

Biases influence not only the way we design a study, but additional biases are inserted into field sampling when multiple observers are used to gather data. Ignoring observer-based variability may lead to incorrect conclusions because results include artifacts, spurious relations, or irreproducible trends. The biases associated with estimations of animal abundances should also be considered carefully in habitat studies. This is because many of our analytical procedures correlate animal numbers with features of the environment. Obviously, a study that has low bias among habitat characteristics can be ruined by biased count data, and vice versa.

Questions concerning animal diversity include those related to vegetation types or specific areas, and those related to specific species or groups of species. The primary goal of vegetation- or area-based questions is to determine the species that occur in specific vegetation types or specific areas. Species-based studies may focus on one or more populations across space or time to determine, for example, geographic or ecological distribution. Species-based studies may be conducted as an inventory, but a thorough description of the species present usually requires that multiple techniques be used. The specific technique(s) selected and the intensity of their implementation is, of course, driven by project goals. Additionally, species-specific surveys will be driven by the behavior of the species of interest. Certain techniques, no matter how popular in the literature, may be unsuitable for determining even the presence of various species. I briefly outlined many of the techniques available for the inventory and monitoring of vertebrates, and provided supplemental recommendations for sources of more detailed directions.

Animal ecologists have expended considerable effort in developing rigorous sampling methods for vertebrates. I am often asked what is the *best* sampling method for a particular species. This question can only be answered when placed in the context of the particular goals of the study. That is, the sampling methods and sampling intensity must be matched with study goals. There is no justification for over- or under-sampling. As synthesized in this chapter, restorationists will find a well-developed literature on vertebrate sampling methods to assist with the design of their projects.

Chapter 10

Before this point in the book, we developed the fundamentals of wildlife ecology as applied to restoration. Although each chapter interrelated many concepts, none described how individual chapters could be linked into an overall synthesis. In this chapter, then, I present case studies that are examples of how wildlife restoration plans can initially be constructed. Rather than develop these case studies myself, I thought it would be much more insightful and informative if I requested that other people write the sections for this chapter. To accomplish this task I presented four ongoing wildlife restoration projects that my students, other coworkers, and I are conducting to a combined undergraduate and graduate course in wildlife restoration that I teach at Texas A&M University. The course was composed of advanced undergraduate and graduate (both MS and PhD) students in wildlife and range science and management. Based on class size, the students divided into four restoration teams and selected from the available projects, which focused on the following subjects:

- An endangered songbird in montane meadows in the Sierra Nevada, California
- A multi-species restoration effort in the Lake Tahoe Basin, California, and Nevada.
- Endangered songbirds on private lands in central Texas.
- An endangered kangaroo rat in central California.

These plans are not engineering and construction plans. Rather, the case studies present the steps and rationale used to develop the overall framework—including goals, objectives, and constraints—needed to initiate a wildlife restoration plan. Each plan follows a common overall format for consistency, and includes a monitoring component set in a general adaptive management framework.

These case studies should cause you to consider the many interrelated factors that must be considered when crafting a wildlife restoration plan. At a minimum these plans should be thought provoking and serve as a valuable template for your own initial planning efforts. Because all of the plans presented in this chapter are works in progress and incorporated aspects of adaptive management, they would in reality be considered working documents.

In Closing

It is my hope that this book will enhance the restoration and management of wildlife. As I wrote in the first edition, I consider this book a work in progress—I welcome positive and negative feedback. I invite all readers to contact me with suggestions for revision, including the addition of new or different material, modification of terminology, copies of pertinent references, and anything else that would enhance the product. I am especially interested in hearing from students regarding the adequacy of this material in the classroom. As an instructor, I realize that my perception of course material is often rather different from that held by my students! I developed the majority of this second edition while teaching a combined undergraduate-graduate course in wildlife restoration during 2007 at Texas A&M University.

Finally, I want to acknowledge all of the practitioners, managers, scientists, and students who are working diligently in the fields of wildlife and restoration. My hope is that this book will assist in fostering communication among all workers and lead to the growth of the field of restoration.

Key terms used throughout this book are listed below along with a brief definition and the primary chapter (in parentheses) where you will find a more
thorough discussion of the topic.

Adaptive management (or **adaptive resource management** [ARM]) is centered primarily on monitoring the effects of land-use activities on key resources, and then using the monitoring results as a basis for modifying
those activities to achieve project goals. (9)

After-only impact designs apply to planned and unplanned treatments but
for which no direct pretreatment data are available. (8)

Alternative stable states model is intermediate between the *deterministic*
and *stochastic* models and asserts that communities are structured and restricted to a certain extent but can develop into numerous stable states,
because of an element of randomness inherent in all ecosystems. (5)

Assembly rules describe broad patterns of species co-occurrence; the study of
how to assemble species encompasses various approaches to finding rules
that govern how ecological assemblages develop. (5)

Artificial insemination is used in captive breeding programs to guide the
mating of individual animals and also reduces potentially unwanted behavioral interactions between mating animals. (3)

Audio strip transects are used to count calling animals (e.g., frogs). (9)

Before-after/control-impact(BACI) framework was developed for application to environmental impact situations where little or no replication was
possible. (8)

Biologically significant means that there is enough difference or a strong enough relationship exists to make us believe that it matters biologically. (8)

Blocking is an extension of pairing to >2 groups. The goal of blocking is to make individual plots as similar as possible within a block; and for the blocks to be as dissimilar as possible. (8)

Bottleneck or population bottleneck (or **genetic bottleneck**) is an evolutionary event in which a significant percentage of a population or species is killed or otherwise prevented from reproducing. (3)

Box traps are the most commonly used device for live trapping small mammals. The most popular traps are manufactured by H.B. Sherman. (9)

Census or census population is the actual number of individuals (complete count) that occur in the population. (3, 9)

Cloning is usually conducted by removing the nucleus from a donor egg cell of the animal that will carry the cloned embryo, and then injecting into the carrier's egg cell the nucleus from a cell of the animal to be cloned. (3)

Coarse-filter (approach) describes managing generally defined or broadscale habitat conditions, such as managing forests within historic ranges of natural disturbances (e.g., wildfire) will provide for the needs of all associated native species. (6)

Controls in study design are required to account for spatial and temporal variation in biological systems. (8)

Community describes the co-occurrence of individuals of several species in time and space, and includes the role of interdependences among the species (populations) under study. (5)

Community species pool is the set of species present in a site within the target community. (5)

Completely randomized design allocates sampling units in a completely randomized manner across a study area. (8)

Compliance monitoring is done when mandated by law or statute to ensure that the project meets legal requirements. (9)

Connectivity refers to the extent to which a species or population can move among landscape elements in a mosaic of habitat. This definition necessitates linkages among species at appropriate spatial and temporal scales. (7)

Conservation breeding is the general effort to manage the breeding of plant and animal species that does not strictly involve captivity, such as moving (relocating) animals to a new location to avoid predators or disease. (3)

Corridor is a linear habitat, embedded in a matrix of dissimilar habitats that connects two or more larger blocks of habitat. (7)

Cover board is an artificial wooden cover used to sample herpetofauna. (9)

Cross-over design can be applied where the treatment is not permanent, but can be crossed-over between control and treatment plots. (8)

Cryopreservation is the freezing and storage of sperm, ova, embryos, or tissues to manage and safeguard against loss of genetic variation. (3)

Deme is defined as individuals of a species that have a high likelihood of interbreeding. (3)

Deterministic model describes a community's development as the inevitable consequence of physical and biotic factors. (5)

Direct measures are variables that link clearly and directly to the question of interest. (9)

Direct roost counts are done by one or more individuals who systematically count all visible bats in a diurnal roost. (9)

Dispersal is the one-way movement, typically of young, away from natal areas. (3)

Disturbance counts are used to overcome the problem of bats roosting out of sight or moving between counts. (9)

Disturbances affecting temporal variance are those that do not alter the mean abundance, but change the magnitude of the occillations between sampling periods. (8)

Drift fence is a short (e.g., 50 cm tall) fence often located to radiate out from a pitfall or funnel trap to help direct animals into the trap. (9)

Dynamic equilibrium incorporates both temporal and spatial variation, where natural factors and levels of resources usually differ between two or more areas being compared, but the differences between mean levels of the resource remain similar over time. (8)

Ecoclines are discontinuities in environmental conditions that occur as relatively broad gradations in conditions over areas of great geographic extent. (7)

Ecological filters form one of the main approaches in assembly rules theory, whereby only those species that are adapted to the abiotic and biotic conditions present at a site will be able to establish themselves successfully. (5)

Ecological indicators are surrogates of the ecological state of a resource of interest. (9)

Ecological restoration is the practice of restoring degraded ecological systems. (2)

Ecological trap is a location that appears to be suitable to an animal because reliable environmental or behavioral cues are mismatched with the actual consequences for the individual's fitness. (3)

Ecosystem engineer is a species that, via morphology or behavior, modifies, maintains, and creates habitat for itself and other organisms. (6)

Ecotones are discontinuities in environmental conditions that occur as relatively sharp breaks in environmental conditions. (7)

Effect size is the difference between the null hypothesis and a specific alternative hypothesis. (7)

Effective population size (N_e) is the number of individuals that contribute genes to succeeding generations. (3)

Effectiveness monitoring is used to evaluate if the stated action met its objective. (9)

Element is an item on which some measurement is made, such as an animal, roost site, snag, or other item of interest. (8)

Error is the difference between the true value and our estimate of the value of a parameter; error is composed of natural variation and (observer) bias. (9)

Factorial designs examine several possible interactions among factors, thus examining the influence of potential factors on your response variable. (8)

Faunal relaxation occurs when wildlife species are lost from isolated environments. (7)

Flagship species are charismatic species that serve as symbols to generate conservation awareness and action. (4)

Focal species (approach) is an attempt to devise ways of simplifying how we study and manage species, including a focus on a selected list of animal species. Focal species are also used to help understand, manage, or conserve ecosystem composition, structure, or function. (6)

Founder effect occurs when a few individuals drive the direction of genetic composition in a population. (3)

Fragmentation refers to the degree of heterogeneity of habitats across a landscape; of particular emphasis is the isolation and size of resource patches. (7)

Functional response of a population refers to changes in behavior of organisms, such as selecting different prey or using different substrates for resting or reproduction. (3)

Genetic drift is the accumulation of random events that change the makeup of a gene pool slightly, but often compound over time. More precisely termed allelic drift, the process of change in the gene frequencies of a population due to chance events determine which alleles (variants of a gene) will be carried forward while others disappear. (3)

Genome banking is the storage of sperm, ova, embryos, tissues, or DNA. (3)

Habitat is defined as the resources and conditions present in an area that produce occupancy—including survival and reproduction—by an organism; habitat is organism-specific. (4)

Habitat abundance refers only to the quantity of resources in the habitat, irrespective of the organisms present in the habitat. (4)

Habitat asssessments are used in monitoring to quantify changes in the amount and condition of habitat, and to predict the impacts of land-use practices, over time and space. (9)

Habitat availability refers to the accessibility and procurability of physical and biological components by animals. (4)

Habitat heterogeneity is the degree of discontinuity in environmental conditions across a landscape for a particular species. (7)

Habitat preference is the consequence of the process, resulting in the disproportional use of some resources over others. (4)

Habitat quality refers to the ability of the environment to provide conditions appropriate for individual and population persistence. (4)

Habitat selection is a hierarchical process involving a series of innate and learned behavioral decisions made by an animal about which habitat it would use at different scales of the environment; the process by which an animal chooses which habitat components to use. (4)

Habitat type refers only to the type of vegetation association in an area, or the potential of vegetation to reach a specified climax stage. (4)

Habitat use is the way an animal uses (or consumes, in a generic sense) a collection of physical and/or biological components (i.e., resources) in a habitat. (4)

Habituation is the waning of behavioral responsiveness by an animal. (9)

Hard release occurs when animals are transported from the capture site (or captive-rearing site) and released into the wild without any conditioning to the release site environment. (3)

Heterozygosity can refer to (1) diversity within individuals, and (2) diversity among individuals in a population. (3)

Hibernaculum counts are done in midwinter to quantify hibernating bats. (9)

Hierarchical designs have levels of one factor occurring in combination with the levels of one or more other factors. (8)

Home range is movement throughout a more or less definable and known space over the course of a day to weeks or months, to locate resources. (3)

Hypothetico-deductive (H-D) method compliments *retroduction* in that it starts with an hypothesis—usually developed through retroduction—and then makes testable predictions about other classes of facts that should be true if the research hypothesis is true. (8)

Impact is a general term used to describe any change that perturbs the environment, whether it is planned or unplanned, human induced, or an act of nature. (8)

Implementation monitoring is used to assess whether or not a directed activity has been carried out as designed. (9)

Independence refers to the probability of one event occurring that is not affected by whether or not another event has or has not occurred. (8)

Indicator species are those whose distribution, abundance, or population dynamics can serve as substitute measures of the status of other species or environmental attributes. (6)

Indirect measures attempt to establish a surrogate for the direct, causal link between variables of interest. (7)

Induction applies to the finding of associations between classes of facts. (8)

Interspersion (of study plots) controls for regular spatial variation in the experimental units. (8)

Intraobserver reliability is a measure of the ability of a specific observer to obtain the same data when measuring the same behavior on different occasions, and thus measures the ability of an observer to be precise in his or her measurements. (7)

Interobserver reliability measures the ability of two or more observers to obtain the same results on the same occasion and often inserts substantial variability into the data set. (9)

Inventory is used to assess the state or status of one or more resources. It can provide information on environmental characteristics such as the distribution, abundance, and composition of wildlife and wildlife habitats. (9)

Keystone species are those that significantly affect one or more key ecological processes or elements to an extent that greatly exceeds what would be predicted from the keystone species' abundance or biomass (6)

Level of biological organization refers to the biological dimension of scale, and whether a study or plan pertains to ecosystems, communities, assemblages, species, or more finely defined entities such as gene pools or ecotypes. (4)

Line transects are a commonly used technique for assessing the distribution and abundance of animals over large areas (i.e., >20 ha) of relatively uniform terrain. (9)

Local species pool is the set of species occurring in a subunit of the biogeographic region, such as a valley segment. (5)

Passage is travel via a corridor by individual animals from one habitat patch to another. (7)

Point counts can be viewed as line transects of zero length conducted while mostly stationary; point counts were initially developed for use in rough terrain where simultaneously walking and observing birds was difficult and potentially dangerous. (9)

Macrohabitat is used to refer to large spatial extent (i.e., landscape-scaled) features, such as seral stages or zones of specific vegetation associations. (4)

Manipulative experiments have the advantage of being able to test treatments and evaluate the response of something to varying degrees of perturbation and do so in a manner usually within your control. (8)

Mensurative studies, also termed **observational studies,** measure differences in treatments without implementing a designed treatment (or experiment). (8)

Metapopulation occurs when a species whose range is composed of more or less geographically isolated patches, interconnected through patterns of gene flow, extinction, and recolonization and has been termed *a population of populations.* (3)

Microhabitat refers to fine-scaled habitat features such as foraging and nest sites. (4)

Migration is a seasonal, cyclic movement typically across latitudes or elevations to track resources or to escape harsh conditions. (3)

Minimum viable population (MVP) is the smallest size population (typically measured in absolute number of organisms rather than density or distribution of organisms) that can sustain itself over time, and below which extinction is inevitable. (3)

Mist nets are the most commonly used method to capture bats and birds. Mist nets are made of light nylon or polyester thread in 2–3 m × 6–12 m panels that are strung between two poles. (9)

Models are conceptualizations of how an ecological process might behave under various scenarios. (9)

Monitoring or **monitoring study** is usually defined as a repeated assessment of the status of some quantity, attribute, or task within a defined area over a specified time period, with the purpose of detecting a change from the present conditions. (9, 10)

Multiyear studies within impact assessment reduce the effects of temporal and spatial variation by removing (subtracting out) naturally varying temporal effects. (8)

Night driving is a form of nonrandom line transect in which the transect is a paved road. The technique is usually used to sample herpetofauna in the evening that are attracted to the heat radiating from the road as the evening air cools. (8)

Nightly emergence counts (or **nightly dispersal counts**) are conducted as bats leave diurnal roosts. (9)

Nuisance species are nonindigenous species that were accidentally or purposefully released into the marine, freshwater, or terrestrial environment. If such species become established and thrive, they will influence the native flora and fauna and their habitats, and may out-compete, prey upon, or bring diseases or parasites to economically and ecologically valuable native species, often adversely changing the ecosystem in the process. (9)

Numerical response of a population refers to absolute changes in abundance of individuals through changes in recruitment. (3)

Operationalization means that ecological concepts such as habitat should have operational definitions, which are the practical, measurable specifications of the ranges of specific phenomena the terms represent. (4)

Optimal study designs are characterized by the availability of pretreatment data, random selection of sites, and appropriate control sites. (8)

Paired designs depict an ability to match the environmental conditions of treatment and control plots. (8)

Pedigree analysis is defined as the genetic study of a multigenerational population with ancestral linkages that are known, knowable, or can be reasonably assumed or modeled. (3)

Physiological niche represents the (usually unknown) potential distribution of a species. (4)

Pilot studies are needed when one has little concrete notion of the best sampling design, sampling intensity, or even sampling methods. (8)

Pitfall trap is a container placed in the ground so that its open end is flush with the surface; animals are captured when they fall into the trap. (9)

Population is a collection of organisms of the same species that interbreed. A broader definition is where a population is considered a collection of individuals of a species in a defined area that might or might not breed with other groups of that species elsewhere. (3)

Population demography is the proximate expression of a host of factors influencing individual fitness and population viability. (3)

Population monitoring is a large and varied area of interest used to monitor trends in abundance or various demographic parameters over time. (9)

Population viability is the likelihood of persistence of well-distributed populations to a specified future time period, typically a century or longer. (3)

Power in study design is the likelihood that we will falsely reach a conclusion of no effect due to a treatment (or impact). (8)

Precision measures how close our estimates are to one another, regardless of how closely they approximate the true value. (9)

Press disturbances are those sustained beyond the initial disturbance. (8)

Pulse disturbances are those not sustained after the initial disturbance; the effects of the disturbance, however, may be long lasting. (8)

Quadrat or **plot sampling** uses square or rectangular plots to sample animals or their sign. (9)

Randomization controls for (reduces) experimenter biases in the assignment of experimental units to treatments. (8)

Randomized complete block design incorporates blocking into the completely randomized design. (8).

Realized niche represents the observed abundance distribution of a species, which usually will not include the full range of conditions under which the species could potentially be found. (4)

Recovery occurs in impact assessment when the dynamics of the impacted areas once again parallel those of the reference (control) area. (8)

Reference conditions are determined by analyzing the data obtained for each variable chosen for the historic range of variation study. (2)

Regional species pool is the set of species occurring in a certain biogeographic or climatic region which are potential members of the target assemblage. (5)

Replication controls for among-replicates variability inherent in any study. (8)

Restoration ecology is the scientific process of developing theory to guide restoration and using restoration to advance ecology. (2)

Restore is understood to mean *to bring back into existence or use*, and *restoration* means the act of restoring. (2)

Retroduction refers to the development of research hypotheses about processes that are explanations or reasons for observed relationships. (8)

Retrospective power analysis is conducted after data collection and is not usually recommended. (8)

Retrospective studies provide baseline data for comparison with modern observations. (9)

Sampled population is the number of the *elements* contained in the *sampling frame*. As noted earlier, the sampled population determines the *population of inference* for a study. (8)

Sampling frame is the complete list of *sampling units*. (8)

Sampling unit is a collection of *elements*, usually taken over some defined space and time. (8)

Sensitization occurs when the presence of an observer heightens an animal's awareness and likely alters its behavior. (9)

Sequential design is a sequence of treatments applied to the same element or plot, with the caveat that the treatment does not cause a permanent change or that the change that occurs is planned. (8)

Sequential sample size analysis is a primarily nonstatistical, graphical method of evaluating sample size during data collection, as well as justifying the collected sample size. (8)

Single-factor designs are those in which one type of treatment or classification factor is applied to all experimental units, such as all elements (e.g., animals, plants) in a unit. (8)

Single-year studies within impact assessment compare impact and control areas within a single year. (8)

SLOSS refers to a debate in ecology and conservation biology during the 1970s and 1980s as to whether Single Large or Several Small (SLOSS) reserves were a superior means of conserving biodiversity in a fragmented habitat. (7)

Snap traps are used for killing rodents during rapid assessment of species richness. (9)

Soft release is when captured animals are held in captivity for an extended period of time (days to months) for a variety of behavioral and physiological reasons, prior to release. (3)

Spatial equilibrium occurs when two or more sampling areas, such as impact and reference, have similar natural factors and, thus, similar levels of a resource. (8)

Species assemblage denotes the group of species present and potentially interacting within a project area. Such an assemblage could be part of a larger community, but there is no need to even invoke the community concept per se with regard to my treatment of assemblage rules for restoration. (5)

Species pool (concept) describes the specific species that occur within different spatial scales (i.e., *regional, local,* and *community species pools*). (5)

Species richness is the number of different species present in an area during a defined period of time. (9)

Split-plot designs are a form of nested factorial design where the study area is divided into blocks, the blocks are further subdivided in large plots, and these large plots are then subdivided into smaller plots, called split plots. (8)

Spot mapping (or **territory mapping**) uses bird behavior to establish rough territory boundaries, thus obtaining an estimate of the density and location of birds in the study area. (9)

Steady-state system is typified by levels of resources, and the natural factors controlling them, showing a constant mean through time. (8)

Stepping stone is used in restoration to denote habitat patches that are used to promote dispersal of individuals between subpopulations. (3)

Stochastic model describes community composition and structure as essentially a random process, depending only on the availability of vacant niches and the order of arrival of organisms. (5)

Strip transects differ from the line transect in assuming that all animals (or animal sign) within the strip are seen. (9)

Suboptimal study designs are characterized by lack of random selection of sites, lack of adequate replication of treatments, lack of pretreatment data, and inadequate control sites. (8)

Subpopulation is used to refer variously to a *deme* or to a portion of a population in a specific geographic location or as delineated by nonbiological criteria. (3)

Substitution of space for time is achieved by finding samples that represent the range of variation for the parameter(s) of interest in order to infer long-term trends. (9)

Surrogate species include concepts such as umbrella species, indicator species, keystone species, ecosystem engineers, flagship species, and focal species. (6)

Target population is the specific area, location, group of animals, or other defined item that is the focus of study. (8)

Temporal fragmentation refers to the degree to which a particular environment, such as an old forest, occupies a specific area through time. (7)

Thresholds (population biology) are used in population viability modeling to indicate conditions of the environment that, when changed slightly past particular values, cause populations to crash. (3)

Thresholds (adaptive management) are specific levels of key resources established prior to initiation of a project and are a central component of adaptive management. (9)

Translocation is the physical movement of individuals from one population to another population. (3)

Trial-and-error approaches to restoration lack plans to fix a problem after implementation, which seldom result in a successful project. (9)

Triggers cause implementation of a specific management action and are a central component of adaptive management. (9)

Type I error, termed alpha, is rejection of the null hypothesis when it is actually true. (8)

Type II error, termed beta, is the failure to reject the null hypothesis when it is in fact false. (3)

Ultrasonic sound detectors are a useful technique for detecting the presence of bats based on echolocation calls. (9)

Umbrella species are those whose conservation confers a protective umbrella to numerous co-occurring species. (6)

Unpaired designs depict an inability to match the environmental conditions of the treatment plots with similar reference plots. (8)

Validation monitoring is used to determine whether established management direction provides guidance to meet its stated objectives. (9)

Visual encounter surveys (VES) are those in which observers walk through an area for a prescribed period of time (i.e., time constrained) systematically searching for animals. (9)

Voucher specimens are an archival record that physically and permanently document data. (9)

Wildlife is a term given to animals and plants that live on their own without taming or cultivation by people. However, in terms of traditional wildlife management, wildlife has been defined as mammals and birds that are hunted (game animals) or trapped (fur-bearing animals). (2)

Wildlife-habitat restoration is focused on providing for the survival and protection of individual organisms in sufficient numbers and locations to maximize the probability of long-term persistence. (3)

LITERATURE CITED

Abate, T. 1992. Which bird is the better indicator species for old-growth forest? *BioScience* 42:8–9.

Ahlering, M. A., and J. Faaborg. 2006. Avian habitat management meets conspecific attraction: If you build it, will they come? *Auk* 123:301–21.

Allendorf, F. W., and G. Luikart. 2007. *Conservation and the genetics of populations.* Oxford: Blackwell Publishing.

American Ornithologists' Union (AOU). 1983. *Check-list of North American birds.* 7th ed. Washington, DC: American Ornithologists' Union.

Anderson, M. K. 1996. Tending the wilderness. *Restoration and Management Notes* 14:154–66.

Andren, H. 1995. Effect of landscape composition on predation rates at habitat edges. Pages 225–55 in L. Hansson, L. Fahrig, and G. Merriam (eds.), *Mosaic landscape and ecological processes.* London: Chapman and Hall.

Angelstam, P. 1996. The ghost of forest past: Natural disturbance regimes as a basis for reconstruction of biologically diverse forests in Europe. Pages 287–337 in R. M. DeGraaf and R. I. Miller (eds.), *Conservation of faunal diversity in forested landscapes.* London: Chapman and Hall.

Ashmole, N. P., and M. J. Ashmole. 1997. The land fauna of Ascension Island: New data from caves and lava flows, and a reconstruction of the prehistoric ecosystem. *Journal of Biogeography* 24:549–89.

Askins, R. A. 2000. *Restoring North America's birds.* New Haven: Yale University Press.

Bailey, R. G. 2005. Identifying ecoregion boundaries. *Environmental Management* 34(1): S14–S26.

Balmford, A., G. M. Mace, and N. Leader-Williams. 1996. Designing the ark: Setting priorities for captive breeding. *Conservation Biology* 10:719–27.

Batzli, G. O. 1992. Dynamics of small mammal populations: A review. Pages 831–50 in D. R. McCullough and R. H. Barrett (eds.), *Wildlife 2001: Populations.* New York: Elsevier Applied Science.

Beazley, K., and N. Cardinal. 2004. A systematic approach for selecting focal species for conservation in the forests of Nova Scotia and Maine. *Environmental Conservation* 31:91–101.

Beesley, D. 1996. Reconstructing the landscape: An environmental history, 1820–1960. Pages 1–24 in D. C. Erman, ed., *Sierra Nevada ecosystem project: Final report to Congress* Vol. 2. *Assessments and scientific basis for management options.* U.S. Geological Survey Digital Data Series DDS-43, Davis, CA.

Beier, P., and R. F. Noss. 1998. Do habitat corridors provide connectivity? *Conservation Biology* 12:1241–52.

Bekoff, M. 1977. *Canis latrans. Mammalian Species* 79:1–9.

Belyca, L. R. 2004. Beyond ecological filters: Feedback networks in the assembly and restoration of community structure. Pages 115–32 in V. M. Temperton, R. J. Hobbs, T. Nuttle, and S. Halle, eds., *Assembly rules and restoration ecology: Bridging the gap between theory and practice.* Washington, DC: Island Press.

Bender, E. A., T. J. Case, and M. E. Gilpin. 1984. Perturbation experiments in community ecology: Theory and practice. *Ecology* 65:1–13.

Bennett, A. F. 1990a. Habitat corridors and the conservation of small mammals in a fragmented forest environment. *Landscape Ecology* 4:109–22.

———. 1990b. Land use, forest fragmentation and the mammalian fauna at Naringal, south-western Victoria, Australia. *Wildlife Restoration* 17:325–47.

Berger, J., P. B. Stacey, L. Bellis, and M. P. Johnson. 2001. A mammalian predator-prey imbalance: Grizzly bear and wolf extinction affect avian neotropical migrants. *Ecological Applications* 11:947–60.

Bibby, C. J., N. D. Burgess, and D. A. Hill. 1992. *Bird census techniques.* 2nd ed. London: Academic Press.

Bibby, C. J., N. D. Burgess, D. A. Hill, and S. H. Mustoe. 2000. *Bird census techniques.* 2nd ed. London: Academic Press.

Bleich, V. C., J. D. Wehausen, and S. A. Holl. 1990. Desert-dwelling mountain sheep: Conservation implications of a naturally fragmented distribution. *Conservation Biology* 4:383–90.

Block, W. M., and L. A. Brennan. 1993. The habitat concept in ornithology: Theory and applications. *Current Ornithology* 11:35–91.

Block, W. M., A. B. Franklin, J. P. Ward Jr., J. L. Ganey, and G. C. White. 2001. Design and implementation of monitoring studies to evaluate the success of ecological restoration on wildlife. *Restoration Ecology* 9:293–303.

Block, W. M., M. L. Morrison, J. Verner, and P. N. Manley. 1994. Assessing wildlife-habitat-relationships models: A case study with California oak woodlands. *Wildlife Society Bulletin* 22:549–61.

Block, W. M., K. A. With, and M. L. Morrison. 1987. On measuring bird habitat: Influence of observer variability and sample size. *Condor* 72:182–89.

Blomberg, S., and R. Shine. 2006. Reptiles. Pages 297–307 in W. J. Sutherland (ed.), *Ecological Census Techniques.* 2nd ed. Cambridge: Cambridge University Press.

Blouin-Demers, G., and P. J. Weatherhead. 2002. Implications of movement patterns for gene flow in black rat snakes (*Elaphe obsoleta*). *Canadian Journal of Zoology* 80:1162–72.

Bogaert, J., A. Farina, and R. Ceulemans. 2005. Entropy increase of fragmented habitats: A sign of human impact? *Ecological Indicators* 5(3): 207–12.

Bolger, D. T., T. A. Scott, and J. R. Rotenberry. 1997. Breeding bird abundance in an urbanizing landscape in coastal southern California. *Conservation Biology* 11:406–21.

Bombay, H. L. 1999. Scale perspectives in habitat selection and reproductive success for willow flycatchers (*Empidonax traillii*) in the central Sierra Nevada, California. MS thesis, California State University, Sacramento.

Bombay, H. L., M. L. Morrison, D. E. Taylor, and J. W. Cain. 2001. *2001 Annual report and preliminary demographic analysis for willow flycatcher monitoring in the central Sierra Nevada*, in partial fulfillment of contract RFQ-IBET-17-01-053 between University of California, San Diego; White Mountain Research Station; and USDA Forest Service, Tahoe National Forest, CA, Nevada City.

Bombay, H. L., M. L. Morrison, and L. S. Hall. 2003. Scale perspectives in habitat selection and animal performance for willow flycatchers (*Empidonax traillii*) in the central Sierra Nevada, California. *Studies in Avian Biology* 26:60–72.

Bombay, H. L., T. M. Ritter, and B. E. Valentine. 2000. *A willow flycatcher survey protocol for California*. 6 June 2000. Pacific Southwest Region, Vallejo, CA: USDA Forest Service.

Bonham, C. D. 1989. *Measurements for terrestrial vegetation*. New York: Wiley.

Bookhout, T. A., ed. 1994. *Research and management techniques for wildlife and habitats*. 5th ed. Bethesda, MD: Wildlife Society.

Bowles M. L., and C. J. Whelan. 1994. Conceptual issues in restoration ecology. Pages 1–7 in M. L. Bowles and C. J. Whelan (eds.), *Restoration of endangered species: Conceptual issues, planning, and implementation*. Cambridge: Cambridge University Press.

Braun, C. E. 2005. *Techniques for wildlife investigations and management*. 6th ed. Bethesda, MD: Wildlife Society.

Brennan, L. A., R. T. Engstrom, and W. E. Palmer. 1998. Whither wildlife without fire? Pages 402–14 in *Transactions of the 63rd North American wildlife and natural resources conference*. Wildlife Management Institute.

Brito, D., and F. A. S. Fernandez. 2000a. Dealing with extinction is forever: Understanding the risks faced by small populations. *Ciencia e Cultura* (Sao Paulo) 52:161–70.

———. 2000b. Metapopulation viability of the marsupial *Micoureus demerarae* in small Atlantic forest fragments in south-eastern Brazil. *Animal Conservation* 3:201–9.

Brook, A., M. Zint, and R. De Young. 2003. Landowners' responses to an endangered species act listing and implications for encouraging conservation. *Conservation Biology* 17:1638–49.

Brown, J. H., D. W. Mehlman, and G. C. Stevens. 1995. Spatial variation in abundance. *Ecology* 76:2028–43.

Browning, M. R. 1993. Comments on the taxonomy of *Empidonax traillii* (willow flycatcher). *Western Birds* 24:241–57.

Bull, E. 1997. *Trees and logs important to wildlife in the interior Columbia River Basin*. USDA Forest Service General Technical Report PNW-391, Portland, OR.

Bury, R. B., D. J. Major, and D. Pilliod. 2002. Responses of amphibians to fire disturbance in Pacific Northwest forests: A review. Pages 34–42 in *The role of fire in nongame wildlife management and community restoration: Traditional uses and new directions*. General Technical Report NE-288. USDA Forest Service, Newtown Square, PA.

Cabe, P. R. 1993. European starling (*Sturnus vulgaris*). Account 48 in A. Poole and F. Gill (eds.), *Birds of North America*. Philadelphia: Academy of Natural Sciences, and Washington, DC: American Ornithologists' Union.

Cain III, J. W. 2001. Nest success of yellow warblers (*Dendroica petechia*) and willow flycatchers (*Empidonax traillii*) in relation to predator activity in montane meadows of the central Sierra Nevada, California. MS thesis, California State University, Sacramento.

Cain III, J. W., M. L. Morrison, and H. L. Bombay. 2003. Predator activity and nest success of willow flycatchers and yellow warblers. *Journal of Wildlife Management* 67(3): 600–610.

Campomizzi, A. J., J. A. Butcher, S. L. Farrell, A. G. Snelgrove, B. A. Collier, K. J. Gutzwiller, M. L. Morrison, and R. N. Wilkins. 2008. Conspecific attraction is a missing component in wildlife habitat modeling. *Journal of Wildlife Management* 72:331–36.

Caprio, A. C., and T. W. Swetnam. 1995. Historic fire regimes along an elevational gradiant on the west slope of the Sierra Nevada, California. In Brown, J. K., R. W. Mutch, C. W. Spoon, and R. H. Wakimoto, technical coordinators, *Proceedings: Symposium on fire in wilderness and park management: Past lessons and future opportunities.* USDA Forest Service General Technical Report INT-320, Ogden, UT.

Carbyn, L. N., H. J. Armbruster, and C. Mamo. 1994. The swift fox reintroduction program in Canada from 1983 to 1992. Pages 247–71 in M. L. Bowles and C. J. Whelan (eds.), *Restoration of endangered species: Conceptual issues, planning, and implementation.* Cambridge: Cambridge University Press.

Carlton, D. K. Berger, S. Holl, and M. Finney. 2000. Wildland fire susceptibility analysis. Pages A-1–A-8 in D. D. Murphy and C. M. Knopp (eds.), *Lake Tahoe watershed assessment.* Vol. 2. USDA Forest Service General Technical Report PSW-175, Berkeley, CA.

Chambers, J. C., and J. R. Miller. 2004. Restoring and maintaining sustainable riparian ecosystems: The Great Basin ecosystem management project. Pages 1–23 in J. C. Chambers and J. R. Miller (eds.), *Great Basin riparian ecosystems: Ecology, management, and restoration.* Washington, DC: Island Press.

Cherry, S. 1998. Statistical tests in publications of The Wildlife Society. *Wildlife Society Bulletin* 26:947–53.

Cimprich, D. A. 2005. *Monitoring of the black-capped vireo during 2005 on Fort Hood.* 2005 Annual Report. Fort Hood, TX: The Nature Conservancy.

Clements, F. E. 1916. *Plant succession.* Washington, DC: Carnegie Institution Publication No. 242.

Clements, F. E. 1920. *Plant indicators.* Washington, DC: Carnegie Institution.

Coblentz, B. E. 1990. Exotic organisms: A dilemma for conservation biology. *Conservation Biology* 4:261–65.

Connor, E. F., and D. Simberloff. 1979. The assembly of species communities: Chance or competition? *Ecology* 60:1132–1340.

Converse, S. J., W. M. Block, and G. C. White. 2006. Small mammal population and habitat responses to forest thinning and prescribed fire. *Forest Ecology and Management* 228:263–73.

Cook, C. W., and J. Stubbendieck, eds. 1986. *Range research: Basic problems and techniques.* Denver: Society for Range Management.

Cooperrider, A. Y., R. J. Boyd, and H. R. Stuart, eds. 1986. *Inventory and monitoring of wildlife habitat.* Denver: USDA Bureau of Land Management Service Center.

Coppolillo, P., H. Gomez, F. Maisels, and R. Wallace. 2004. Selection criteria for suites of landscape species as a basis for site-based conservation. *Biological Conservation* 115:419–30.

Corn, P. S. 1994. Terrestrial amphibian communities in the Oregon Coast Range. Pages 304–17 in L. F. Ruggiero et al., technical coordinators, *Wildlife vegetation of unmanaged Douglas-fir forests.* General Technical Report PNW-GTR-285. Washington, DC: USDA Forest Service.

Cox, C. B., and P. D. Moore. 1993. *Biogeography: An ecological and evolutionary approach.* 5th ed. Boston: Blackwell.

Craig, D., and P. L. Williams. 1998. Willow flycatcher (*Empidonax traillii*). In *The Riparian bird conservation plan: A strategy for reversing the decline of riparian-associated birds in California*. California Partners in Flight. http://www.prbo.org/calpif/htmldocs/riparian_v-2.html

Crump, M. L., and N. J. Scott Jr. 1994. Visual encounter surveys. Pages 84–92 in W. R. Heyer, M. A. Donnelly, R. W. McDiarmid, L. C. Hayek, and M. S. Foster (eds.), *Measuring and monitoring biological diversity: Standard methods for amphibians*. Washington, DC: Smithsonian.

Curson, D. R., C. B. Goguen, and N. E. Mathews. 2000. Long-distance commuting by brown-headed cowbirds in New Mexico. *Auk* 117:795–99.

Daly, M., M. Wilson, P. R. Behrends, and L. F. Jacobs. 1990. Characteristics of kangaroo rats, *D. merriami*, associated with differential predation risk. *Animal Behavior* 40:380–89.

Daubenmire, R. 1968. *Plant communities: A textbook of plant synecology*. New York: Harper and Row.

Davidson, C. 1998. Issues in measuring landscape fragmentation. *Wildlife Society Bulletin* 26:32–37.

Davis, R., C. Dunford, and M. V. Lomolino. 1988. Montane mammals of the American Southwest: The possible influence of post-Pleistocene colonization. *Journal of Biogeography* 15:841–48.

DeBenedetti, S. H., and D. J. Parsons. 1979. Natural fire in subalpine meadows: A case description from the Sierra Nevada. *Journal of Forestry* 77(8): 477–79.

DeBoer, L.E.M. 1992. Current status of captive breeding programmes. Pages 5–16 in H. D. M. Moore, W. V. Holt, and G. M. Mace (eds.), *Biotechnology and the conservation of genetic diversity*. Symposia of the Zoological Society of London, No. 64. Oxford: Oxford University Press.

DeBoer, T. S., and D. D. Diamond. 2006. Predicting presence-absence of the endangered golden-cheeked warbler. *Southwestern Naturalist* 51:181–90.

Diamond, J. M. 1975. The assembly of species communities. Pages 342–44 in M. L. Cody and J. M. Diamond (eds.), *Ecology and evolution of communities*. Cambridge: Harvard University Press.

Dickson, J. G. 2002. Fire and bird communities in the south. Pages 52–57 in *The role of fire in nongame wildlife management and community restoration: Traditional uses and new directions*. USDA Forest Service General Technical Report NE-288, Newtown Square, PA.

Diefenbach, D. R., L. A. Baker, W. E. James, R. J. Warren, and M. J. Conroy. 1993. Reintroducing bobcats to Cumberland Island, Georgia. *Restoration Ecology* 1:241–47.

Dull, R. A. 1999. Palynological evidence for 19th century grazing-induced vegetation change in the southern Sierra Nevada, California, USA. *Journal of Biogeography* 26:899–912.

Dueser, R. D., and H. H. Shugart Jr. 1978. Microhabitats in forest-floor small mammal fauna. *Ecology* 59:89–98.

Eckrich, G. H., T. E. Koloszar, and M. D. Goering. 1999. Effective landscape management of brown-headed cowbirds at Fort Hood, Texas. *Studies in Avian Biology* 18:267–74.

Egan, D., and E. A. Howell. 2001. Introduction. Pages 1–23 in D. Egan and E. A. Howell (eds.), *Historical ecology handbook*. Washington, DC: Island Press.

Ehrlich, P. R., D. S. Dobkin, and D. Wheye. 1988. *The birder's handbook*. New York: Simon & Schuster.

Elias, S. A. 1992. Late Quaternary zoogeography of the Chihuahuan Desert insect fauna, based on fossil records from packrat middens. *Journal of Biogeography* 19:185–97.

Elzinga, C. L., D. W. Salzer, J. W. Willoughby, and J. P. Gibbs. 2001. *Monitoring plant and animal populations.* Malden, MA: Blackwell Science.

Falk, D. A., C. M. Richards, A. M. Montalvo, and E. E. Knapp. 2006. Population and ecological genetics in restoration ecology. Pages 14–41 in D. A. Falk, M. A. Palmer, and J. B. Zedler (eds), *Foundations of restoration ecology.* Washington, DC: Island Press.

Farrell, S. L. 2007. Brown-headed cowbird parasitism on endangered species: Relationships with neighboring avian assemblages. MS thesis, Texas A&M University, College Station.

Feinsinger, P. 2001. *Designing field studies for biodiversity conservation.* Washington, DC: Island Press.

Fellers, G. M., and C. A. Drost. 1994. Sampling with artificial cover. Pages 146–50 in W. R. Heyer, M. A. Donnelly, R. W. McDiarmid, L. C. Hayek, and M. S. Foster (eds.), *Measuring and monitoring biological diversity: Standard methods for amphibians.* Washington, DC: Smithsonian.

Flaspohler, D. J., S. A. Temple, and R. N. Rosenfield. 2001. Species-specific edge effects on nest success and breeding bird density in a forested landscape. *Ecological Applications* 11:32–46.

Flather, C. H., K. R. Wilson, D. J. Dean, and W. C. McComb. 1997. Identifying gaps in conservation networks: Of indicators and uncertainty in geographic-based analyses. *Ecological Applications* 7:531–42.

Fleishman, E., J. B. Dunham, D. D. Murphy, and P. F. Brussard. 2004. Explanation, prediction, and maintenance of native species richness and composition. Pages 232–60 in J. C. Chambers and J. R. Miller (eds.), *Great Basin riparian ecosystems: Ecology, management, and restoration.* Washington, DC: Island Press.

Flett, M. A., and S. D. Sanders. 1987. Ecology of a Sierra Nevada population of willow flycatchers. *Western Birds* 18:37–42.

Ford, W. M., M. A. Menzel, D. W. McGill, J. Larem, and T. S. McCay. 1999. Effects of a community restoration fire on small mammals and herpetofauna in the southern Appalachians. *Forest Ecology and Management* 114:233–43.

Forman, R. T. T. 1995. *Land mosaics: The ecology of landscapes and regions.* Cambridge: Cambridge University Press.

Fox, B. J. 1999. The genesis and development of guild assembly rules. Pages 23–57 in E. Weiher and P. Keddy (eds.), *Ecological assembly rules: Perspectives, advances, and retreats.* Cambridge: Cambridge University Press.

Franklin, J. 1989. Importance and justification of long-term studies in ecology. Pages 3–19 in G. E. Likens (ed.), *Long-term studies in ecology: Approaches and alternatives.* New York: Springer-Verlag.

Freemark, K. E., J. B. Dunning, S. J. Hejl, and J. R. Probst. 1995. A landscape ecology perspective for research, conservation, and management. Pages 381–427 in T. E. Martin and D. M. Finch (eds.), *Ecology and management of neotropical migratory birds: A synthesis and review of critical issues.* New York: Oxford University Press.

Freudenberger, D., and L. Brooker. 2004. Development of the focal species approach for biodiversity conservation in the temperate agricultural zones of Australia. *Biodiversity and Conservation* 13:253–74.

Friedmann, H. 1929. *The cowbirds: A study in the biology of parasitism.* Springfield, IL: Charles C. Thomas.

Fule, P. Z., A. E. Cocke, T. A. Heinlein, and W. W. Covington. 2004. Effects of an intense prescribed forest fire: Is it ecological restoration? *Restoration Ecology* 12:220–30.

Fule, P. Z., W. W. Covington, and M. M. Moore. 1997. Determining reference conditions for ecosystem management of southwestern ponderosa pine forests. *Ecological Applications* 7:895–908.

Fuller, M. R., J. J. Millspaugh, K. E. Church, and R. E. Kenward. 2005. Wildlife radiotelemetry. Pages 377–417 in C. E. Braun (ed.), *Techniques for wildlife investigation and management*. 6th ed. Bethesda, MD: Wildlife Society.

Garrett, K. L., M. G. Raphael, and R. D. Dixon. 1996. White-headed woodpecker (*Picoides albolarvatus*). Account 252 in A. Poole and F. Gill (eds.), *Birds of North America*. Academy of Natural Sciences, Philadelphia, PA, and American Ornithologists' Union, Washington, DC.

Gauss, C. F. 1934. *The struggle for existence*. Baltimore, MD: Williams and Wilkins.

Gavin, T. A. 1989. What's wrong with the questions we ask in wildlife research? *Wildlife Society Bulletin* 17:345–50.

Gawlik, D. E. 2006. The role of wildlife science in wetland ecosystem restoration: Lessons from the Everglades. *Ecological Engineering* 26:70–83.

Gibbons, D. W., and R. D. Gregory. 2006. Pages 308–50 in W. J. Sutherland (ed.), *Ecological census techniques*. 2nd ed. Cambridge: Cambridge University Press.

Gilpin, M. E. 1987. Minimum viable populations: A restoration perspective. Pages 301–5 in W. R. Jordan III, M. E. Gilpin, and J. D. Aber (eds.), *Restoration ecology: A synthetic approach to ecological research*. Cambridge: Cambridge University Press.

Gilpin, M. E., and M. E. Soulé. 1986. Minimum viable populations: Processes of species extinction. Pages 19–34 in M. E. Soulé (ed.), *Conservation biology: The science of scarcity and diversity*. Sunderland, MA: Sinauer Associates.

Gleason, H. A. 1917. The structure and development of the plant association. *Bulletin of the Torrey Botany Club* 44:463–81.

———. 1926. The individualistic concept of the plant association. *Bulletin of the Torrey Botany Club* 53:7–26.

Gogan, P. J. P., and J. F. Cochrane. 1994. Restoration of woodland caribou to the Lake Superior region. Pages 219–42 in M. L. Bowles and C. J. Whelan (eds.), *Restoration of endangered species: Conceptual issues, planning, and implementation*. Cambridge: Cambridge University Press.

Goguen, C. B., and N. E. Mathews. 1999. Review of the causes and implications of the association between cowbirds and livestock. *Studies in Avian Biology* 18:10–17.

Good, S. V. 1997. Population structure of *Dipodomys ingens* (Heteromyidae): The role of spatial heterogeneity in maintaining genetic diversity. *Evolution* 51:1296–1310.

Goodwin, H. T. 1995. Pliocene-Pleistocene biogeographic history of prairie dogs, *Cynomys* (Sciuridae). *Journal of Mammology* 76:100–122.

Goodyear, N. C., and J. Lazell. 1994. Status of a relocated population of endangered *Iguana pinguis* on Guana Island, British Virgin Islands. *Restoration Ecology* 2:43–50.

Gorman, L., and D. K. Rosenberg. 2000. Reassessment of temporal and spatial patterns of population size of San Joaquin kangaroo rats at NAS Lemoore. Unpublished report to U.S. Navy, Engineering Field Activities West, San Diego, CA.

Gotelli, N. J. 1999. How do communities come together? *Science* 286:1684–85.

Gotfryd, A., and R. I. C. Hansell. 1985. The impact of observer bias on multivariate analyses of vegetation structure. *Oikos* 45:223–34.

Govindarajulu, P., R. Altwegg, and B. R. Anholt. 2005. Matrix model investigation of invasive species control: Bullfrogs on Vancouver Island. *Ecological Applications* 15:2161–70.

Graber, J. W. 1961. Distribution, habitat requirements, and life history of the black-capped vireo (*Vireo atricapilla*). *Ecological Monographs* 31:313–36.

Green, G. A., H. L. Bombay, and M. L. Morrison. 2003. *Conservation Assessment of the Willow Flycatcher in the Sierra Nevada*. Region 5, Vallejo, CA: USDA Forest Service.

Green, R. H. 1979. *Sampling design and statistical methods for environmental biologists*. New York: Wiley.

Greenberg, C. H., D. G. Neary, and L. D. Harris. 1994. A comparison of herpetofaunal sampling effectiveness of pitfall, single-ended, and double-ended funnel traps used with drift fences. *Journal of Herpetology* 28:319–24.

Greig-Smith, P. 1983. *Quantitative plant ecology*. 3rd ed. Berkeley: University of California Press.

Griffin, A. S., D. T. Blumstein, and C. S. Evans. 2000. Training captive-bred or translocated animals to avoid predators. *Conservation Biology* 14:13–26.

Griffith, B. J., M. Scott, J. W. Carpenter, and C. Reed. 1989. Translocation as a species conservation tool: Status and strategy. *Science* 245:477–80.

Grinnell, J. 1917. The niche-relationships of the California thrasher. *Auk* 34:427–33.

Grover, M. C. 1998. Influence of cover and moisture on abundances of the terrestrial salamanders *Plethodon cinereus* and *Plethodon glutinosus*. *Journal of Herpetology* 32:489–97.

Grumbine, E. W. 1994. What is ecosystem management? *Conservation Biology* 8:27–38.

Grzybowski, J. A. 1995. Black-capped vireo (*Vireo atricapillus*). In A. Poole and F. Gill, (eds.), *Birds of North America*, No. 181. Academy of Natural Sciences, Philadelphia, and American Ornithologists' Union, Washington DC, USA.

Grzybowski, J. A., D. J. Tazik, and G. D. Schnell. 1994. Regional analysis of black-capped vireo breeding habitats. *Condor* 96:512–44.

Gutierrez, D. 1997. Importance of historical factors on species richness and composition of butterfly assemblages (Lepidoptera: Rhopalocera) in a northern Iberian mountain range. *Journal of Biogeography* 24:77–88.

Haddad, N. M., D. R. Bowne, A. Cunningham, B. J. Danielson, D. J. Levey, S. Sargent, and T. Spira. 2003. Corridor use by diverse taxa. *Ecology* 84:609–15.

Hafner, D. J. 1993. North American pika (*Ochotona princeps*) as a late quaternary biogeographic indicator species. *Quaternary Research* 39:373–80.

Hall, L. S., P. R. Krausman, and M. L. Morrison. 1997. The habitat concept and a plea for standard terminology. *Wildlife Society Bulletin* 25:173–82.

Halle, S., and M. Fattorini. 2004. Advances in restoration ecology: Insights from aquatic and terrestrial ecosystems. Pages 10–33 in V. M. Temperton, R. J. Hobbs, T. Nuttle, and S. Halle (eds.), *Assembly rules and restoration ecology: Bridging the gap between theory and practice*. Washington, DC: Island Press.

Halliday, T. 2006. Amphibians. Pages 278–96 in W. J. Sutherland (ed.), *Ecological census techniques*. 2nd ed. Cambridge: Cambridge University Press.

Hamilton, W. T. 2004. *Brush management past, present, future*. College Station, Texas: Texas A&M University Press.

Hanski, I. 1996. Metapopulation ecology. Pages 13–43 in O. E. Rhodes Jr., R. K. Chesser, and M. H. Smith (eds.), *Population dynamics in ecological space and time*. Chicago: University of Chicago Press.

Hansson, L. 1991. Dispersal and connectivity in metapopulations. *Biological Journal of the Linnean Society* 42:89–103.

Harris, A. H. 1993. Wisconsin and pre-pleniglacial biotic changes in southeastern New Mexico. *Quaternary Research* 40:127–33.

Harris, J. H., S. D. Sanders, and M. A. Fleet. 1987. Willow flycatcher surveys in the Sierra Nevada. *Western Birds* 18:27–36.

Hayes, M. P., and M. R. Jennings. 1986. Decline of ranid frog species in western North America: Are bullfrogs (*Rana catesbeiana*) responsible? *Journal of Herpetology* 20:490–509.

Hein, E. W. 1997. Improving translocation programs. *Conservation Biology* 11:1270–71.

Hejl, S. J., and J. Verner. 1990. Within-season and yearly variation in avian foraging locations. *Studies in Avian Biology* 13:202–9.

Herkert, J. R. 1994. The effects of habitat fragmentation on midwestern grassland bird communities. *Ecological Applications* 4:461–71.

Herzog, F., A. Lausch, E. Muller, H. Thulke, U. Steinhardt, and S. Lehmann. 2001. Landscape metrics for assessment of landscape destruction and rehabilitation. *Environmental Management* 27:91–107.

Hess, G. R. 1994. Conservation corridors and contagious disease: A cautionary note. *Conservation Biology* 8:256–62.

Heyer, W. R., M. A. Donnelly, R. W. McDiarmid, L. C. Hayek, and M. S. Foster, editors. 1994. *Measuring and monitoring biological diversity: Standard methods for amphibians.* Washington, DC: Smithsonian

Hilty, J. A., W. Z. Lidicker Jr., and A. M. Merenlender. 2006. *Corridor ecology: The science and practice of linking landscapes for biodiversity conservation.* Washington, DC: Island Press.

Hobbs, R. J., and D. A. Norton. 1996. Towards a conceptual framework for restoration ecology. *Restoration Ecology* 4:93–110.

——. 2004. Ecological filters, thresholds, and gradients in resistance to ecosystem reassembly. Pages 72–95 in V. M. Temperton, R. J. Hobbs, T. Nuttle, and S. Halle (eds.), *Assembly rules and restoration ecology: Bridging the gap between theory and practice.* Washington, DC: Island Press.

Holmes, R. T. 1981. Theoretical aspects of habitat use by birds. Pages 33–37 in D. E. Capen (ed.), *The use of multivariate statistics in studies of wildlife habitat.* Fort Collins, CO: USDA Forest Service General Technical Report RM-87.

Hunter, M. L. 1991. Coping with ignorance: The coarse-filter strategy for maintaining biodiversity. Pages 266–81 in K. A. Kohm (ed.), *Balancing on the brink of extinction.* Washington, DC: Island Press.

Hunter, P., N. A. Mahony, P. J. Ewins, D. Baird, and M. Field. 1997. Artificial nesting platforms for bald eagles in southern Ontario, Canada. *Journal of Raptor Research* 31:321–26.

Hurlbert, S. J. 1984. Pseodoreplication and the design of ecological field experiments. *Ecological Monographs* 54:187–211.

Huston, M. A. 2002. Introductory essay: Critical issues for improving predictions. Pages 7–21 in J. M. Scott, P. J. Heglund, M. L. Morrison, J. B. Haufler, M. G. Raphael, W. A. Wall, and F. B. Samson (eds.), *Predicting species occurrences: Issues of accuracy and scale.* Washington, DC: Island Press.

Hutchinson, G. E. 1957. Concluding remarks. *Cold Spring Harbor Symposium on Quantitative Biology* 22:415–27.

Hutchinson, K. B. 1995. New habitat plan hides old game. *Austin-American Statesman.* 16 January 1995. A11.

Hutto, R. L. 1985. Habitat selection by nonbreeding, migratory land birds. Pages 455–76 in M. L. Cody (ed.), *Habitat selection in birds.* Orlando: Academic Press.

——. 2006. Toward meaningful snag-management guidelines for postfire salvage logging in North American conifer forests. *Conservation Biology* 20:984–93.

Huxel, G. R., and A. Hastings. 1999. Habitat loss, fragmentation, and restoration. *Restoration Ecology* 7:309–15.

Ibarzabal, J., and A. Desrochers. 2001. Lack of relationship between forest edge proximity and nest predator activity in an eastern Canadian boreal forest. *Canadian Journal of Forest Research* 31:117–22.

Ice, G. G., D. G. Neary, and P. W. Adams. 2004. Effects of wildfire on soils and watershed processes. *Journal of Forestry* 102:16–20.

Immelmann, K., and C. Beer. 1989. *A dictionary of ethology*. Cambridge: Harvard University Press.

International Union for the Conservation of Nature (IUCN). 1998. *Guidelines for re-introductions*. Gland, Switzerland and Cambridge: IUCN/SSC Re-introduction Specialist Group.

Jaeger, R. G. 1994a. Patch sampling. Pages 107–9 in W. R. Heyer, M. A. Donnelly, R. W. McDiarmid, L. C. Hayek, and M. S. Foster (eds.), *Measuring and monitoring biological diversity: Standard methods for amphibians*. Washington, DC: Smithsonian.

——. 1994b. Transect sampling. Pages 103–7 in W. R. Heyer, M. A. Donnelly, R. W. McDiarmid, L. C. Hayek, and M. S. Foster (eds.), *Measuring and monitoring biological diversity: Standard methods for amphibians*. Washington, DC: Smithsonian.

Jaeger, R. G., and R. F. Inger. 1994. Quadrat sampling. Pages 97–102 in W. R. Heyer, M. A. Donnelly, R. W. McDiarmid, L. C. Hayek, and M. S. Foster (eds.), *Measuring and monitoring biological diversity: Standard methods for amphibians*. Washington, DC: Smithsonian.

James, F. C. 1971. Ordinations of habitat relationships among breeding birds. *Wilson Bulletin* 83:215–36.

James, F. C., and H. H. Shugart Jr. 1970. A quantitative method of habitat description. *Audubon Field Notes* 24:727–36.

Jennings, M. R. 1985. Pre-1900 overharvest of California red-legged frogs (*Rana aurora draytonii*). *Herpetologica* 41:94–103.

Johnson, D. H. 1980. The comparison of usage and availability measurements for evaluating resource preference. *Ecology* 61:65–71.

——. 1999. The insignificance of statistical significance testing. *Journal of Wildlife Management* 63:763–72.

Johnson, N. K. 1994. Pioneering and natural expansion of breeding distributions in western North America. *Studies in Avian Biology* 15:27–44.

Jones, C., W. J. McShea, M. J. Conroy, and T. H. Kunz. 1996. Capturing mammals. Pages 115–55 in D. E. Wilson, F. R. Cole, J. D. Nichols, R. Rudran, and M. S. Foster (eds.), *Measuring and monitoring biological diversity: Standard methods for mammals*. Washington, DC: Smithsonian.

Kalinowski, S. T., P. W. Hedrick, and P. S. Miller. 2000. Inbreeding depression in the Speke's gazelle captive breeding program. *Conservation Biology* 14:1375–84.

Kasworm, W. F., M. F. Proctor, C. Servheen, and D. Paetkau. 2007. Success of grizzly bear augmentation in northwest Montana. *Journal of Wildlife Management* 71:1261–66.

Kattlemann, R., and M. Embury. 1996. Riparian areas and wetlands. Pages 201–73 in D. C. Erman (ed.), *Sierra Nevada ecosystem project: Final report to Congress*. Vol. 3. *Assessments and scientific basis for management options*. U.S. Geological Survey Digital Data Series DDS-43, Davis, CA.

Kay, C. E. 1998. Are ecosystems structured from the top-down or bottom-up? A new look at an old debate. *Wildlife Society Bulletin* 26:484–98.

Keane, J. J., and M. L. Morrison. 1999. Temporal variation in resource use by black-throated gray warblers. *Condor* 101:67–75.

Keddy, P., and E. Weiher. 1999. Introduction: The scope and goals of research on assembly rules. Pages 1–20 in E. Weiher and P. Keddy (eds.), *Ecological assembly rules: Perspectives, advances, and retreats.* Cambridge: Cambridge University Press.

Kelly, S. T., and M. E. Decapita. 1982. Cowbird control and its effects on Kirtland's warbler reproductive success. *Wilson Bulletin* 94:363–65.

Kendeigh, S. C. 1944. Measurement of bird populations. *Ecological Monographs* 14:67–106.

Kenward, R. E. 2000. *A manual of wildlife radio tagging.* 2nd ed. San Diego: Academic Press.

Kessel, B., and D. D. Gibson. 1994. A century of avifaunal change in Alaska. *Studies in Avian Biology* 15:4–13.

Kiff, L. F., and D. J. Hough. 1985. *Inventory of bird egg collections of North America.* Norman: American Ornithologists' Union and Oklahoma Biological Survey.

Kissling, M. L., and E. O. Garton. 2006. Estimating detection probability and density from point-count surveys: A combination of distance and double-observer sampling. *Auk* 123:735–52.

Knapp, E. E., J. E. Keeley, E. A. Ballenger, and T. J. Brennan. 2005. Fuel reduction and coarse woody debris dynamics with early season and late season prescribed fire in a Sierra Nevada mixed conifer forest. *Forest Ecology and Management* 208:383–97.

Knapp, R. A., D. M. Boiano, and V. T. Vredenburg. 2007. Removal of nonnative fish results in population expansion of a declining amphibian (mountain yellow-legged frog, *Rana muscosa*). *Biological Conservation* 135:11–20.

Kostecke, R. M., S. G. Summers, G. H. Eckrich, and D. A. Cimprich. 2005. Effects of brown-headed cowbird (*Molothrus ater*) removal on black-capped vireo (*Vireo atricapilla*) nest success and population growth on Fort Hood, Texas. *Ornithological Monographs* 57:28–37.

Krebs, C. J. 2006. Mammals. Pages 351–69 in W. J. Sutherland (ed.), *Ecological census techniques.* 2nd ed. Cambridge: Cambridge University Press.

Krebs, C. J., S. Boutin, R. Boonstra, A. R. Sinclair, J. N. M. Smith, M. T. Dale, K. Martin, and R. Turkington. 1995. Impact of food and predation on the snowshoe hare cycle. *Science* 269:1112–15.

Kreisel, K. J., and S. J. Stein. 1999. Bird use of burned and unburned coniferous forests during winter. *Wilson Bulletin* 111:243–50.

Kreuter, U. P., H. E. Amestoy, M. M. Kothmann, D. N. Ueckert, W. A. McGinty, and S. R. Cummings. 2005. The use of brush management methods: A Texas landowner survey. *Rangeland Ecology and Management* 58:284–91.

Kristoferson, L. L., and M. L. Morrison. 2001. Integrating methods to determine breeding and nesting status of 3 western songbirds. *Wildlife Society Bulletin* 29:688–96.

Kroll, J. C. 1980. Habitat requirements of the golden-cheeked warbler: Management implications. *Journal of Range Management* 33:60–65.

Kuenzi, A. J., and M. L. Morrison. 2003. Temporal patterns of bat activity in southern Arizona. *Journal of Wildlife Management* 67:52–64.

Kuenzi, A. J., G. T. Downward, and M. L. Morrison. 1999. Bat distribution and hibernacula use in west-central Nevada. *Great Basin Naturalist* 59:213–20.

Kunz, T. H., editor. 1988. *Ecological and behavioral methods for the study of bats.* Washington, DC: Smithsonian.

Kunz, T. H., D. W. Thomas, G. C. Richards, C. R. Tidemann, E. D. Pierson, and P. A. Racey. 1996. Observational techniques for bats. Pages 105–14 in D. E. Wilson, F. R. Cole, J. D. Nichols, R. Rudran, and M. S. Foster (eds.), *Measuring and monitoring biological diversity: Standard methods for mammals.* Washington, DC: Smithsonian.

Kus, B. E. 1999. Impacts of brown-headed cowbird parasitism on productivity of the endangered Bell's vireo. *Studies in Avian Biology* 18:160–66.

Lacava, J., and J. Hughes. 1984. Determining minimum viable population levels. *Wildlife Society Bulletin* 12:370–76.

Lacy, R. C. 1994. Managing genetic diversity in captive populations of animals. Pages 63–89 in M. L. Bowles and C. J. Whelan (eds.), *Restoration of endangered species: Conceptual issues, planning, and implementation.* Cambridge: Cambridge University Press.

Lacy, R. C., J. D. Ballou, F. Princee, A. Starfield, and E. A. Thompson. 1995. Pedigree analysis for population management. Pages 57–75 in J. D. Ballou, M. Gilpin, and T. J. Foose (eds.), *Population management for survival and recovery: Analytical methods and strategies in small population conservation.* New York: Columbia University Press.

Ladd, C., and L. Gass. 1999. Golden-cheeked warbler (*Dendroica chrysoparia*). Account 420 in A. Poole, and F. Gill (eds.), *Birds of North America.* Academy of Natural Sciences, Philadelphia, PA, and American Ornithologists' Union, Washington, DC, USA.

Lahti, D. C. 2001. The "edge effect on nest predation" hypothesis after twenty years. *Biological Conservation* 99:365–74.

Lambeck, R. J. 1997. Focal species: A multi-species umbrella for nature conservation. *Conservation Biology* 11:849–56.

———. 2002. Focal species and restoration ecology: Response to Lindenmayer et al. *Conservation Biology* 16:549–51.

Lancia, R. A., and J. W. Bishir. 1996. Removal methods. Pages 210–17 in D. E. Wilson, F. R. Cole, J. D. Nichols, R. Rudran, and M. S. Foster (eds.), *Measuring and monitoring biological diversity: Standard methods for mammals.* Washington, DC: Smithsonian.

Lancia, R. A., C. E. Braun, M. W. Collopy, R. D. Dueser, J. G. Kie, C. J. Martinka, J. D. Nichols, T. D. Nudds, W. R. Porath, and N. G. Tilghman. 1996. ARM! For the future: Adaptive resource management in the wildlife profession. *Wildlife Society Bulletin* 24:436–42.

Lande, R. 1987. Extinction thresholds in demographic models of territorial populations. *American Naturalist* 30(4): 624–35.

Lande, R., and G. F. Barrowclough. 1987. Effective population size, genetic variation, and their use in population management. Pages 87–123 in M. E. Soulé (ed.), *Viable populations.* New York: Cambridge University Press.

Landres, P. B., J. Verner, and J. W. Thomas. 1988. Ecological uses of vertebrate indicator species: A critique. *Conservation Biology* 2:316–28.

Langpap, C. 2004. Conservation incentives programs for endangered species: An analysis of landowner participation. *Land economics* 80:375–81.

———. 2006. Conservation of endangered species: Can incentives work for private landowners. *Ecological Economics* 57:558-572–72.

LaPolla, V. N., and G. W. Barrett. 1991. Effects of corridor width and presence on the population dynamics of the meadow vole (*Microtus pennsylvanicus*). *Landscape Ecology* 8:25–37.

Laymon, S. A. 1987. Brown-headed cowbirds in California: Historical perspectives and management opportunities in riparian habitats. *Western Birds* 18:63–70.

Laymon, S. A., and R. H. Barrett. 1986. Developing and testing habitat-capability models: Pitfalls and recommendations. Pages 87–91 in J. Verner, M. L. Morrison, and C. J. Ralph (eds.), *Wildlife 2000: Modeling habitat relationships of terrestrial vertebrates*. Madison: University of Wisconsin Press.

Leopold, A. 1933. *Game management*. New York: Scribners.

Levey, D. J., B. M. Bolker, J. J. Tewksbury, S. Sargent, and N. M. Haddad. 2005. Effects of landscape corridors on seed dispersal by birds. *Science* 309:146–48.

Levins, R. 1969. Some demographic and genetic consequences of environmental heterogeneity for biological control. *Bulletin of the Entomological Society America* 15:237–40.

———. 1970. Extinction. *Lectures on Mathematics in the Life Sciences* 2:75–107.

Li, W. H. 1978. Maintenance of genetic variability under the joint effect under selection, mutation and random drift. *Genetics* 90:349–82.

Lindenmayer, D. B., and J. F. Franklin. 2002. *Conserving forest biodiversity: A comprehensive multiscaled approach*. Washington, DC: Island Press.

Lindenmayer, D.B., and R.J. Hobbs, editors. 2007. *Managing and designing landscapes for conservation: Moving from perspectives to principles*. Washington, DC: Island Press.

Lindenmayer, D. B., A. D. Manning, P. L. Smith, H. P. Possingham, J. Fischer, I. Oliver, and M. A. McCarthy. 2002. The focal-species approach and landscape restoration: A critique. *Conservation Biology* 16:338–45.

Lindstrom, S., P. Rucks, and P. Wigand. 2000. A contextual overview of human use and environmental conditions. Pages 23–127 in *Lake Tahoe watershed assessment*. Vol. 1. USDA Forest Service General Technical Report PSW-GTR-175, Berkeley, CA.

Litvaitis, J. A., K. Titus, and E. M. Anderson. 1994. Measuring vertebrate use of terrestrial habitats and foods. Pages 254–74 in T. A. Bookhout (ed.), *Research and management techniques for wildlife and habitats*. 5th ed. Bethesda, MD: Wildlife Society.

Lockwood, J. L., and S. L. Pimm. 1999. When does restoration succeed? Pages 363–92 in E. Weiher and P. Keddy (eds.), *Ecological assembly: Advances, perspectives, and retreats*. Cambridge: Cambridge University Press.

Lowther, P. E. 1993. Brown-headed cowbird (*Molothrus ater*). Account 47 in A. Poole and F. Gill (eds.), *Birds of North America*. Academy of Natural Sciences, Philadelphia, PA, and American Ornithologists' Union, Washington, DC, USA.

Mace, G. M., J. M. Pemberton, and H. F. Stanley. 1992. Conserving genetic diversity with the help of biotechnology: Desert antelope as an example. Pages 123–34 in H. D. M. Moore, W. V. Holt, and G. M. Mace (eds.), *Biotechnology and the conservation of genetic diversity*. Symposia of the Zoological Society of London, No. 64. Oxford: Oxford University Press.

MacKenzie, D. I., J. D. Nichols, J. A. Royle, K. H. Pollock, L. L. Bailey, and J. E. Hines. 2006. *Occupancy estimation and modeling*. San Diego: Academic Press.

Magness, D. R, R. N. Wilkins, and S. J. Hejl. 2006. Quantitative relationships among golden-cheeked warbler occurrence and landscape size, composition, and structure. *Wildlife Society Bulletin* 34:473–79.

Manley, P. N., J. A. Fites-Kaufmann, M. E. Barbour, M. D. Schlesinger, and D. M. Rizzo. 2000. Biological integrity. Pages 403–598 in D. D. Murphy and C. Knopp (eds.), *Lake Tahoe watershed assessment*. USDA Forest Service General Technical Report PSW-GTR-175, Berkeley, CA.

Marcot, B. G., R. E. Gullison, and J. R. Barborak. 2001. Protecting habitat elements and natural areas in the managed forest matrix. Pages 523–58 in R. A. Fimbel, A. Grajal, and J. G. Robinson (eds.), *The cutting edge: Conserving wildlife in logged tropical forests.* New York: Columbia University Press.

Margules, C. R., and M. B. Usher. 1981. Criteria used in assessing wildlife conservation potential: A review. *Biological Conservation* 21:79–109.

Margules, C. R., A. O. Nicholls, and R. L. Pressey. 1988. Selecting networks of reserves to maximize biological diversity. *Biological Conservation* 43:63–76.

Marion, W. R., P. A. Quincy, C. G. Cutlip Jr., and J. R. Wilcox. 1992. Bald eagles use artificial nest platform in Florida. *Journal of Raptor Research* 26:266.

Marsh, D. M., and P. C. Trenham. 2001. Metapopulation dynamics and amphibian conservation. *Conservation Biology* 15:40–49.

Martin, P., and P. Bateson. 1993. *Measuring behavior.* 2nd ed. Cambridge: Cambridge University Press.

Martin, T. E. 1993. Nest predation among vegetation layers and habitats: Revising the dogmas. *American Naturalist* 141:897–913.

Martin, T. E., and G. R. Geupel.1993. Nest-monitoring plots: Methods for locating nests and monitoring success. *Journal of Field Ornithology* 64:507–19.

Marzluff, J. M., J. J. Millspaugh, and M. S. Handcock. 2004. Relating resources to a probabilistic measure of space use: Forest fragments and Steller's jays. *Ecology* 85:1411–27.

Maschinski, J. 2006. Implications of population dynamics and metapopulation theory for restoration. Pages 59–87 in D. A. Falk, M. A. Palmer, and J. B. Zedler (eds.), *Foundations of restoration ecology.* Washington, DC: Island Press.

Mathewson, H. A., H. L. Loffland, and M. L. Morrison. 2006a. 2005 Annual report and preliminary demographic analysis for willow flycatcher monitoring in the central Sierra Nevada. Unpubl. report, USDA Forest Service, Region 5, Vallejo, CA.

———. 2006b. 2006 Annual report and preliminary demographic analysis for willow flycatcher monitoring in the central Sierra Nevada. Unpubl. report, USDA Forest Service, Region 5, Vallejo, CA.

Matthies, D., I. Brauer, W. Maiborn, and T. Tscharntke. 2004. Population size and the risk of local extinction: Empirical evidence from rare plants. *Oikos* 105:481–88.

Mayfield, H. F. 1977. Brown-headed cowbird: Agent of extinction? *American Birds* 31:107–13.

McCabe, R. A. 1991. *The little green bird: Ecology of the willow flycatcher.* Madison, WI: Rusty Rock Press.

McDowell, C., R. Sparks, J. Grindley, and E. Moll. 1989. Persuading the landowner to conserve natural ecosystems through effective communication. *Journal of Environmental Management* 28:211–25.

Mech, S. G., and J. G. Hallett. 2001. Evaluating the effectiveness of corridors: A genetic approach. *Conservation Biology* 15:467–74.

Meffe, G. K., and C. R. Carroll. 1997. *Principles of conservation biology.* 2nd ed. Sunderland, MA: Sinauer Associates.

Merriam, G., M. Kozakiewicz, E. Tsuchiya, and K. Hawley. 1989. Barriers as boundaries for metapopulations and demes of *Peromyscus leucopus* in farm landscapes. *Landscape Ecology* 2:227–35.

Miller, J. R. 2007. Habitat and landscape design: Concepts, constraints and opportunities. Pages 81–95 in D. B. Lindenmayer and R.J. Hobbs (eds), *Managing and designing landscapes for conservation: Moving from perspectives to principles.* Washington, DC: Island Press.

Mills, L. S. 2007. *Conservation of wildlife populations: Demography, genetics, and management.* Oxford: Blackwell Publishing.

Millspaugh, J. J., and J. M. Marzluff, editors. 2001. *Radio tracking and animal populations.* San Diego: Academic Press.

Mitsch, W. J., and J. G. Gosselink. 1993. *Wetlands.* 2nd ed. New York: Van Nostrand Reinhold.

Monroe, M. E. and S. J. Converse. 2006. The effects of early season and late season prescribed fires on small mammals in a Sierra Nevada mixed conifer forest. *Forest Ecology and Management* 236:229–40.

Monty, A. M., E. J. Heist, E. R. Wagle, R. E. Emerson, E. H. Nicholson, and G. A. Feldhamer. 2003. Genetic variation and population assessment of eastern woodrats in southern Illinois. *Southeastern Naturalist* 2:243–60.

Moore, H. D. M., W. V. Holt, and G. M. Mace, editors. 1992. *Biotechnology and the conservation of genetic diversity.* Symposia of the Zoological Society of London, No. 64. Oxford: Oxford University Press.

Morgan, P., G. H. Aplet, J. B. Haufler, H. C. Humphries, M. M. Moore, and W. D. Wilson. 1994. Historical range of variability: A useful tool for evaluating ecosystem change. Pages 87–111 in R. N. Sampson, D. L. Adams, and M. J. Enzer (eds.), *Assessing forest ecosystem health in the inland west.* New York: Haworth Press.

Morrison, M. L. 1984a. Influence of sample size and sampling design on analysis of avian foraging behavior. *Condor* 86:146 50.

———. 1984b. Influence of sample size and sampling design on discriminant function analysis of habitat use by birds. *Journal of Field Ornithology* 55:330–35.

———. 1986. Birds as indicators of environmental change. *Current Ornithology* 3:429–51.

———. 1997. Experimental design for plant removal and restoration. Pages 104–16 in J. O. Lucken and J. W. Thieret (eds.), *Assessment and management of plant invasions.* New York: Springer-Verlag.

———. 2001. Techniques for discovering historic animal assemblages. Pages 295–316 in D. Egan and E. A. Howell (eds.), *The historical ecology handbook.* Washington, DC: Island Press.

———. 2002. *Wildlife restoration: Techniques for habitat analysis and animal monitoring.* Washington, DC: Island Press.

Morrison, M. L., and A. Averill-Murray. 2002. Evaluating the efficacy of manipulating cowbird parasitism on host nesting success. *Southwestern Naturalist* 47:236–43.

Morrison, M. L., W. M. Block, D. R. Strickland, B. A. Collier, and M. J. Peterson. 2008. *Wildlife study design.* 2nd edition. New York: Springer-Verlag.

Morrison, M. L., W. M. Block, M. D. Strickland, and W. L. Kendall. 2001. *Wildlife Study design.* New York: Springer-Verlag.

Morrison, M. L., W. M. Block, and J. Verner. 1991. Wildlife-habitat relationships in California's oak woodlands: Where do we go from here? Pages 105–9 in *Proceedings of the symposium on California's oak woodlands and hardwood rangeland.* Washington, DC: USDA Forest Service General Technical Report PSW-126.

Morrison, M. L., and L. S. Hall. 1998. Responses of mice to fluctuating habitat quality. 1: Patterns from a long-term observational study. *Southwestern Naturalist* 43:123–36.

———. 1999. Habitat relationships of amphibians and reptiles in the Inyo-White mountains, California and Nevada. Pages 233–37 in S. B. Monsen and R. Stevens (eds.), *Proceedings: Ecology and management of pinyon-juniper communities within the interior west.* Proceedings RMRSP-9. Ogden, UT: USDA Forest Service, Rocky Mountain Research Station.

——. 2002. Standard terminology: Toward a common language to advance ecological understanding and applications. Pages 43–52 in J. M. Scott et al. (eds.), *Predicting species occurrences: Issues of scale and accuracy.* Washington, DC: Island Press.

Morrison, M. L., L. S. Hall, S. K. Robinson, S. I. Rothstein, D. C. Hahn, and T. D. Rich (eds.). 1999. Research and management of the brown-headed cowbird in western landscapes. *Studies in Avian Biology* 18:204–17.

Morrison, M. L., R. C. Heald, and D. L. Dahlsten. 1990. Can incense-cedar be managed for birds? *Western Journal of Applied Forestry* 5:28–30.

Morrison, M. L., C. M. Kuehler, T. A. Scott, A. A. Lieberman, W. T. Everett, R. B. Phillips, C. E. Koehler, P. A. Aigner, C. Winchell, and T. Burr. 1995. San Clemente loggerhead shrike: Recovery plan for an endangered species. *Proceedings of the Western Foundation of Vertebrate Zoology* 6:293–95.

Morrison, M. L., and B. G. Marcot. 1995. An evaluation of resource inventory and monitoring programs used in national forest planning. *Environmental Management* 19:147–56.

Morrison, M. L., B. G. Marcot, and R. W. Mannan. 1992. *Wildlife-habitat relationships: Concepts and applications.* Madison: University of Wisconsin Press.

——. 1998. *Wildlife-habitat relationships: Concepts and applications.* 2nd ed. Madison: University of Wisconsin Press.

——. 2006. *Wildlife-habitat relationships: Concepts and applications.* 3rd edition. Washington, DC: Island Press.

Morrison, M. L., and E. C. Meslow. 1984. Response of avian communities to herbicide-induced vegetation changes. *Journal of Wildlife Management* 48:14–22.

Morrison, M. L., T. A. Scott, and T. Tennant. 1994a. Wildlife-habitat restoration in an urban park in southern California. *Restoration Ecology* 2:17–30.

——. 1994b. Laying the foundation for a comprehensive program of restoration for wildlife habitat in a riparian floodplain. *Environmental Management* 18:939–55.

Mueller-Dombois, D., and H. Ellenberg. 1974. *Aims and methods of vegetation ecology.* New York: Wiley.

Murphy, D. D., and C. M. Knopp, technical editors. 2000. *Lake Tahoe Watershed Assessment.* Vol.1. USDA Forest Service General Technical Report PSW-GTR-175, Berkeley, CA.

Nei, M., T. Maruyama, and R. Chakraborty. 1975. The bottleneck effect and genetic variability in populations. *Evolution* 29:1–10.

Newman, D., and D.A. Tallmon. 2001. Experimental evidence for beneficial fitness effects of gene flow in recently isolated populations. *Conservation Biology* 15:1054–63.

Newmark, W. D. 1986. Mammalian richness, colonization, and extinction in western North American national parks. PhD dissertation, University of Michigan, Ann Arbor.

——. 1995. Extinction of mammal populations in western North American national parks. *Conservation Biology* 9:512–26.

Nice, M. M. 1957. Nesting success in altricial birds. *Auk* 74:305–21.

Nichols, J. D., and M. J. Conroy. 1996. Techniques for estimating abundance of species richness. Pages 177–92 in D. E. Wilson, F. R. Cole, J. D. Nichols, R. Rudran, and M. S. Foster (eds.), *Measuring and monitoring biological diversity: Standard methods for mammals.* Washington, DC: Smithsonian.

Niemi, G. J., J. M. Hanowski, A. R. Lima, T. Nicholls, and N. Weiland. 1997. A critical analysis on the use of indicator species in management. *Journal of Wildlife Management* 61:1240–52.

Noon, B. R. 1981. Techniques for sampling avian habitats. Pages 42–52 in D. E. Capen (ed.), *The use of multivariate statistics in studies of wildlife habitat*. Fort Collins, CO: USDA Forest Service General Technical Report RM-87.

Noss, R. F. 1985. On characterizing presettlement vegetation: How and why. *Natural Areas Journal* 5:5–19.

Noss, R. F., M. A. O'Connell, and D. M. Murphy. 1997. *The science of conservation Planning: Habitat conservation under the Endangered Species Act*. Washington, DC: Island Press.

Nuttle, T., R. J. Hobbs, V. M. Temperton, and S. Halle. 2004. Assembly rules and ecosystem restoration: Where to from here? Pages 410–22 in V. M. Temperton, R. J. Hobbs, T. Nuttle, and S. Halle (eds.), *Assembly rules and restoration ecology: Bridging the gap between theory and practice*. Washington, DC: Island Press.

O'Connor, R. J. 2002. The conceptual basis of species distribution modeling: Time for a paradigm shift? Pages 25–33 in J. M. Scott, P. J. Heglund, M. L. Morrison, J. B. Haufler, M. G. Raphael, W. A. Wall, and F. B. Samson (eds.), *Predicting species occurrences: Issues of accuracy and scale*. Washington, DC: Island Press.

Odum, E. 1971. *Fundamentals of ecology*. 3rd ed. Philadelphia: Saunders.

O'Farrell, M. J., B. W. Miller, and W. L. Gannon. 1999. Qualitative identification of free-flying bats using the Anabat detector. *Journal of Mammalogy* 80.11–23.

Ollice of Technology and Assessment (OTA). 1993. *Harmful non-indigenous species in the United States*. OTA-F-565. 2 vols. Washington, DC: U.S. Congress, Office of Technology Assessment.

Olson, S. L. 1977. Additional notes on subfossil bird remains from Ascension Island. *Ibis* 119:37–43.

Orchard, S. A. 2005. Bullfrog control web site. <http://www.bullfrogcontrol.com>. Accessed 20 March 2007.

Orr, R. T. 1949. *Mammals of Lake Tahoe*. San Francisco: California Academy of Sciences.

Orr, R. T., and J. Moffitt. 1971. *Birds of the Lake Tahoe region*. San Francisco: California Academy of Sciences.

Padoa-Schioppa, E., M. Baietto, R. Massa, and L. Bottoni. 2006. Bird communities as bioindicators: The focal species concept in agricultural landscapes. *Ecological Indicators* 6:83–93.

Palmer, M. A., D. A. Falk, and J. B. Zedler. 2006. Ecological theory and restoration ecology. Pages 1–10 in D. A. Falk, M. A. Palmer, and J. B. Zedler (eds.), *Foundations of restoration ecology*. Washington, DC: Island Press.

Parker, K. R., and J. A. Wiens. 2005. Assessing recovery following environmental accidents: Environmental variation, ecological assumptions, and strategies. *Ecological Applications* 15:2037–51.

Patoski, J. N. 1997. The war on cedar. *Texas Monthly*. December:114–22.

Patton, D. R. 1992. *Wildlife habitat relationships in forested ecosystems*. Portland, OR: Timber Press.

Peters, R. H. 1991. *A critique for ecology*. Cambridge: Cambridge University Press.

Pilliod, D. S., R. B. Bury, E. J. Hyde, C. A. Pearl, and P. S. Corn. 2003. Fire and amphibians in North America. *Forest Ecology and Management* 178:163–81.

Powell, R. A. 2000. Animal home ranges and territories and home range estimators. Pages 65–110 in L. Boitani and T. K. Fuller (eds.), *Research techniques in animal ecology: Controversies and consequences*. 2nd ed. New York: Columbia University Press.

Power, D. M. 1994. Avifaunal change on California's coastal islands. *Studies in Avian Biology* 15:75–90.

Pulich, W. M. 1976. The golden-cheeked warbler: A bioecological study. Austin, TX: Texas Parks and Wildlife Department.

Pulliam, H. R. 1988. Sources, sinks, and population regulation. American Naturalist 132:652–61.

Ralph, C. J., G. R. Geupel, P. Pyle, T. E. Martin, and D. F. DeSante. 1993. Handbook for field methods for monitoring landbirds. General Technical Report PSW-144. Washington, DC: USDA Forest Service.

Ralph, C. J., J. R. Sauer, and S. Droege. 1995. Monitoring bird populations by point counts. General Technical Report PSW-GTR-149. Berkely: USDA Forest Service, Pacific Southwest Research Station.

Ralph, C. J., and J. M. Scott. 1981. Estimating numbers of terrestrial birds. Studies in Avian Biology 6:1–630.

Ramey III, R. R., G. Luikart, and F. J. Singer. 2000. Genetic bottlenecks results from restoration efforts: The case of bighorn sheep in Badlands National Park. Restoration Ecology 8:85–90.

Ratliff, R. D. 1982. A meadow site classification for the Sierra Nevada, California. USDA Forest Service, Pacific Southwest Forest and Range Experimental Station, Albany, CA: USDA Forest Service Gen. Tech. Report. GTR-PSW-60.

———. 1985. Meadows in the Sierra Nevada of California: State of knowledge. USDA Forest Service, Pacific Southwest Forest and Range Experimental Station, Albany, CA: USDA Forest Service Gen. Tech. Report. GTR-PSW-84.

Reinert, H. K. 1984. Habitat separation between sympatric snake populations. Ecology 65:478–86.

Reynolds, R. P., R. I. Crombie, R. W. McDiarmid, and T. L. Yates. 1996. Voucher specimens. Pages 63–69 in D. E. Wilson, F. R. Cole, J. D. Nichols, R. Rudran, and M. S. Foster (eds.), Measuring and monitoring biological diversity: Standard methods for mammals. Washington, DC: Smithsonian.

Reynolds, R. T., J. M. Scott, and R. A. Nussbaum. 1980. A variable circular-plot method for estimating bird numbers. Condor 82:309–13.

Ricklefs, R. E. 1969. An analysis of nesting mortality in birds. Smithsonian Contributions to Zoology 9:1–48.

Robbins, C. S., D. Bystrak, and P. H. Geissler. 1986. Breeding bird survey: Its first fifteen years, 1965–1979. Research Publication 157. Washington, DC: U.S. Fish and Wildlife Service.

Roberson, E. 1996. Impacts of livestock grazing on soils and recommendations for management. San Diego, CA: California Native Plant Society.

Roberts, C. W., B. L. Pierce, A. W. Braden, R. R. Lopez, N. J. Silvy, D. Ransom Jr. 2006. Comparison of camera and road survey estimates for white-tailed deer. Journal of Wildlife Management 70:263–67.

Roberts, L. 1988. Hard choices ahead on biodiversity. Science 241:1759–61.

Robinson, S. K., J. A. Gryzbowski, S. I. Rothstein, M. C. Brittingham, L. J. Petit, and F. R. Thompson III. 1993. Management implications of cowbird parasitization on neotropical migrant songbirds. Pages 93–102 in D. M. Finch and P. W. Stangel (eds.), Status and management of neotropical migratory birds. USDA Forest Service, Rocky Mountain Forest and Range Experiment Station, Fort Collins, CO: USDA Forest Service Gen. Tech. Rep. RM-229.

Robinson, S. K., S. I. Rothstein, M. C. Brittingham, L. J. Petit, and J. A. Grzybowski. 1995. Ecology and behavior of cowbirds and their impact on host populations. Pages 428–60 in T. E. Martin and D. M. Finch (eds.), Ecology and management of neotropical mi-

gratory birds: A synthesis and review of critical issues. New York: Oxford University Press.

Romesburg, H. C. 1981. Wildlife science: Gaining reliable knowledge. *Journal of Wildlife Management* 45:293–313.

Rood, S. B., C. R. Gourley, E. M. Ammon, L. G. Heki, J. R. Klotz, M. L. Morrison, D. Mosley, G. G. Scoppettone, S. Swanson, and P. L. Wagner. 2003. Flows for floodplain forests: A successful riparian restoration. *BioScience* 53:647–56.

Roques, D. and T. Gaman. 2002. A report on the economics of forest restoration in the Sierra Nevada. East-West Forestry Associates. Inverness, CA. http://www.forestdata .com/sierrall.htm. Accessed 26 Sept 2007.

Roth, R. R. 1976. Spatial heterogeneity and bird species diversity. *Ecology* 57:773–82.

Rothstein, S. I., J. Verner, and E, Stevens. 1984. Radio-tracking confirms a unique diurnal pattern of spatial occurrence in the parasitic brown-headed cowbird. *Ecology* 65:77–88.

Rudran, R., T. H. Kunz, C. Southwell, P. Jarman, and A. P. Smith. 1996. Observational techniques for nonvolant mammals. Pages 81–104 in D. E. Wilson, F. R. Cole, J. D. Nichols, R. Rudran, and M. S. Foster (eds.), *Measuring and monitoring biological diversity: Standard methods for mammals.* Washington, DC: Smithsonian.

Ruggiero, L. F., R. S. Holthausen, B. G. Marcot, K. B. Aubry, J. W. Thomas, and E. C. Meslow. 1988. Ecological dependency: The concept and its implication for research and management. *North American Wildlife and Natural Resources Conference* 53:115–26.

Ryan, T. J., T. Philippi, Y. A. Leiden, M. E. Dorcas, T. B. Wigley, and J. W. Gibbons. 2001. Monitoring herpetofauna in a managed forest landscape: Effects of habitat types and census techniques. *Forest Ecology and Management* 167:83–90.

Saenz, D., K. A. Baum, R. N. Conner, D. C. Rudolph, and R. Costa. 2002. Large-scale translocation strategies for reintroducing red-cockaded woodpeckers. *Journal of Wildlife Management* 66:212–21.

Sanders, J. C. 2005. Relationships among landowner and land ownership characteristics and participation in conservation programs in central Texas. MS thesis, Texas A&M University, College Station.

Sanders, S. D. and M. A. Flett. 1989. *Ecology of a Sierra Nevada population of willow flycatchers (Empidonax traillii), 1986–1987.* Sacramento, CA: California Department of Fish and Game, Wildlife Management Division, Nongame Bird and Mammal Section.

Schelhaas, M. J., G. J. Nabuurs, M. Sonntag, and A. Pussinen. 2002. Adding natural disturbances to a large-scale forest scenario model and a case study for Switzerland. *Forest Ecology and Management* 167(1–3): 13–26.

Schooley, R. L. 1994. Annual variation in habitat selection: Patterns concealed by pooled data. *Journal of Wildlife Management* 58:367–74.

Schreuder, H. T., T. G. Gregoire, and G. B. Wood. 1993. *Sampling methods for multiresource forest inventory.* New York: Wiley.

Scott, J. M., F. Davis, B. Csuti, R. Noss, B. Butterfield, C. Grives, H. Anderson, et al. 1993. Gap analysis: A geographical approach to protection of biodiversity. *Wildlife Monographs* 123:1–41.

Scott, J. M., P. J. Heglund, M. L. Morrison, J. B. Haufler, M. G. Raphael, W. A. Wall, and F. B. Samson. 2002. Introduction. Pages 1–5 in J. M. Scott, P. J. Heglund, M. L. Morrison, J. B. Haufler, M. G. Raphael, W. A. Wall, and F. B. Samson (eds.), *Predicting species occurrences: Issues of accuracy and scale.* Washington, DC: Island Press.

Scott Jr. N. J. 1994. Complete species inventories. Pages 78–84 in W. R. Heyer, M. A. Donnelly, R. W. McDiarmid, L. C. Hayek, and M. S. Foster (eds.), *Measuring and monitoring biological diversity: Standard methods for amphibians.* Washington, DC: Smithsonian. *Journal of Wildlife Management* 60:363–68.

Serena, M. 1982. *Status and distribution of the willow flycatcher (Empidonax traillii) in selected portions of the Sierra Nevada, 1982.* California Department of Fish and Game, Wildlife Management Division, Sacramento, CA: Administrative Report 82-5.

Sergio, F., and I. Newton. 2003. Occupancy as a measure of territory quality. *Journal of Animal Ecology* 72:857–65.

Sheremet'ev, I. S. 2004. Stability of isolated populations of terrestrial mammals inhabiting the islands of the Peter the Great Bay, Sea of Japan. *Russian Journal of Ecology* 35:171–75.

Sherry, A. L., D. M. Engle, D. M. Leslie Jr., J. S. Fehmi, J. Kretzer. 2003. Comparison of vegetation sampling procedures in a disturbed mixed-grass prairie. *Proceedings of Oklahoma Academy of Science* 83:7–15.

Shump, K. A. Jr., and A. U. Shump. 1982. *Lasiurus borealis: Mammalian Species* 183:1–6.

Simberloff, D. 2007. Individual species management: Threatened taxa and invasive species. Pages 295–310 in D. B. Lindenmayer and R. J. Hobbs (eds), *Managing and designing landscapes for conservation: Moving from perspectives to principles.* Washington, DC: Island Press.

Simberloff, D., and J. Cox. 1987. Consequences and costs of conservation corridors. *Conservation Biology* 1:63–71.

Simberloff, D. S., J. A. Farr, J. Cox, and D. W. Mehlman. 1992. Movement corridors: conservation bargains or poor investments. *Conservation Biology* 6:493–504

Simberloff, D., L. Stone, and T. Dayan. 1999. Ruling out a community assembly rule: The method of favored states. Pages 58–74 in E. Weiher and P. Keddy (eds.), *Ecological assembly rules: Perspectives, advances, and retreats.* Cambridge: Cambridge University Press.

Simon, T. L., 1980. An ecological study of the marten in Tahoe National Forest, California. MS thesis, California State University, Sacramento.

Singer, F. J., C. M. Papouchis, and K. K. Symonds. 2000. Translocations as a tool for restoring populations of bighorn sheep. *Restoration Ecology* 8:6–13.

Skalski, J. R., and D. S. Robson. 1992. *Techniques for wildlife investigations: Design and analysis of capture data.* San Diego: Academic Press.

Smallwood, K. S. 2001. Linking habitat restoration to meaningful units of animal demography. *Restoration Ecology* 9:253–61.

Smallwood, K. S., and M. L. Morrison. 1999. Estimating burrow volume and excavation rate of pocket gophers (Geomyidae). *Southwestern Naturalist* 44:173–83.

———. 2006. San Joaquin kangaroo rat (*Dipodomys n. nitratoides*): Conservation research in resource management Area 5, Naval Air Station, Lemoore. Unpubl. report, U.S. Navy Engineering Command, San Diego, CA.

Smeins, F., S. Fuhlendofr, and C. Taylor Jr. 1997. Environmental and land use changes: A long term perspective. In C. A. Taylor (ed.), *Juniper symposium proceedings.* San Angelo, TX: Texas Agriculture Experiment Station Technical Report.

Smith, B. D. 2007. The ultimate ecosystem engineers. *Science* 315:1797–98.

Smith, R. J. 1999. Repeal the endangered species act. *Human Events.* 12 March 1999.

Smucker, K. M., R. L. Hutto, and B. M. Steele. 2005. Changes in bird abundance after wildfire: Importance of fire severity and time since fire. *Ecological Applications* 15:1535–49.

SonoBat. 2007. Products and set-up of bat monitoring via the use of bat detectors, recording devices, and sonogram analysis. <http://www.sonobat.com>. Accessed 15 May 2007.

Soulé, M. E. 1980. Thresholds for survival: Maintaining fitness and evolutionary potential. Pages 151–70 in M. E. Soulé and B. A. Wilcox (eds.), *Conservation biology: An evolutionary-ecological perspective*. Sunderland, MA: Sinauer.

———. 1990. The onslaught of alien species, and other challenges in the coming decades. *Conservation Biology* 4:233–39.

Stafford, M. D. and B. E. Valentine. 1985. A preliminary report on the biology of the willow flycatcher in the central Sierra Nevada. CAL-NEVA Wildlife Transactions 1985:66–77.

Stake, M. M., and D. A. Cimprich. 2003. Using video to monitor predation at black-capped vireo nests. *Condor* 105:348–57.

Stake, M. M., J. Faaborg, and F. R. Thompson III. 2004. Video identification of predators at golden-cheeked warbler nests. *Journal of Field Ornithology* 75:337–44.

Stebbins, R. C. 2003. *Field guide to western reptiles and amphibians*. 3rd ed. New York: Houghton and Mifflin.

Stefani, R. A., H. L. Bombay, and T. M. Benson. 2001. Willow flycatcher. Pages 143–95 in USDA Forest Service, Sierra Nevada Forest Plan Amendment Final Environmental Impact Statement, vol. 3, part 4.4. Sacramento, CA: USDA Forest Service, Pacific Southwest and Intermountain Regions.

Steidl, R. J., and R. G. Anthony. 2000. Experimental effects of human activity on breeding bald eagles. *Ecological Applications* 10:258–68.

Steidl, R. J., and L. Thomas. 2001. Power analysis and experimental design. Pages 14–36 in S. M. Scheinerand J. Gurevitch (eds.), *Design and analysis of ecological experiments*. New York: Oxford University Press.

Stein, B. A., L. L. Master, and L. E. Morse. 2002. Taxonomic bias and vulnerable species. *Science* 297:1807.

Steinhart, P. 1990. *California's wild heritage: Threatened and endangered animals in the golden state*. San Francisco, CA: California Academy of Science.

Stevens, R. 1999. *Property tax exemption for wildlife management*. Noble Foundation. <http://www.noble.org/Ag/Wildlife/TaxExempt/index.html/>. Accessed 9 February 2007.

Stewart-Oaten, A., J. R. Bence, and C. W. Osenberg. 1992. Assessing the effects of unreplicated perturbations: No simple solutions. *Ecology* 73:1396–1404.

Storer, T. I. and R. L. Usinger. 1963. *Sierra Nevada natural history: An illustrated handbook*. Berkeley, CA: University of California Press.

Strayer, D., J. S. Glitzenstein, C. G. Jones, J. Kolasa, G. E. Likens, M. J. McDonnell, G. G. Parker, and S.T.A. Pickett. 1986. *Long-term ecological studies: An illustrated account of their design, operation, and importance to ecology*. Occasional Paper 2. Washington, DC: Institute of Ecosystem Studies.

Summers, S. G., R. M. Kostecke, and G. L. Norman. 2006. Efficacy of trapping and shooting in removing breeding brown-headed cowbirds. *Wildlife Society Bulletin* 34:1107–12.

Summers, S. G., M. M. Stake, G. H. Eckrich, R. M. Kostecke, and D. A. Cimprich. 2006. Reducing cowbird parasitism with minimal-effort; shooting: A pilot study. *Southwestern Naturalist* 51:409–11.

Sutherland, J. P. 1974. Multiple stable states in natural communities. *American Naturalist* 108:859–73.

Swetnam, T. W., C. D. Allen, and J. L. Betancourt. 1999. Applied historical ecology: Using the past to manage for the future. *Ecological Applications* 9:1189–1206.

Szacki, J., J. Babinska-Werka, and A. Liro. 1993. The influence of landscape spatial structure on small mammal movements. *Acta Theriologica* 38:113–23.

Szewczak, J. M. 2004. Advanced analysis techniques for identifying bat species. Pages 121–27 in R. M. Brigham, E. K. V. Kalko, G. Jones, S. Parsons, and H. J. G. A. Limpens (eds.), *Bat echolocation research: Tools, techniques, and analysis*. Austin, TX: Bat Conservation International.

Szewczak, J. M., S. M. Szewczak, M. L. Morrison, and L. S. Hall. 1998. Bats of the White and Inyo mountains of California-Nevada. *Great Basin Naturalist* 58:66–75.

Taylor, A. H. 2007. *Forest changes since Euro-American settlement and ecosystem restoration in the Lake Tahoe Basin, USA*. USDA Forest Service General Technical Report PSW-203, Berkeley, CA.

Temperton, V. M., and R. J. Hobbs. 2004. The search for ecological assembly rules and its relevance to restoration ecology. Pages 34–54 in V. M. Temperton, R. J. Hobbs, T. Nuttle, and S. Halle (eds.), *Assembly rules and restoration ecology: Bridging the gap between theory and practice*. Washington, DC: Island Press.

Temperton, V. M., R. J. Hobbs, T. Nuttle, and S. Halle. 2004. *Assembly rules and restoration ecology: Bridging the gap between theory and practice*. Washington, DC: Island Press.

Texas Environmental Profiles. 2000. *Public land and public recreation. Sunset commission report*. <http://www.texasep.org/html/lnd/lnd_5pub.html>. Accessed 10 February 2007.

Texas Parks and Wildlife Department (TPWD). 2007. Texas Parks and Wildlife Department Homepage. <http://www.tpwd.state.tx.us/landwater/land/maps/gis/map_downloads/>. Accessed 1 April 2007.

Thompson, J. R., V. C. Bleich, S. G. Torres, and G. P. Mulcahy. 2001. Translocation techniques for mountain sheep: Does the method matter? *Southwestern Naturalist* 46:87–93.

Thompson, W. L., G. C. White, and C. Gowan. 1998. *Monitoring vertebrate populations*. San Diego: Academic Press.

Trammell, M. A., and J. L. Butler. 1995. Effects of exotic plants on native ungulate use of habitat. *Journal of Wildlife Management* 59:808–16.

Trulio, L. A. 1995. Passive relocation: A method to preserve burrowing owls on disturbed sites. *Journal of Field Ornithology* 66:99–106.

Tscharntke, T., I. Steffan-Dewenter, A. Kruess, and C. Thies. 2002. Contribution of small habitat fragments to conservation of insect communities of grassland-cropland landscapes. *Ecological Applications* 12:354–63.

Turchin, P. 1998. *Quantitative analysis of movement: Measuring and modeling population redistribution in animals and plants*. Sunderland, MA: Sinauer.

Tyser, R. W., and C. A. Worley. 1992. Alien flora in grasslands adjacent to road and trail corridors in Glacier National Park, Montana (U.S.A.). *Conservation Biology* 4:251–60.

Underwood, A. J. 1994. On beyond BACI: Sampling designs that might reliably detect environmental disturbances. *Ecological Applications* 4:3–15.

USDA Forest Service. 2005. *Taylor and Tallac Watersheds Environmental Assessment Report. Lake Tahoe, CA*.

USDA Forest Service. 2006a. *Lake Tahoe Basin Management Unit, Recommended Projects for Funding* (online). Updated April 2006. <www.fs.fed.us/r5/ltbmu/documents/ltfac/ recommend/Capital-Projects-Primary-Secondary-Lists.pdf>. Accessed 20 March 2007.

USDA Forest Service. 2006b. *Lake Tahoe Basin management unit, 2005/2006 monitoring program annual report.* Ecosystem Conservation Department, Adaptive Management Monitoring Program, South Lake Tahoe, CA.

United States Department of Agriculture (USDA) Plant Database. 2007. http://www.fs .fed.us/database/feis/plants/shrub/sallem/all.html, http://www.fs.fed.us/database/feis/ plants/shrub/salgey/all.html. Accessed 26 April 2007.

United States Department of the Interior/United States Department of Agriculture (USDI/USDA). 1996. *Sampling vegetation attributes: Interagency technical reference.* Cooperative Extension Service. BLM/RS/ST-96/002+1730, Denver, CO.

U.S. Fish and Wildlife Service (USFWS). 1981. *Standards for the development of habitat suitability index models.* Ecological Services Manual 103. Washington, DC: Government Printing Office.

U.S. Fish and Wildlife Service (USFWS). 1991. *Black-capped vireo (Vireo atricapillus) recovery plan.* Austin, TX: Ecological Services.

U.S. Fish and Wildlife Service (USFWS). 1992. *Golden-cheeked warbler (Dendroica chrysoparia) recovery plan.* Albuquerque, NM: Ecological Services.

U.S. Fish and Wildlife Service (USFWS). 1995. Final rule determining the status of the southwestern willow flycatcher. *Federal Register* 10,694-10,715.

Unitt, P. 1987. *Empidonax truillii extimus:* An endangered subspecies. *Western Birds* 18(3): 137–62.

Valentine, B. E., T. A. Roberts, S. D. Boland and A. P. Woodman. 1988. Livestock management and productivity of willow flycatchers in the central Sierra Nevada. *Transactions of the Western Section of the Wildlife Society* 24:104–14.

Van Andel, J. 2006. Populations: Intraspecific interactions. Pages 70–81 in J. van Andel and J. Aronson (eds.), *Restoration ecology.* Oxford: Blackwell.

Van Andel, J., and A. P. Grootjans. 2006. Concepts in restoration ecology. Pages 16–28 in J. van Andel and J. Aronson (eds.), *Restoration ecology.* Oxford: Blackwell.

Van Horne, B. 1983. Density as a misleading indicator of habitat quality. *Journal of Wildlife Management* 47:893–901.

———. 2002. Approaches to habitat modeling: The tensions between pattern and process and between specificity and generality. Pages 63–72 in J. M. Scott, P. J. Heglund, M. L. Morrison, J. B. Haufler, M. G. Raphael, W. A. Wall, and F. B. Samson (eds.), *Predicting species occurrences: Issues of accuracy and scale.* Washington, DC: Island Press.

Van der Valk, A. G., R. L. Pederson, and C. B. Davis. 2004. Restoration and creation of freshwater wetlands using seed banks. *Wetlands Ecology and Management* 1:191–97.

Van Wieren, S. E. 2006. Populations: Re-introductions. Pages 82–92 in J. van Andel and J. Aronson (eds.), *Restoration ecology.* Oxford: Blackwell.

VanderWerf, E. A. 1993. Scales of habitat selection by foraging 'elepaio in undisturbed and human-altered forests in Hawaii. *Condor* 95:980–89.

Vander Zanden, M. J., J. D. Olden, and C. Gratton. 2006. Food-web approaches in restoration ecology. Pages 165–89 in D. A. Falk, M. A. Palmer, and J. B. Zedler (eds.), *Foundations of restoration ecology.* Washington, DC: Island Press.

Varner III, J. M. D. R. Gordon, F. E. Putz, and J. K. Hiers. 2005. Restoring fire to long-unburned *Pinus palustris* ecosystems: Novel fire effects and consequences for long-unburned ecosystems. *Restoration Ecology* 13:536–44.

Verner, J. 1984. The guild concept applied to management of bird populations. *Environmental Management* 8:1–14.

Verner, J., M. L. Morrison, and C. J. Ralph. 1986. *Wildlife 2000: Modeling habitats of terrestrial vertebrates.* Madison: University of Wisconsin Press.

Vickery, P. D., M. L. Hunter Jr., and J. V. Wells. 1992. The use of a new reproductive index to evaluate relationships between habitat quality and breeding success. *Auk* 109:697–705.

Vreeland, J. K., and W. D. Tietje. 2002. Numerical response of small vertebrates to prescribed fire in California oak woodland. Pages 100–110 in W. M. Ford, K. R. Russell, and C. E. Moorman (eds.), *The role of fire in nongame wildlife management and community restoration: Traditional uses and new directions.* USDA Forest Service General Technical Report NE-288, Newtown Square, PA.

Walters, C. 1986. *Adaptive management of renewable resources.* New York: MacMillan.

Ward, M. P., and S. Schlossberg. 2004. Conspecific attraction and the conservation of terrestrial songbirds. *Conservation Biology* 18:519–25.

Wecker, S. C. 1964. Habitat selection. *Scientific American* 211:109–16.

Weiher, E., and P. Keddy. editors. 1999. *Ecological assembly rules: Perspectives, advances, retreats.* Cambridge: Cambridge University Press.

Weixelman, D. A., D. C. Zamudio, and K. A. Zamudio. 1999. *Eastern Sierra Nevada riparian field guide.* USDA Forest Service, Humboldt-Toiyabe National Forest, Sparks, NV: Intermountain Region Report R-4-ECOL-99-01.

Weller, T. J., S. A. Scott, T. J. Rodhouse, P. C. Ormsbee, and J. M. Zinck. 2007. Field identification of the cryptic verpertilionid bats, *Myotis lucifugus* and *M. yumanensis. Acta Chiropterologica* 9:133–47.

Welsh Jr., H. H. and A. J. Lind. 1995. Habitat correlates of Del Norte salamander, *Plethodon elongates* (Caudata: Plethodontidae), in northwestern California. *Journal of Herpetology* 29:198–210.

———. 2002. Multiscale habitat relationships of stream amphibians in the Klamath-Siskiyou region of California and Oregon. *Journal of Wildlife Management* 66:581–602.

Wemmer, C., T. H. Kunz, G. Lundie-Jenkins, and W. J. McShea. 1996. Mammalian Sign. Pages 157–76 in D. E. Wilson, F. R. Cole, J. D. Nichols, R. Rudran, and M. S. Foster (eds.), *Measuring and monitoring biological diversity: Standard methods for mammals.* Washington, DC: Smithsonian.

Wenny, D. G., R. L. Clawson, J. Faaborg, and S. L. Sheriff. 1993. Population density, habitat selection, and minimum area requirements of three forest-interior warblers in central Missouri. *Condor* 95:968–79.

Whisson, D. A. 1999. Modified bait stations for California ground squirrel control in endangered kangaroo rat habitat. *Wildlife Society Bulletin* 27:172–77.

White, G. C., and R. A. Garrott. 1990. *Analysis of wildlife radio-tracking data.* San Diego: Academic Press.

White, L. D., and C. W. Hanselka. 1991. *Prescribed range burning in Texas.* Texas Agricultural Extension Service, Texas A&M University System, College Station, TX: PWD-BK-W7000-0196-7/91.

White, P.S., and A. Jentsch. 2004. Disturbance, succession, and community assembly in terrestrial plant communities. Pages 342–66 in V. M. Temperton, R. J. Hobbs, T. Nuttle, and S. Halle (eds.), *Assembly rules and restoration ecology: Bridging the gap between theory and practice.* Washington, DC: Island Press.

Whitefield, M. J., K. M. Enos, and S. P. Rowe. 1999. Is brown-headed cowbird trapping effective for managing populations of the endangered southwestern willow flycatcher? *Studies in Avian Biology* 18:260–66.

Wiens, J. A. 1972. Anuran habitat selection: Early experience and substrate selection in Rana cascadae tadpoles. *Animal Behavior* 20:218–20.

————. 1984. The place of long-term studies in ornithology. *Auk* 101:202–3.

————. 1989a. *The ecology of bird communities.* Vol. 1. *Foundations and patterns.* Cambridge: Cambridge University Press.

———— 1989b. *The ecology of bird communities.* Vol. 2. *Processes and variations.* Cambridge: Cambridge University Press.

Wiens, J. A., and K. R. Parker. 1995. Analyzing the effects of accidental environmental impacts: Approaches and assumptions. *Ecological Applications* 5:1069–83.

Wilkins, R. N., R. A. Powell, A. A. T. Conkey, and A. G. Snelgrove. 2006. *Population status and threat analysis for the black-capped vireo.* Region 2, Albuquerque, NM: US-FWS.

Willis, K. J., and H. J. B. Birks. 2006. What is natural? The need for a long-term perspective in biodiversity conservation. *Science* 314:1261–65.

Willson, M. F. 1974. Avian community organization and habitat structure. *Ecology* 55:1017–29.

Wilson, D. E., F. R. Cole, J. D. Nichols, R. Rudran, and M. S. Foster (eds.). 1996. *Measuring and monitoring biological diversity: Standard methods for mammals.* Washington, DC: Smithsonian.

Wing, L. 1947. Christmas census summary, 1900–1939. Pullman: State College of Washington. Mimeograph

Wolf, C. M., B. Griffith, C. Reed, and S. A. Temple. 1996. Avian and mammalian translocations: Update and reanalysis of 1987 survey data. *Conservation Biology* 10:1142–54.

Wolton, R. J., and J. R. Flowerdew. 1985. Spatial distribution and movements of wood mice, yellow-necked mice, and bank voles. *Symposia of the Zoological Society of London* 55:249–75.

Wood, P. B., M. W. Collopy, and C. M. Sekerak. 1998. Postfledging nest dependence period for bald eagles in Florida. *Journal of Wildlife Management* 62:333–39.

Wootton, J. T., and D. A. Bell. 1992. A metapopulation model of the peregrine falcon in California: Viability and management strategies. *Ecological Applications* 2:307–21.

Wright, S. 1931. Evolution in Mendelian populations. *Genetics* 16:97–159.

Zeng, Z., and J. H. Brown. 1987. Population ecology of a desert rodent: *Dipodomys merriami* in the Chihuahuan desert. *Ecology* 68:1328–40.

Zimmerman, B. L. 1994. Audio strip transects. Pages 92–97 in W. R. Heyer, M. A. Donnelly, R. W. McDiarmid, L. C. Hayek, and M. S. Foster (eds.), *Measuring and monitoring biological diversity: Standard methods for amphibians.* Washington, DC: Smithsonian.

Zwartjes, M. 1999. *Conservation plan for the Barton Creek habitat preserve. Travis County, TX:* The Nature Conservancy of Texas.